# RECONFIGURING GLOBAL SOCIETIES IN THE PRE-VACCINATION PHASE OF THE COVID-19 PANDEMIC

*Edited by Jack Fong*

*Reconfiguring Global Societies in the Pre-Vaccination Phase of the COVID-19 Pandemic* examines lived experiences of the COVID-19 pandemic in communities and societies around the world before the arrival of vaccines. This collection presents analyses of scholars from eight countries, all of whom were engaged in the unfolding crisis of social forces across the world.

This timely volume conveys valuable insights about how public officials, the state, healthcare workers, and, ultimately, citizens responded to consequences of the pandemic upon not only the body but also social relations in community, city, and society. The contributing scholars document how state apparatuses, urban configurations, places of employment, legal structures, and ways of life responded to crisis-altered social conditions during the pandemic. The book investigates what societies experiencing crisis around the world reveal about the state's efficacy and inefficacy in fulfilling its social contract for its citizens, especially on unresolved issues related to social relations based on politics, race, ethnicity, gender, and crime.

This collection brings together a cross section of scholars experiencing the same temporal moment of crisis together, watching and observing how the pandemic of their age uncoiled itself into the fabric of community, onto the institutions and bureaucracies of society, and into the most intimate confines of the home.

JACK FONG is a professor of sociology at California State Polytechnic University, Pomona.

# Reconfiguring Global Societies in the Pre-Vaccination Phase of the COVID-19 Pandemic

EDITED BY JACK FONG

UNIVERSITY OF TORONTO PRESS
Toronto Buffalo London

© University of Toronto Press 2024
Toronto Buffalo London
utorontopress.com

ISBN 978-1-4875-2707-5 (cloth)     ISBN 978-1-4875-2710-5 (EPUB)
ISBN 978-1-4875-2708-2 (paper)     ISBN 978-1-4875-2709-9 (PDF)

**Library and Archives Canada Cataloguing in Publication**

Title: Reconfiguring global societies in the pre-vaccination phase of the
   COVID-19 Pandemic / edited by Jack Fong.
Names: Fong, Jack, 1970– editor.
Description: Includes bibliographical references and index.
Identifiers: Canadiana (print) 20230578144 | Canadiana (ebook) 20230578152 |
   ISBN 9781487527082 (paper) | ISBN 9781487527075 (cloth) |
   ISBN 9781487527099 (PDF) | ISBN 9781487527105 (EPUB)
Subjects: LCSH: COVID-19 Pandemic, 2020- – Social aspects – Case studies. |
   LCGFT: Case studies.
Classification: LCC RA644.C67 R43 2024 | DDC 362.1962/4144–dc23

Cover design: Lara Minja/ Lime Design
Cover image: People from an Anglo-Indian community sit maintaining social distance in an open-air area; AFP via Getty Images/Dibyangshu Sarkar

Excerpt(s) from COLLECTED FICTIONS: VOLUME 3 by Jorge Luis Borges, translated by Andrew Hurley, copyright © 1998 by Maria Kodama, translation copyright © 1998 by Penguin Random House LLC. Used by permission of Viking Books, an imprint of Penguin Publishing Group, a division of Penguin Random House LLC. All rights reserved.

We wish to acknowledge the land on which the University of Toronto Press operates. This land is the traditional territory of the Wendat, the Anishnaabeg, the Haudenosaunee, the Métis, and the Mississaugas of the Credit First Nation.

University of Toronto Press acknowledges the financial support of the Government of Canada, the Canada Council for the Arts, and the Ontario Arts Council, an agency of the Government of Ontario, for its publishing activities.

# Contents

*List of Figures, Images, and Tables*  ix

*Acknowledgments*  xi

*Preface*  xv

Introduction  3

**The Far East**

1 Reconsidering the Third Place: Social Distancing and Inequality in South Korea during the Era of Coronavirus  33
   KELLY HUH AND HYEJIN YOON

2 The Anthropocene, Zoonotic Diseases, and the State – Japan's Response to the COVID-19 Pandemic: Criminal Negligence or Crimes against Humanity?  56
   HIROSHI FUKURAI

**South and Southeast Asia**

3 Witnessing amidst Distancing: Structural Vulnerabilities and the Researcher's Gaze in Pandemic Times in Relation to Migrant Workers of India and Singapore  85
   AMRITORUPA SEN AND JUNBIN TAN

4 Social Distancing? "No Problem!": Explaining Thailand's Successful Containment of COVID-19  110
   PIYA PANGSAPA

5 Trust in Numbers? The Politics of Zero Deaths
   and Vietnam's Response to COVID-19   134
   AMY DAO

## A Global Address

6 An "Unseen Enemy" and the "Shadow Pandemic":
  The COVID-19 Pandemic and Domestic Violence   157
  SHWETA ADUR AND ANJANA NARAYAN

## The United States

7 An Investigation into the Economic, Social, and
  Psychological Dimensions of COVID-19   177
  KEVIN MCCAFFREE AND ANONDAH SAIDE

8 Unsettling Contact: The Collapse of Emotional Distance
  at a COVID-19 Medical Frontline   200
  JUNBIN TAN AND PHU TRAN

9 A Spatial Snapshot of the Relationship between the
  COVID-19 Pandemic and Selected Crimes in California   221
  GABRIELE PLICKERT AND EMILY COOPER

10 Employing Lyn Lofland's and Ray Oldenburg's Urban
   Sociology to "Read" the Emptying of Los Angeles' Publics   245
   JACK FONG

## European Union

11 Trust between Citizens and State as a Strategy to Battle
   the Pandemic: Were Senior Citizens Collateral Damage
   in the Swedish Government's Plan to Flatten the Curve?   275
   ANN-CHRISTINE PETERSSON HJELM

12 The German Reaction to Corona: The Interplay of Care,
   Control, and Personal Responsibility within the Welfare State   301
   ALBERT SCHERR

## South Pacific

13 The Benefits and Drawbacks of Social Distancing:
   Lessons from New Zealand   327
   MARIA ARMOUDIAN AND BERNARD DUNCAN

Conclusion   361

*Contributors*   381

*Index*   383

# List of Figures, Images, and Tables

**Figures**

| | | |
|---|---|---|
| 0.1. | The Disaster Impact Model | 8 |
| 1.1. | Public transportation usage on weekends in metropolitan areas between 18 April and 11 October 2020, South Korea | 46 |
| 1.2. | Cell phone–based movement on weekends between 18 April and 11 October 2020, South Korea | 47 |
| 1.3. | Credit card usage on weekends between 18 April and 11 October 2020, South Korea | 47 |
| 4.1. | Mask wearing and hand sanitizer use, Southeast Asia | 119 |
| 7.1. | Visual depiction of conceptual relations | 183 |
| 7.2. | Visual depiction of the pared-down significant mediation effects among Republicans | 189 |
| 7.3. | Visual depiction of the pared-down significant mediation effect among Democrats | 190 |
| 9.1. | California COVID-19 progression and implemented health measures at the time of crime analysis, March–June 2020 | 230 |
| 9.2. | Percent change of crimes by selected California cities, March–June 2019 and 2020 | 233 |
| 9.3. | Selected crimes, Los Angeles, 2019–20 | 235 |
| 9.4. | Selected crimes, San Francisco, 2019–20 | 236 |
| 9.5. | Selected crimes, San Diego, 2019–20 | 236 |

## Images

| | | |
|---|---|---|
| 0.1. | Japan's Prime Minister Shinzo Abe at the first Novel Coronavirus Expert Meeting | 13 |
| 0.2. | Migrant workers returning to their villages, India | 14 |
| 0.3. | COVID-19 testing registration desk, Hanoi | 16 |
| 0.4. | Refrigerated semi-truck trailers, New York | 20 |
| 0.5. | Swedish elderly celebrating after receiving vaccines, Linköping, Sweden | 22 |
| 0.6. | Masked passengers in a subway light rail tram, Düsseldorf | 24 |
| 1.1. | Mandatory self-quarantine app and public notice, Incheon International Airport | 44 |
| 3.1. | Migrant workers outside their dormitory rooms, Singapore | 95 |
| 4.1. | "Cupboard of Happiness," Thailand | 123 |
| 4.2. | Visiting the cupboard | 124 |
| 4.3. | Items inside Cupboard of Happiness | 124 |
| 6.1. | Protest against domestic violence, Tel Aviv | 165 |
| 10.1. | Empty 60 and 710 freeway interchange, Los Angeles | 261 |
| 10.2. | Rally against anti-Asian hate crimes, Manhattan | 266 |
| 13.1. | COVID-19 testing facility, Central Wellington, NZ | 334 |
| 13.2. | At the entrance of an appliance store | 345 |

## Tables

| | | |
|---|---|---|
| 1.1. | Three-level social distancing in South Korea | 48 |
| 7.1. | Self-reported Republicans | 186 |
| 7.2. | Self-reported Democrats | 188 |
| 14.1. | Lockdown durations around the world, 16 January 2020–15 January 2021 | 371 |

# Acknowledgments

Our volume was begun and completed during a very surreal and challenging period of our shared humanity, one whose lingering effects will be, in my view, seen historically as a bona fide pandemic saeculum. My gratitude must thus be directed to all my contributors, who exhibited much patience with my updates, queries, and deadlines. Many among you had to contend with existential predicaments in your respective lifeworlds as well as respond to ongoing demands of your profession, all combined with a further need to ensure some semblance of quotidian certainty and quality of life.

In alphabetical order, I express my heartfelt thanks to Associate Professor Dr. Shweta Adur at California State University, Los Angeles' Sociology Department; Dr. Maria Armoudian, Senior Lecturer at the University of Auckland's Politics and International Relations Department, New Zealand; Emily Cooper, Criminology Master's Degree candidate at George Mason University; Associate Professor Dr. Amy Dao in the Geography and Anthropology Department at California State Polytechnic University, Pomona; Bernard Duncan, Communications and Engagement Lead of the Ministry of Education of New Zealand; Dr. Hiroshi Fukurai, Professor of Sociology and Legal Studies and President of the Asian Law and Society Association in the Sociology Department at the University of California, Santa Cruz; Dr. Kelly Huh, Department Chair and Associate Professor of Geography in the Geography and Anthropology Department at California State Polytechnic University, Pomona; Dr. Kevin McCaffree, Associate Professor of Sociology at the University of North Texas's Department of Sociology; Dr. Anjana Narayan, Department Chair and Professor of Sociology in the Sociology Department at California State Polytechnic University, Pomona; Dr. Piya Pangsapa Assistant Professor and Assistant Dean for Research and International Affairs, School of Global Studies, Thammasat University, Thailand; Dr. Ann-Christine Petersson Hjelm, Associate Professor and Senior Lecturer of Jurisprudence and Commercial Law, Department of

Business Studies, at Uppsala University, Sweden; Dr. Gabriele Plickert, Professor of Sociology in the Sociology Department at California State Polytechnic University, Pomona; Assistant Professor Dr. Anondah Saide of the Educational Psychology Department at the University of North Texas; Dr. Albert Scherr, Senior Professor at the Institute of Sociology of the Freiburg University of Education, Germany, and Research Fellow at University of the Free State, QwaQwa Campus, South Africa; Dr. Amiritorupa Sen, Research Scholar at the National University of Singapore; Junbin Tan, PhD candidate in the Anthropology Department at Princeton University; Dr. Phu Tran, General Anesthesiology and Nocturnist (MD) at New York Presbyterian-Queens Hospital; and Dr. Hyejin Yoon, Associate Professor and Geographer in the Department of Geography at the University of Wisconsin–Milwaukee.

I express my sincere appreciation to Dr. Matthew Bowker at the University of Buffalo, New York, for tackling issues related to the pandemic through a political and psychoanalytic lens. He will be part and parcel of, I believe, a "school" of pandemic studies surely to emerge in the years to come. My gratitude also extends to Dr. Michelle Stack, Professor of Educational Studies at the University of British Columbia, Vancouver, for giving me important insights on how to be a better editor. Her suggestions were conveyed to me while she was in Vancouver, Canada, and I was under a mandatory 14-day quarantine in Thailand when the country was devastated by the UK's alpha variant of COVID-19 in late summer 2021. I also extend my gratitude to Dr. James Andre at the University of California, Riverside, Director of the Granite Mountains Desert Research Center at the Mojave Natural Preserve, for providing tips on where to conduct astronomy as a means of stress relief while the volume was being assembled. I express my heartfelt thanks to the University of Toronto Press's Jodi Lewchuk and its production team. I am especially grateful to Jodi, Judy Williams, Barbara Porter, and Stephanie Mazza for their thorough support and flexibility in ensuring that our project come to fruition. All of us established some semblance of community while undertaking this effort in spite of the lockdowns. Lastly, I thank my son, Pattarakorn, and my spouse, Thasanee, for their wonderful support, one that entailed regularly reminding me about the need for balance while the volume was being composed, assembled, and brought to fruition.

"Occasionally a few birds, a horse perhaps, have saved the ruins of an amphitheater."
– Jorge Luis Borges, from *Tlön, Uqbar, Orbis Tertius*
(Borges, 1998, p. 78)

# Preface

With a humanitarian lens inflected by cultures and worldviews from different countries, the social thinkers in our volume offer their illuminations on the disruptions caused by pandemic distancing due to COVID-19, undertaken and framed with their own articulations and positionalities derived from their domiciles or area studies. Through data, theory, investigation, analysis, reflection, and description, the material consequences of pandemic distancing, along with its stressors, are given structural form and their iterations discerned. As the effects of the pandemic progressed from spring 2020 until the end of the year, which for many of the world's population was the pre-vaccination phase of the crisis, our panoply of authors offers readers unique cross-sectional insights into a historical saeculum. In what Nassim Nicholas Taleb (2010) would term a "Black Swan" event, the COVID-19 pandemic has drastically reconfigured community and society in a variety of international and interactional contexts.

Taleb's reference is highly appropriate, yet aside from my use of the term in an earlier work about Friedrich Nietzsche's philosophy as one that can inspire actors to overcome such Black Swans, never did it cross my mind that such an event would offer twenty-first-century humanity its first global existential crisis, one more immediately palpable and visceral than climate change. Taleb describes a Black Swan event as a rare event that appears from an unexpected realm to create significant impacts and liminalities upon humanity, one that ultimately compels us in hindsight to interpret its historical impact. The armamentarium assembled from information about societies experiencing the COVID-19 pandemic will help *Reconfiguring Global Societies* make this Black Swan visible. Although we have approached the horizon where we may begin to have a retrospective on its corporeal health impacts, we are far from understanding COVID-19's impact on the health of drastically reconfigured communities and societies. How this volume informs further understanding of our dramatically altered lifeworlds, a term made popular by *fin de siècle* philosopher Edmund Husserl and by contemporary sociologist and philosopher Jürgen Habermas, thus propels the energy of this edited collection.

During a period defined by a graphic mortality, interspersed with health and political narratives advocating or mandating a variety of pandemic mitigation measures, many members of society sought alternative forms of existence through communities that found sublimation, while others demanded that their social institutions offer effective emergency responses *or* respite from additional duress. The sentiments born from the emptying of communities and publics during different types of lockdowns, followed by discontent with such strategies expressed by many, ushered in new social contexts that emerged to fill society's breached interactions and spaces. Such breaching seen around the world reveals how communities, societies, and states attempted to fulfil their social contracts in ways that urgently responded not only to their own society's humanity but also to our shared humanity.

The COVID-19 pandemic is being experienced in an unprecedented manner. When the SARS-1 pandemic erupted between 2002 and 2004, much of the world remained reliant on official or cultural channels to be informed; the proliferation of online media as an instrument of citizen journalism, consumerism, and cultural production had yet to become a conventional means for knowledge dissemination of information and, in unfortunate cases, misinformation. Whether advised by political leaders, journalists, or individuals with social standing, the citizenry of any one country had yet to be exposed to a multitude of simultaneous communiqués and cultural narratives, sometimes contradicting one another, emanating from their own country as well as from other countries during the same period. In the context of experiencing the COVID-19 pandemic, however, the expansive and quick flow of pandemic information and misinformation revealed unfiltered angst and discontent, along with perseverance, and hope as experienced across the various countries where our volume's contributors live. Indeed, one can forward the view that the COVID-19 pandemic is history's first affectively *shared humanity* health crisis at the global and local levels – it is not Ebola, contextualized in sub-Saharan Africa, mad cow disease (not readily transmissible to humans), contextualized in the United Kingdom, or SARS-1, which wreaked havoc in East Asia twenty years ago, the early iteration of COVID-19 (Slack, 2021).[1] Our edited volume thus captures the social dynamics of pandemic distancing as COVID-19 made its first appearance in many countries around the world in early 2020. It is a non-longitudinal "snapshot" which should be seen as offering an understanding of the pandemic during its outset. As a snapshot, our work captures the social responses during the pre-vaccination period of different countries before they generated a comparatively more coherent global and local narrative for pandemic mitigation by the end of December 2020, when the rollout of vaccines began (with the UK administering its first vaccination in early December 2020).[2] The pre-vaccination phase under examination in this volume is a significant period because of the culture shock of COVID-19 as it arrived at our doorsteps with no "medicine" to thwart its advance. It is this non-longitudinal component that we hope can reveal our attempts at illuminating some semblance of order emerging from disorder. Indeed, I am of the view that when we have enough temporal distance in the future for clear hindsight, analysts will be able to look at the effects of the pandemic saeculum in two distinct historical phases: the paroxysmic pre-vaccination period and the post-vaccination period that hints at COVID-19's endemicity. It is my hope that this volume, by focusing on the merits of this snapshot/cross-sectional method, will be exhibiting some sort of yet to be appreciated prescience.

Our researchers have decided to undertake this project because the desiccation of civil society and public spheres during the pandemic motivated us to examine the social contexts and "insides" of what were traditionally assumed to be stable social structures, structures that then became subjected to acute duress, to destructive forces *and* forces of renewal as they churned out alternative narratives of survival upon the human condition. During our discussions, it became clear that a community's or country's path towards social revitalization occurred in stages depending on public policy, interspersed with additional intervening variables "on the ground" that are specific to a country. It is in this context that our contributors were inspired and committed to understanding the gravity of the pandemic and its resultant social dislocations. During this period, a variety of exchanges between our contributors exhibited aspirational desires to wrestle with the consequences of the pandemic beyond its corporeal effects – and from this realization emerged the need to assemble and compose this edited volume.

Such an undertaking could not have been timelier, since our scholars, in their respective countries, have generated a repository of knowledge born from "local researchers" who know the "best local contexts to study local disasters" (Gaillard, 2019, p. S14). As such, the power of analysis has been transferred "to local scholars to take up the lead" in examining systemic crises (Gaillard, 2019, p. S15). Around the world in their respective geographical wombs, if not having travelled back to them to frequently conduct research, our authors have been entangled in their social contexts, experiencing varying degrees of state regulation, community discontent, social angst, and even cautious optimism. Indeed, the coterie of contributors seen in this volume experienced varying degrees of culture shock shaped greatly by the pandemic and its reconfiguration of community and society. Yet in spite of the nuances of different cultures and transnational voices communicating from or about their geographical wombs, our thinkers agree that pandemic distancing, especially if enacted through the quarantining of its citizens, is no Balinese *nyepi*.[3]

By offering prescient, hopeful, critical, and even pessimistic insights, our contributors give content and contour, through their respective observations and analyses, to the existential angsts and social changes experienced by many during the pandemic. Our contributors do not believe that we are too socio-historically close to the current phenomenon to offer a clear view of societies around the world experiencing the consequences of reconfigured communities. Marshall McLuhan presciently noted as far back as the 1960s that we live in a "global village" where the global is local, a condition not lost among our writers as they embarked on their analyses. It should thus be noted that our contributors examined the consequences of the 2020 pandemic upon the state apparatus, social life, social interaction, and sense of community in close temporal proximity to the culture shocks generated by the enormity of the pandemic during the pre-vaccination phase. The pandemic generated a foundational existential awakening for our writers as it affected not only particular lifeworlds in resource-rich nations but also, to a great degree, the lifeworlds of social inequalities that remain embedded and unresolved across many countries affected by COVID-19.

Illuminating the reconfiguring of community and society through its emaciated publics is more complicated than approaching a society writhing in the context of militarized conflict. In the latter context, the destruction of infrastructure and lives is viscerally felt and, in undignified fashion insofar as casualties are concerned, quantifiable. Although the same arithmetic that captures mortality can be seen across the planet when measuring outcomes of the coronavirus pandemic, the social implications are more amorphous and varied. This facticity obligates us to observe how societies suffer in a new light, since rebuilding and reassembling community and society at the collateral and infrastructural levels – as in redeveloping social spaces after a militarized conflict – will still not quell nihilistic or metaphysical concerns expressed by pandemic-fatigued members of society. Instead, as actors are tucked away in their more intimate confines, alternative spatialities are conjured, impatient anticipation of state responses builds, alternative modes of interaction are relied upon, and introspective authoring of solitude is engaged to respond to deformed social environments. Also surfacing are the percolating, unresolved dynamics and conflict between intimates and kin – ignited by unemployment, absolute deprivation, quotidian disruption, death, fear, and stress – dynamics that can, under normal circumstances, allow public resolution of the private to occur. Our work thus demonstrates that there are novel social structures and responses, positive and negative, with and without state apparatuses, that force upon actors the confrontation of their human condition, their mortality, while the public empties its civil society and drains its cities of resources, embedding numerous members of community in their own as well as in state-mandated liminalities. Some countries exhibited a noticeably better outcome than others during our period of research, and our edited volume intends to reveal the consequences and intricacies of how public policy, spatiality, culture, and political climate greatly influence the well-being of those exposed to pandemic conditions and their accompanying systemic crises.

At this juncture, we need to ask: why should social scientists study a health crisis that wreaks havoc so intimately on our bodies and our personal mental states? Sociologist Robert Stallings offers some important dialectical insights: observers and analysts of society can, through what crisis makes visible, begin to examine "aspects of social structures and processes that are hidden in everyday life," by "studying the 'exception' in order to better understand the 'rule'" (2002, p. 283). He celebrates how sociologists such as Anthony Giddens once noted that the study of "critical situations" offers privileged insight into "the accustomed routines of daily life," especially when they are "dramatically disrupted" (Giddens, 1979, cited in Stallings, 2002, p. 284), while noting how the great theorist of social systems Niklas Luhmann (see 1982, 1989, 1993) also examined such disruptions, orientations that inspired us to employ the term "systemic crisis/es" as our preferred analogue to the term disaster/s. Gary Kreps argues how information derived from crises "provide[s] rich data for addressing basic questions about social organization – its origins, adaptive practices, and survival," and that these questions were actually considered "fundamental by the classic figures of sociology" (1984, p. 310). For Dynes and Drabek (1994) and Fritz (1961, 1996), systemic crises create a sociological laboratory for new considerations and conceptualizations for one's sociological imagination and theorizing processes. Scholars in our collection also heed insights offered by Geyer and Schweitzer's 1976 classic, *Theories of Alienation*. Of particular importance is the passage below, one that describes well our contributors' efforts to counter

> the "overpsychologization" of the alienation concept, further reinforced by the virtual absence of truly macrosociological research methods ... Even when alienation is viewed as a subjective state, research should encompass the objective environmental determinants of this subjective state. Unfortunately, this rarely happens in empirical alienation studies. Consequently, the researcher is unable to make a research-based judgement about the structure and conditions which exist in society at large; at issue are only the individual's feelings, perceptions, and attitudes. By focusing on the subjective states of individuals, social structural problems which lie at the root of alienation are by definition excluded. (Geyer & Schweitzer, 1976, pp. xxi–xxii)

By examining different countries' responses to pandemic distancing at different times during the spread of COVID-19 in 2020, our volume offers important considerations and insights into how members of society around the world respond to macro-level emergencies. Yet it is also a collection that conveys important analyses by social thinkers who passionately feel a sense of urgency to highlight how unpredictable changes to community and society affect the human condition. The changes are very visceral and real. In addition to the loss of loved ones during the pandemic, major disruptions to our quotidian lifeworlds over many months have dramatically altered and exhausted lives even when deaths were not involved. Our volume thus conveys the concerns and analyses of 19 scholars residing in the seven countries of Germany, India, New Zealand, Singapore, Sweden, Thailand, and the United States of America. In turn, analyses span 10 countries and one global address: Germany, India, Japan, New Zealand, Singapore, South Korea, Sweden, Thailand, Vietnam, and the United States, all of which offer us insights into pandemic mitigation efforts of the Far East, South and Southeast Asia, the United States of America, the European Union, and the South Pacific, as well as around the world. Moreover, our chapters are sequenced with these regional classifications to allude to the geopolitics involved in pandemic mitigation that inflect the most localized of ideological and cultural politics. The chapters are written by social thinkers working from a variety of disciplinary perspectives born from their anthropological, geographical, political, legal, medical, social work, and sociological backgrounds, perspectives that will document how our state apparatuses, urban configurations, places of employment, and ways of life responded to societies during the pandemic. We hope that this volume about the human condition experiencing systemic crisis will convey valuable insights for cautious extrapolations and inferences that can allow public officials, the state, healthcare workers, and, most importantly, the people, to shape public policy in ways that respond to the consequences of the pandemic not only upon the body but also upon reconfigured communities and societies.

NOTES

1 The distinguished historian Paul Slack, in *Plague: A very short introduction*, concedes that Western discourse on historicizing pandemics as ultimately stemming from Asia "is one created by Western eyes," and that the "global history of plague might look very different if Chinese and Indian accounts of ancient epidemics, of which there are many, had been fully examined through modern medical spectacles; or if there were any records at all of similar outbreaks in sub-Saharan Africa before Europeans arrived" (2021, pp. 20–1).
2 The BBC reported in early December 2020 that British citizens were sanguine about the prospects of vaccination. Margaret Keenan, at 90 years, was the first to receive the vaccine and proclaimed that it was the "best early birthday present" and that she felt "privileged to be the first person vaccinated against COVID-19" (BBC, 2020).
3 The New Year holiday celebrated in the Balinese calendar where its citizens willingly observe a day of silence, reflection, and contemplation. Civic functions and public facilities cease operations. Members of the community remain indoors, a stipulation also mandated of tourists. Road travel is prohibited unless one is a first responder or law enforcement official. Community patrols known as the *pecalang* ensure compliance.

RECONFIGURING GLOBAL SOCIETIES IN THE
PRE-VACCINATION PHASE OF THE COVID-19 PANDEMIC

# Introduction

Although the COVID-19 pandemic was dispersed around the world in dramatic fashion, further amplified by the mass media's expediency in conveying graphic imagery of its consequences, namely death, social disarray, and social upheaval, one might presuppose that an introductory chapter about how pandemic social environments have reconfigured community and society can analytically begin and end in a global context. However, major concerns exhibited by your American editor and his international colleagues about the dysfunctions seen in our respective societies during the pandemic motivated us, instead, to analyse at the local context, but in a manner that could be understood *within* the larger framework of the global confronting its latest morbidity.

At the time our volume was first drafted during the conclusion of 2020, the United States had the world's highest death toll of COVID-19 with over 330,000 fatalities. By the time our work was close to completion, the United States had recorded over one million fatalities, while around the world over six million had perished. In the United States, the reckless politicization and Sinophobia exhibited by then President Donald Trump during the crisis, along with his enabling of beliefs in pseudo-science, such as Dr. Stella Immanuel's claim that hydroxychloroquine can "cure COVID," if not the pure rejection of scientific facts altogether (Andrews & Paquette, 2020), further exacerbated the conditions of the pandemic. For example, anti-Asian American violence increased exponentially, some American citizens' obstinate refusal to wear masks, owing to beliefs that such practices infringed on their individual rights, resulted in verbal and physical confrontations between the masked and unmasked in public spaces, shortages in protective professional equipment for healthcare workers, along with emotional burnout, stymied public efforts at providing healthcare, and deurbanization through the emptying of publics and its deleterious effects upon walled confines of the home resulted in the rise of domestic violence. Furthermore, the death of George Floyd in Minneapolis, Minnesota on 25 May 2020, at the hands of law enforcement, sparked nationwide protests at a time when physical distancing was called for across the United States, the dynamics of which were coeval with a pandemic-generated economic downturn. The aforementioned social discontents under burgeoning pandemic constraints thus convinced me to reach out to my colleagues domestically and internationally to ascertain social dynamics in their respective lifeworlds. Such a gesture was undertaken if only for the sake of having alternative viewpoints for understanding how communities and societies elsewhere have been affected by COVID-19 during the pre-vaccination period of 2020.

During this period, an opportunity to begin establishing some semblance of community surfaced first with our panoply of writers: our contributors were primarily researching within their home environs; many of us were relatively productive without the demands of long commutes to work via personal or public transportation, a privilege of which we are mindful, since many workers around the world suffered more severe outcomes in their employment situation. Yet it would be misplaced to envision the birth of this project as one exclusively based on convenience, for it was our need to make visible the changing contours of our affected communities, families, and ourselves, along with our deep sense of compassion and empathy for an injured humanity, that constituted the genesis of this volume. It came into being because we have witnessed an epic event that has affected humanity around the world, an event that also unmasked the fragility and vulnerability of our lifeworlds, within which we customarily thrive. In this context, I begin my introduction by offering a condensed summary on the discourse of disaster studies, followed by chapter summaries highlighting our international and transnational contributors writing about Germany, India, Japan, New Zealand, Singapore, South Korea, Sweden, Thailand, Vietnam, the United States, and the world, grouped into their regions of examination: the Far East, South and Southeast Asia, the United States of America, the European Union, and the South Pacific. Yet regardless of place, it should be emphasized that the ethos propelling this volume is that we care deeply about one another's lifeworlds, how to understand them during systemic crises, and how the health of our lifeworlds, not just our bodies, has the ability to affect each and every one of us. Moreover, although we celebrate our cultural diversity, with COVID-19's obliviousness to country or creed, we are also situationally aware that we live in a global village of a shared humanity that inflects itself at the local.

In the context of the USA, the field of "disaster studies" can be traced back to the 1950s during exigencies generated by the Cold War.[1] E.L. Quarantelli, a foremost scholar in the area, notes how social science research during this period approached crises as concerns about outcomes resulting from militarized conflict. Research institutions that were active during this period, specifically the National Opinion Research Center (NORC), and universities such as the University of Chicago, University of Maryland, and University of Oklahoma, along with the Committee on Disaster Studies, which later transformed into the Disaster Research Center (DRC) that is still operating from its intellectual womb at the University of Delaware, thus tended to focus on civilian and soldier reactions to war, as well as how to maintain order during such crises. In the early iteration of disaster studies, the emphasis on peacetime disasters was relegated to less importance if it did not fit the foci of various federal agencies that harboured Cold War concerns. Indeed, peacetime crises were mistakenly seen to have "the closest parallel to a wartime situation" (Quarantelli, 1987, p. 290), and findings were designed with "the ultimate aim of extrapolation to military situations" (1987, p. 292),[2] although NORC and the DRC would later tackle other forms of crisis, such as earthquakes, floods, and tornadoes, along with aircraft accidents, fireworks explosions, and hospital fires. In spite of a later shift towards concerns affecting the general population, the discourse on disasters was not so easily separated from how it would affect the state security apparatus. Sociologist of collective behaviour Lewis Killian noted at the time that data from such research "were intended almost exclusively for use by the Army" (cited in Quarantelli, 1987, p. 292). In this regard, the DRC in the first decade or so of disaster studies had to contend with funding agencies concerned with "wartime or military organization, [with] extrapolations that could be made from peacetime or civilian groups" (1987, p. 297).

Much has changed since the Cold War period. With this in mind, we can begin to appreciate the many social scientific disciplines that have attended to social crises. Indeed, sociology, psychology, and social psychology all offered important narratives on the consequences of systemic crisis, although these oriented themselves towards applied research so as to attend to emergencies. Many of these studies thus focused on planning rather than management, a trend that continued to the period of Quarantelli's writing in 1987. In the last 40 years, sociological concerns and the discipline's greater autonomy in maintaining a more civilian perspective became the foundational approach towards disaster studies, with "primarily positive functional consequences" for the development of the field (Quarantelli, 1994, p. 26). In addition to its collective behavioural approach, the area of disaster studies also frequently employed legal/public policy perspectives that harnessed symbolic interactionism to provide cues for their methodologies. Early researchers on disasters such as Charles Fritz's examination of disasters on mental health (1996) adopted Weberian ideal types to address crises. Yet it would be Quarantelli's prescience on examining systemic crisis that is embodied by our pandemic volume and its international and local contributors:

> A fundamental reconceptualization of the phenomena being studied is necessary if a qualitative improvement in the research and its application is to happen. Furthermore, our view is that the impetus is more likely to come out of a consideration of basic questions reinforced by the progressive *internationalization* of disaster studies, than from practical concerns. (1994, p. 26; emphasis added)

Quarantelli's concerns were made to address how funding by federal agencies still gravitated towards civilian studies of crises that could be extrapolated to wartime situations. Quarantelli explains that it would again be sociologists who were instrumental in harnessing their sociological imaginations to respond to civilian crises, a process that draws on an older tradition in sociology where the "great majority of other researchers who did disaster studies in the middle and late 1950s were … sociologists" focusing on collective behaviour (Quarantelli, 1994, p. 27).

In an observation that predates Nassim Nicholas Taleb's anti-fragility thesis by decades (2010, 2012), sociologists interested in systemic crises held the rather optimistic view that "crisis situations represented occasions allowing for the emergence and creation of new behaviors rather than the breakdown of social order or the presence of unfavorably or deviantly viewed social behavior (and they continue to do so today …)" (Quarantelli, 1994, p. 28). Nonetheless, it should be emphasized that sociologists in this field are indebted to how disaster studies evolved from a variety of other disciplinary perspectives. For example, geography's foci on natural hazards and risk studies that attended to accident research contributed greatly to sociological readings of disasters, with the NSF in the 1960s fielding a research grant to study matters such as "coastal erosion, frost and high wind, snow hazards, and water quality" (Quarantelli 1987, p. 303; see also Burton et al., 1978; Mitchell, 1990),[3] further informed by pioneering research from Canada (Tyhurst, 1950), France (Chandessais, 1966), and Japan (Okabe & Hirose, 1985). In the case of the United States, Quarantelli celebrates how, along with sociologists, "geographers … anthropologists … political scientists" and "public administration specialists as well as risk analysts … have tended to form their own subprofessional formal disaster groupings or associations" (1994, p. 29).

Given our discussion, it becomes clear that a variety of access points can be employed to enter a discussion of different disasters or systemic crises. In Stallings's approach to studying earthquake crises, he makes the distinction between actual earthquake disasters and risks of earthquake disasters, and how these are conceptualized through time (1997). Brenda Z. Guiberson's condensed history of natural and human-generated crises across the centuries can serve as another example (2010). Guiberson's body of knowledge, derived from examining disasters historically (more specifically the period between the smallpox pandemic that affected Indigenous peoples of North America in the seventeenth century and the 2005 Hurricane Katrina destruction of New Orleans in the state of Louisiana), can be seen in contrast to an epistemology that can emerge based on speculating about the risk (or threat) of disasters which have yet to occur. In this regard, Guiberson echoes Stallings's concern about the risks associated with *future* crises based on technological failures, especially through the interrelationship between nature and technology. Although peer-reviewed articles offer hope of generating a more immediate snapshot of systemic crises, the social sciences still tend to concern themselves with justifying certain disciplinary approaches towards crises or defining the purpose and mission of disaster studies. One can see this attribute in Baum and Fleming's emphasis on continuing a psychological approach towards disaster studies, especially in how it can attend to crises caused by technological systems – in their case, systems that in their malfunction have created disasters "involving toxic substances" – which were designed to "tame the hostile world of our ancestors" yet have now "turned" against us and "bitten the hand that runs them" (1993, p. 665),[4] and in Block, Silber, and Perry's 1956 examination of the psychiatry of how children coped in the wake of tornados that wreaked havoc in the southern United States of America.

Furthermore, only a few writers "land" their theories on the material outcomes of crises through the scholarship of authors who have assessed their consequences in close temporal proximity. Those who do tend to extrapolate from the events of one nation-state and/or one territorial unit. For example, Comfort et al. (2011) examined disaster public policy in terms of its efficacy via local, regional, national, and international organizations that attended to the massive earthquake that struck Haiti on 12 January 2010. Harry Estill Moore's seminal sociological study *Tornadoes over Texas* (1958) examined two cities in Texas, Waco and San Angelo, affected by the 1953 "Waco Tornado," still the most destructive tornado recorded in Texas since 1900. It remains a mandatory read if only to ascertain how "nearly every type of data and data analysis currently found in disaster studies" was expertly employed across "its 14 empirical chapters" (cited in Stallings 2002, p. 285).[5] Recent offerings continue such a trend, as in Adam Higginbotham's *Midnight in Chernobyl* (2020) on Russia's nuclear accident, Robert Kerbeck's *Malibu Burning* documenting the devastating wildfire that struck Los Angeles County in Malibu, California (2019), and *Fukushima: The Story of a Nuclear Disaster* on Japan's 2011 Tōhoku quake and the subsequent tsunami, by David Lochbaum et al. (2014).

Michael Lindell (2013) offers excellent discernments on the discourse of disaster studies, highlighting how research agendas include definitional discussions of disasters, the significance of understanding casualties, their impact upon individuals' mental health, and how such disruptions severely impact the essential functions of society. Lindell and Prater (2003) also note how the disaster studies discourse can be further enhanced by research on emergency management and preparedness as well as on disaster recovery efforts. Their formulation of the Disaster Impact Model highlights how "the physical impacts" of disasters "in turn, cause the disaster's *social impacts*," all of which "can be reduced by *community recovery resources* and *extra-community assistance*" (2003, p. 176). Such a model encapsulates many of our contributors' foci (see Figure 0.1).

8  Introduction

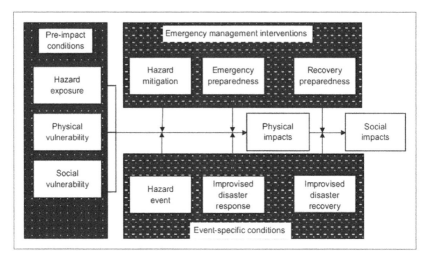

Figure 0.1. The Disaster Impact Model; adapted from Lindell and Prater (2003), in Lindell (2013, p. 799).

Others such as Brenda Phillips (1997) approach disaster studies by advocating for continued use of qualitative methods. She celebrates arguably the greatest, if not earliest, work on disasters, Samuel Henry Prince's *Catastrophe and Social Change: Based upon a Sociological Study of the Halifax Disaster* (2011). First published in 1920, Prince's book adopted a qualitative approach to ascertain the massive 1917 maritime explosion in Halifax, Canada, which he described as "one of the greatest catastrophes in history" (2011, p. 7). Indeed, our volume is driven by sentiments seen in and inspired by the architecture of Prince's monograph, shaped by "the shock and disintegration as the writer first observed it," followed by documenting many instances of what Prince made visible: "individual and group reactions ... examined in the light of sociological theory" (1920, p. 8).[6] Phillips argues that continuing disaster researchers need to celebrate this legacy of qualitative research, further enhanced by the continued use of non-obtrusive measures and longitudinal research. Phillips even provides a "wish list" of research that can uphold the qualitative tradition, a well-intentioned typology that would, however, draw the ire of disaster researchers such as J.C. Gaillard, who insisted that such researchers be mindful of their own Western technocratic approach towards responding to systemic crises as well as the Orientalisms that include exploitation of subaltern intellectual "workers" to assist in knowledge production defined by a Western gaze. Here, Sarah Turner's 2010 study of "silenced assistants" or "ghost workers" about rarely recognized Vietnamese and Chinese research assistants associated with Western scholars gives us reflexive pause.

Harnessing Fanon (2005), Freire (1970), and Said (1978) to inform his critical approach, Gaillard's political critique of the epistemology needed to understand crisis is, in my view, untenable in acute crises where social change needs to be rearticulated to mitigate the damages done to the social, especially if material consequences result in absolute deprivations. In this regard, our other aforementioned scholars have expedited their responses to such acute situations, a process that cannot in its entirety be placed in the matrix of Gaillard's prescriptions. Nonetheless, Gaillard's and Turner's elaborations have important relevance for ongoing disaster studies, as they remind thinkers to be fully reflexive of their positionalities as they serve as epistemic authorities in the crisis scenario under scrutiny. That said, Gaillard's critique and Phillips's "wish list" for a more rigorous qualitative methods are not an approach that is exclusively employed in our work, because 1) many of our contributors could not gather extensive field data because of pandemic regulations that mandated indoor confinement, one that compelled us to carefully manage our exposure to the public – that is, no fieldwork or engagement with research assistants was possible for many of our contributors; and 2) without primary *field* data, many of our scholars have relied on a grounded theory approach to assemble a variety of data derived from public policy communiqués, government mandates, and news reports, a process that renders our work as primarily inductive, descriptive, theoretical, yet explanatory through cautious extrapolations. Certainty is relegated to less importance while we analyse and highlight what is probable under the conditions of what can be deemed a worldwide emergency.

Given the circumstances that shape our epistemological practices, did our scholars trained in upholding the Enlightenment ideals of reason and rationality debase a deeper approach towards understanding indigenous conceptions of stress? We are of the view that this did not happen, for at least in this volume, we feel that we have met the concerns of staunch critics of disaster studies and the visionaries of a more robust disaster methodology halfway: although trained in the process of data acquisition defined by a Western discourse, our authors are localized and have thus been compelled to report about their lifeworld environments. Moreover, many of our contributors, examining from within the USA how other countries have fared, are nonetheless intimately linked as members of diasporic groups to their demotic and the professional class residing in their respective geographical wombs. Our epistemologies, then, have been birthed from each contributor's social context, which, because of the exigencies involved, has allowed us to engage in a localized response while accommodating a symbiosis of data and practices that can be acquired from the global. Such an approach remains primarily consistent with the vulnerability paradigm, one that approaches crises as a social issue affecting culture, demographics, economics, and politics, rather than envisioning them through the hazard paradigm that views crises as able to "overwhelm people and societies," a view that removes agency from persons who can remain stronger than circumstance, who are able to overcome (Hewitt, 1983).

A key issue – and we believe this is a glaring one – that affects our examination of community and urban dysfunctions during systemic crisis is that theorists of social systems rarely illuminate systemic malfunction in a manner that makes visible the existential conditions of those actors they examine. Many write with the assumption that social institutions are primarily invulnerable when spatially embedded in society. One sees this tendency in the otherwise excellent edited volumes *Urban Ills: Twenty-First-Century Complexities of Urban Living in Global Contexts* by Carol Camp Yeakey, Vetta L. Sanders Thompson, and Anjanette Wells (2014a, 2014b). By examining a variety of social, economic, and political outcomes that detract from community, culture, and social life, Yeakey et al. do effectively lay bare the complexities and diversity of community and urban issues. However, although the chapters adopt a global perspective, social conditions and a variety of urban publics are still assumed to be amendable; that is, the conditions within which their contributors conduct research are contextualized within community and state apparatuses that have not been rendered inoperative by a global systemic crisis, as was the case with our contributors during the pandemic. Although poverty, low wages, political malfeasance, racism, substandard health care, and resource-deprived schools remain crucially important topics for social analyses of community, Yeakey et al. see these urban ills within social systems that have not been acutely transformed in a short period of time. The volumes did not yield an epistemology from researchers who were experiencing what the global population was experiencing at the time of their writing, one that could simultaneously offer a localized understanding of what is seen as an existential threat to the planet's population.

The voices of our contributors may, on initial glance, appear cacophonous, akin to Friedrich Nietzsche's fictitious town of Motley Cow in *Thus Spoke Zarathustra*, an environment where divergent and noisy viewpoints define a community's communicative temperament. Nonetheless, committed analyses from our diverse panoply of scholars reveal how, in spite of the many moving parts in our volume, there remains a centre of gravity that coheres our ideas: to explain through our respective disciplines how our *social* pandemic experiences have been affected by changing systemic conditions under a variety of stressors that include the deleterious process Thornburg, Knottnerus, and Webb describe as deritualization, the systemic disruption, if not abandonment, during disasters of social scripts that conventionally ensured harmony in quotidian life (2007). Moreover, by highlighting indigenous conceptions of stress, there is indeed adhesion and coherence between our contributors as they attend to their respective lifeworld disruptions in ways that nonetheless preserve their disciplinary language and autonomy of analysis. Thus, what might be interpreted as a somewhat frenetic pace of our work is by design, given that all our contributors began their writings in urgent contexts of community death and renewal at a time when the world had no vaccines to respond to COVID-19. We now segue to pertinent summaries of our authors' chapters, grouped by regions around the world.

**The Far East**

*South Korea*

Kelly Huh and Hyejin Yoon in chapter 1 document the interplay of South Korea's state apparatus, its corporate sector, and customary practices by those in civil society as effective mechanisms in reducing the initial onslaught of COVID-19. In Huh and Yoon's chapter, titled "Reconsidering the Third Place: Social Distancing and Inequality in South Korea during the Era of Coronavirus," our authors document through the lens of geography how the pandemic prompted a government-led corporatocratic ethos that effectively mobilized all spaces of the Korean lifeworld. The process, however, generated tensions between the country's social institutions that had to balance governmental mandates for pandemic mitigation and corporate imperatives that desired to maintain economic production and mass employment. Huh and Yoon, however, note that the existential threat of the pandemic forced these divergent forces back together in détente, in a manner where state, public, and corporate enterprises found coherence once more through technological approaches towards crisis management. Our authors argue that this was achieved when South Korea's civil society and state harnessed the power of social media for health practices: by tapping into resources offered by an online culture, amalgamating it with state mandates, and combining these with cultural practices such as the donning of face masks seen in the many third places, that is, places of interactive recreation, of South Korea. As such, Huh and Yoon note that a high degree of public compliance was established for the nation. By revealing the contours of how the South Korean state has responded to the crisis, Huh and Yoon conclude that the main impetus driving the policies and practices of the government during the pandemic was inspired by a variety of historical issues, mainly South Korea's relationship with North Korea, as well as current demographic factors affecting the country's population. Drawing from the country's experiences with wartime urgencies, Huh and Yoon argue that the South Korean population's compliance with government mandates reflects their respect for state approaches towards crisis management.

*Japan*

Hiroshi Fukurai's chapter 2, titled "The Anthropocene, Zoonotic Diseases, and the State – Japan's Response to the COVID-19 Pandemic: Criminal Negligence or Crimes against Humanity?" offers a critical examination of the country's responses to the pandemic. Fukurai outlines how the Japanese government first declared a nationwide state of emergency on 16 April 2020, more than three months after the initial discovery of the novel strain of coronavirus in Wuhan, China in December 2019, as well as the World Health Organization (WHO)'s announcement of the danger of the zoonotic virus outbreak in early January 2020. Fukurai assesses different countries' pandemic mitigation practices, all discussed in a manner that offers cues for further understanding Japan's public policies towards thwarting the spread of COVID-19. Fukurai's chapter critiques Japan's delayed response and cites how the initial tepid reaction by its government was due to its desire to move forward with the 2020 Tokyo Summer Olympic Games. Japan's admission of the pandemic would have required the postponement, if not total cancellation, of the highly lucrative event. Indeed, since 2012, Prime Minister Shinzo Abe had been promoting the 2020 Summer Olympic Games as a multibillion-dollar economic stimulus which would help with recovery from multiple economic setbacks and crises during Japan's lost decades of the Heisei era. In hopes of critically responding to the aforementioned issues, Fukurai's chapter discusses successful approaches of neighbouring countries in responding to the pandemic, as well as exploring the international humanitarian contributions of Cuba, who sent doctors to Vietnam, China, and other countries in Asia and beyond. His chapter also highlights the significance of the collaboration between China and Cuba that resulted in wider distribution of Cuba's anti-viral drugs to multiple developing countries in Asia, Africa, Latin America, and the Caribbean. In this regard, Fukurai moves towards other horizons to tease out indigenous conceptions of stress that prioritize the population rather than economic imperatives, employing these to critically assess the Abe Administration's failure to protect the public from the deadly virus. With Taiwan's, Venezuela's, and New Zealand's pandemic mitigation efforts as additional examples, Fukurai reveals the global expanse of pandemic mitigation strategies in ways that offer insights into how Japan could have improved its pandemic response. Fukurai's chapter thus highlights rich transnational collaborations between countries attending to pandemic mitigation rather than highly politicized approaches that exacerbate international division in mitigation efforts.

Image 0.1. Japan's Prime Minister Shinzo Abe joins the first Novel Coronavirus Expert Meeting on 16 February 2020 (Cabinet Secretariat, Cabinet Public Relations Office, 16 February 2020).

## South and Southeast Asia

*India and Singapore*

In chapter 3, Amritorupa Sen and Junbin Tan examine how the migrant underclasses, non-permanent residents of their host communities who live in congested urban environments, were among the most affected by the COVID-19 outbreak. Their chapter, "Witnessing amidst Distancing: Structural Vulnerabilities and the Researcher's Gaze in Pandemic Times in Relation to Migrant Workers of India and Singapore," documents the challenges of accessing migrant workers' underclass experiences and the structural conditions that frame their difficulties in India and Singapore. In Singapore, migrant workers were disproportionate victims of COVID-19. That the spread of COVID-19 among migrant workers clusters had not affected wider Singaporean society attests to a racial/class/occupational divide that is further compounded by state efforts to segregate all migrant workers from the wider population after infections emerged in the group. As such, Sen and Tan note how migrant workers have become a category that is derived from and in turn induces segregational

Image 0.2. Migrant workers walk along a road to return to their villages, during a 21-day nationwide lockdown to limit the spreading of COVID-19, in New Delhi, India, 26 March 2020 (Reuters/Danish Siddiqui).

effects; these effects made ethnographic research difficult if not impossible. However, compared to migrant workers in Singapore, India's migrant workers are neither clearly segregated nor easily identifiable. These attributes are further compounded by their work in the informal sector that is unregulated by government. Increase in joblessness and lack of pay had forced thousands to ply the streets as the homeless while sinking deeper into poverty. The pandemic prompted many Indian migrant workers to plan their return home, while the most unfortunate found it impossible to leave, remaining trapped in their host cities. Within this context, from April to June 2020, Sen and Tan followed the flow of events in different media channels, reflecting on their own privileged positions in relation to the migrant underclass, as well as confronting questions of gazing, of complicity, of guilt, of responsibility, and of their incapacity to act while observing social distancing. Our authors thus interrogate the meaning of observing from a distance, a gaze in a mediated form that protects observers from the dangers to which they are exposed. They are inspired by Judith Butler's (2012) inquiry into questions of ethical obligation to suffering from a distance, and "what it means for our ethical obligations when we are up against another person or group, find ourselves invariably joined to those we never chose?" Employing Jacques Lacan's (1979) concept of the "gaze" – that a researcher's analysis is situated within, disciplined by, and proceeds from a privileged position due to broader structural relations between ourselves and our subjects/interlocutors – Sen and Tan respond to Butler's question affirmatively: that we have an instinctual capacity, when startled and confronted by the discomforting glare of migrant adversities, "to respond ethically to suffering," allowing ourselves to be "invariably joined to those we never chose."

## Thailand

In chapter 4, titled "Social Distancing? 'No Problem!': Explaining Thailand's 'Successful' Containment of COVID-19," Piya Pangsapa notes how on 13 January 2020, Thailand became the first country outside China to report a coronavirus infection (in a 61-year-old tourist from Wuhan city). By mid-March, there were over 400 confirmed cases in the country, prompting the prime minister to declare a state of emergency on 26 March 2020. In light of such conditions, Thailand imposed a nationwide curfew on 3 April, followed by a nationwide ban on public gatherings, a nationwide ban on the sale of alcohol, a ban on all inbound international flights, and the implementation of lockdown measures across the country. Yet, many community outreach activities surfaced during this period to inspire the Thai population towards compliance for the greater national good. Such efforts meant that by 30 July 2020, Thailand was ranked first out of 184 countries in a global survey for the country's effective handling of COVID-19 and recovery from the crisis, its index score of 82.06 placing it on top of the global rankings as an example of best practices in tackling the virus (Global COVID-19 Index (GCI), 2020).[7] Pangsapa explains how Thailand's "successful" containment of COVID-19 may be attributed to a combination of factors (it was the only country in ASEAN to adopt Universal Precautions in its fight against the virus) but argues that the country's "success" in this respect would not have been possible without full public cooperation. Her chapter thus examines the Thai government's response to the COVID-19 pandemic as well as the Thai public's response to the state's pandemic mitigation policies. Pangsapa argues that Thailand's mitigation practices are noteworthy in that the country shut down effectively, even though it utilized draconian measures (difficult to do in most democracies) that secured citizen compliance in ways where they were willing, ready, and able to acquiesce with the measures. Pangsapa explains how the Thai state's proactive approach in handling the coronavirus may also explain why social distancing was taken seriously from the inception of the pandemic, and thus perhaps Thailand's "success" can be attributed to Thai-ness, that is, the character dispositions and personality of the Thai people, as well as Thai culture itself. For

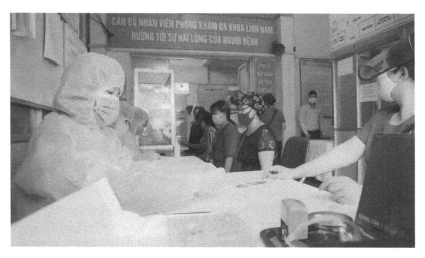

Image 0.3. COVID-19 testing registration desk, Hanoi (Truyền Hình Pháp Luật, 18 April 2020).

Pangsapa, Thailand's success in thwarting the early variant of COVID-19 supports Crispin's contention in the Asia Times (2020) that Thailand's "so far mild COVID-19 experience likely owes to a uniquely Thai mix of factors."

### Vietnam

In chapter 5, titled "Trust in Numbers? The Politics of Zero Deaths and Vietnam's Response to COVID-19," Amy Dao examines the atmosphere of doubt and surprise in news media reporting on Vietnam's response to COVID-19, with particular focus on the country's death count – or lack thereof – during the early months of the pandemic. Vietnam's swift response to the COVID-19 contagion garnered international acclaim from journalists across the globe as well as from the WHO. Indeed, in the month of April 2020, when many societies around the world were experiencing the COVID-19 onslaught, Vietnam reported only 270 confirmed cases, a most remarkable achievement despite its limited healthcare infrastructure, sizable population of 97 million people, and densified urban environments. Dao thus examines how doubt about Vietnam's positive indicators in early news reports raises fundamental questions about how power and privilege operate in global development; specifically, how the country's data is validated by Occidental forces that attempt to colonize Vietnam's narratives on pandemic mitigation. Invoking Edward Said's critique of colonial renderings of other nations and cultures in *Orientalism* (1978) to counter such practices, Dao's chapter assesses which countries have the ability to claim "indelible truth" through the sanctity of data. In the undertaking, Dao successfully decolonizes non-Vietnamese narratives so as to accord Vietnam its rightful place in its approach towards embracing and reporting on indigenous conceptions of health stressors. Through incisive examples of Western incredulity about Vietnam's low death count and its efficacious approach towards mitigating the effects of the pandemic, a journalistic audit of Vietnam's death count, and the controversy between a US-based economist and other scientists involved in Vietnam's reporting of COVID-19, Dao's chapter demonstrates how the "Western gaze of development" continues to exhibit its hubris towards Vietnam's successful containment of COVID-19 during the pre-vaccination period.

**A Global Address**

From Iraq, where a woman was beaten so badly by her husband that she fatally immolated herself (Human Rights Watch, 2020), to India where a certain Heena would be beaten by her husband to a "pulp" in front of their seven-year-old son only to be thrown out of their home at 3 a.m. (Daily Sabah, 2020), to Brazil, Tunisia, France, Nigeria, Australia, the United States, and the United Kingdom, to name but a few countries, the global surge in intimate partner violence amid the pandemic compelled the United Nations Secretary-General António Guterres to proclaim that "violence is not confined to the battlefield" (2020). Shweta Adur and Anjana Narayan's chapter 6, titled, "An 'Unseen Enemy' and the 'Shadow Pandemic': The COVID-19 Pandemic and Domestic Violence," adds to Guterres's concerns, namely that for many women and girls threats "loom largest where they should be safest – in their own homes" (2020). For Adur and Narayan, the global impact of the coronavirus pandemic, its accompanying lockdowns, and the stipulations for maintaining distancing to combat the crisis have laid bare once again two key aspects about gender-based intimate partner violence and domestic violence in general: first, that the private sphere/"home" can be a most violent and oppressive social environment; second, violence in such intimate confines can be reproduced by institutional, systemic, and structural contexts. Drawing upon contemporary scholarship on gender-based violence and recent reporting on the pandemic crisis, Adur and Narayan's chapter examines how state-mandated lockdowns around the world have inadvertently and disproportionately exacerbated violence against women in the home environment, enabling conditions and opportunities for perpetrators to evade scrutiny. With accounts and data that offer readers a panoramic gaze into the intersectionalities of global intimate partner violence and domestic violence in general, Adur and Narayan discuss measures that state apparatuses, multilateral organizations, and community organizations have taken to contend with the scourge of the pandemic and its link to a surge in violence against women. The global gaze of the chapter cites scholars from India and elsewhere around the world, employs opening vignettes, historicizes the 1918 Influenza Pandemic,[8] offers data on Brazil and India as well as narratives from the United Nations' António Guterres, and conveys additional considerations illuminating conditions on the ground (e.g., Canada and the states of Alabama and Texas in the USA). The chapter congregates these themes to frame how women's rights can articulate their global concerns and discontents in pandemic conditions affecting intimate partner violence.

## United States of America

### United States of America – National

When the Black Death began decimating Europe, wealthy political and military elites escaped to the countryside to avoid the dense urban environments that seemed hardest hit by the disease. Peasants, labourers, and other commoners, so dependent on markets in the city, disproportionately stayed behind, and disproportionately died as a result. Kevin McCaffree and Anondah Saide in chapter 7 discuss the degree to which such patterns can be seen in the United States' present-day pandemic responses. In "An Investigation into the Economic, Social, and Psychological Dimensions of COVID-19," McCaffree and Saide more specifically interrogate the degree to which psychological, social, and economic measures – aspects of people's well-being – further mediated by news media, politicize people's understanding of and willingness to comply with pandemic mitigation practices. By employing the most recent COVID-19 pandemic as a case study, our authors consider why resistances to compliance might be easier or more desirable for those who are less dependent on their immediate networks for income or psychological support. McCaffree and Saide's data suggest, among a set of interesting findings, that resisting pandemic mitigation policies such as that based on quarantines is linked to how socially connected and financially secure the group is, a trend seen in Republicans. Democrats and those with healthier psychological well-being prioritized compliance in pandemic mitigation policies. Additionally, McCaffree and Saide document how, when Republicans become more similar to Democrats in their COVID-related attitudes, the less financially or socially secure they feel. Similarly, their data suggest that partisans across parties express concerns about COVID-19 in a manner greatly inflected and mediated by news media content. Our authors identify a blind spot in the "tribal partisanship" model, which needs to account for within-group status hierarchy effects where Republicans are defining themselves in opposition not only to (perceived) Democrat positions but also to those of their political group who have sympathies towards Democrat-inspired pandemic mitigation policies. McCaffree and Saide remind readers that, in spite of a systemic crisis that is existentially threatening, a continuing politicization of public health accompanies such apprehensions, one that requires further analyses.

*United States of America – New York*

Hospitals have become overcrowded and understaffed in New York City, the American city worst affected by the first wave of COVID-19 infections. In chapter 8, "Unsettling Contact: The Collapse of Emotional Distance at a COVID-19 Medical Frontline," anthropologist Junbin Tan and anesthesiologist Phu Tran examine New York City medical frontline workers' encounters with the pandemic and severely ill patients, largely through Tran's lived experiences in the hospital setting when faced with health risks, burgeoning patient numbers, and resource constraints. Tran notes how, as New York hospitals faced equipment shortages and infrastructural constraints during the early phase of the pandemic, and anesthesiologists raced against time to intubate critically ill patients, it became more common for anesthesiologists and respiratory therapists to see patients die before their eyes. The rule of emotional distancing, which healthcare practitioners are trained to adopt, threatened to crumble when the respiratory team was besieged by an overwhelming number of deaths, the risk of catching an infection themselves, and sorrow, remorse, and guilt that they – because of inadequate infrastructure and failed equipment – were somehow complicit in patients' deaths. Tan and Tran argue that in such a process, deep-set boundaries between "us" and "them" became unsettled as they interacted closely with patients, and as they confronted the mortality of patients who, if not for their sudden illness and complications, would have led normal lives. The chapter centres on Tran's accounts of the medical frontline and his thoughts on his respiratory therapist co-worker Macy's emotional breakdown in the face of a patient, "Ms. A," and her death. By analysing his co-author's accounts, Tan unpacks what unsettling encounters at the COVID-19 frontline indicate about changing social relations involving emotional intimacy and distance, and engagement and vulnerabilities among medical workers, as well as between medical workers and their patients.

*United States of America – California*

The geographic association between social distancing and crime in the context of California's pandemic is carefully elaborated in Gabriele Plickert and Emily Cooper's chapter 9, titled "A Spatial Snapshot of the Relationship between the

20  Introduction

Image 0.4. Refrigerated semi-truck trailers hold the dead at Bellevue Hospital Center in Manhattan, New York, on 4 April 2020 (Alamy/Steven Greaves).

COVID-19 Pandemic and Selected Crimes in California." The chapter attempts to answer a difficult question: To what extent does social distancing have a positive, negative, or no effect on the motivations and patterns of crime? In the process, they examine how the unprecedented spatial constraints under California's social distancing mandates, paired with the experience of affective and economic distress in the state, have critically altered the mapping of crime patterns. For example, although social distancing might reduce the occurrence of property crimes in residential areas because these are now populated with more vigilant community members staying at home in compliance with the state, areas of initially low risk such as non-essential businesses have potentially become places for motivating criminal behaviour. By adopting a structural model that is made to be acutely sensitive to the spatialities of social distancing, Plickert and Cooper further enhance their analysis by relating the aforementioned dynamics to geographic boundaries. Such an undertaking highlights the most probable area where offenders can be linked to potential crime targets. While crimes of domestic violence have already risen since the pandemic's arrival in California, Plickert and Cooper's chapter empirically attends to crime motivation and crime pattern effects of property crime (i.e., burglary, robbery, vandalism, and motor vehicle theft), in hopes of analysing changes in criminality that will have great implications for mapping crime beyond the current pandemic.

*United States of America – Los Angeles*

Chapter 10, my chapter titled "Employing Lyn Lofland's and Ray Oldenburg's Urban Sociology to 'Read' the Emptying of Los Angeles' Publics," provides an urban analysis of the city during the early months of the pandemic. I forward the view that, during the period of physical distancing in the city, revitalized ideas by urban sociologists Lyn Lofland (1973, 1998) and Ray Oldenburg (1999) can offer a reading of the dynamics of the public realm and its ability to influence urban residents' interactions with one another. The chapter thus makes operative Oldenburg's concept of "third places," those areas of society designed to enable a controlled release of community tensions through recreative interaction. Drawing from daily news data about Los Angeles, as well as documenting public officials' daily pronouncements during the pandemic, my chapter reveals how the city's third places collapsed, temporarily transforming Los Angeles into, essentially, a warehouse, or what urban sociologists and geographers often refer to as a "break-of-bulk" point. The chapter attempts to draw some rudimentary conclusions about how deurbanization forces appear as the city ceased third place operations during the pandemic. In the undertaking, the chapter also uncovers the complexity of the city's social angst and how its germaphobia is inflected with xenophobia in a manner that reflects the underlying racial and ethnic discontents between identity groups that underlie the American experience. It highlights the limits of technocratic management and favours a greater need for urban centres to express multicultural coexistence, a process that was amorphous during the initial phase of pandemic mitigation efforts but later exhibited traction in reaction to the racialized pandemic-driven communal violence that surfaced. The chapter concludes by noting how social conditions born from urban environments bereft of healthy publics are, nonetheless, epistemological because they reveal how crisis reconfigures social relations under duress in that most idealized environment of modernity, the city.

**The European Union**

*Sweden*

In chapter 11, Ann-Christine Petersson Hjelm highlights Swedish public policy that affected its elderly population during the country's experience with the onset of the pandemic. In "Trust between Citizens and State as a Strategy to

Image 0.5. Swedish elderly in Linköping, Sweden, celebrate after receiving their vaccines at their retirement home (Jeppe Gustafsson, 27 December 2020).

Battle the Pandemic: Were Senior Citizens Collateral Damage in the Swedish Government's Plan to Flatten the Curve?" Petersson Hjelm advocates for the pandemic's most vulnerable population, Sweden's elderly, who reside in the home care service sector and retirement homes. Petersson Hjelm highlights how, during the spring of 2020, Sweden's Public Health Agency, an institution normally tasked with ensuring that the population is protected against communicable diseases, failed to protect its elderly, evincing how its system for elderly care is facing a multitude of challenges with regard to a vast, aging population. Petersson Hjelm's chapter draws attention to how the Swedish Constitution was restrained at the outset of the pandemic, and that historically, in peacetime or in crisis situations, Sweden has had no harsh social regulations designed specifically for a "state of emergency." Yet Petersson Hjelm reports how, by 1 April 2020, the government decided to prohibit visits to all of the country's retirement homes to prevent the spread of COVID-19. In such contexts, Petersson Hjelm's chapter underscores the tradeoff between individual freedoms and societal imperatives when health institutions scrambled to attend to the pandemic crisis as it affected the Swedish elderly. When seen in the context of practice during the pandemic, Petersson Hjelm's chapter underscores the need for greater coordination and interaction between personnel working in Sweden's social services and the country's elderly population. By employing a socio-legal approach that analyses how different municipalities operate to handle the pandemic in the care of Sweden's senior citizens, Petersson Hjelm also highlights how the country's socio-legal institutions, especially its legal system and legal culture, responded to the welfare of the elderly both at the societal and individual levels.

*Germany*

Albert Scherr argues in chapter 12, "The German Reaction to Corona: The Interplay of Care, Control, and Personal Responsibility within the Welfare State," how nation-states of the North, in spite of being subsumed under globalization dynamics, still exhibit tremendous autonomy in enforcement of public policy. Scherr's chapter highlights the continuing high importance of nation-states in a globalized world society, and shows why Germany was relatively successful in overcoming the pandemic crisis in its early phase through state measures. Scherr's chapter makes visible how the pandemic crisis during spring 2020 highlighted the importance of Germany's public healthcare system and its mandate of providing compulsory health insurance for all, a process seen as an effective form of rational crisis management that upholds the sanctity of Germany's welfare state. Additionally, he highlights how the state's mandate that individual rights are not impinged upon as stipulated in the German constitution led to the government's renunciation of strict curfews and surveillance measures at the time. Scherr illuminates how, like the state, a large part of the population deferred to the need to establish balance between restrictions on freedom and individual rights. Scherr thus argues that Germany's handling of the pandemic during the pre-vaccination phase reveals the effectiveness of the established arrangement of the welfare state, democracy, and the implementation of individual civil liberties. He considers how Germany's response to its citizens can be seen as an articulation of Thomas Pogge's notion of "common nationalism," a benevolent expression of the state towards its citizens that, as will be emphatically conveyed in his chapter, renounces an ideological orientation that establishes a superiority of one's own nation in relation to others. As a result, Scherr notes how the social contract as it unfolded upon German citizens during the pandemic operated with comparatively less friction than other state-mandated measures seen elsewhere around the world. Moreover, Scherr documents how the use of public spaces and national borders during the crisis occurred with considerable compliance to restrictions. Scherr's observations of everyday life in Germany at the time of his writing reveal how social compliance with pandemic restrictions was ensured not so much by sanctions and controls as by the personal responsibility of citizens who deferred to the welfare state's policies and practices, all shaped by a scientific discourse that was not seen as subversive by the majority of its citizens.

Image 0.6. Masked passengers sit inside a subway light rail tram in Düsseldorf, Germany (Peeradon Warithkorasuth, 7 August 2020).

## South Pacific

*New Zealand*

Maria Armoudian and Bernard Duncan's chapter 13, "The Benefits and Drawbacks of Social Distancing: Lessons from New Zealand," examines the seeming success of the country's efforts in containing the pandemic at the outset. With a population of 4.8 million people spread out across its two main islands located near the "bottom of the world" in the South Pacific – with no other country within 4,000 kilometres (2,500 miles) – it also has low population density, ranking at 202 among 235 ranked countries. Armoudian and Duncan illuminate in their chapter how these two features factored into New Zealand's success in reducing the impact of the coronavirus in a relatively short period of time. Through their sociopolitical account, Armoudian and Duncan note how access to a global geography of pandemic information gave the island nation time to watch the virus ravage many parts of the world, while acquiring insights and wisdoms by assessing which strategies seen around the world would work best for New Zealand. Armoudian and Duncan's chapter also illuminate how pandemic mitigation policies and outcomes were shaped by political institutions and cultural actors from many sections of New Zealand's lifeworld. The authors' journalistic approach of systematizing sociopolitical issues and events in a manageable chronology highlights how the pandemic functioned as a catalyst in shaping New Zealand's October 2020 election, one that reelected Jacinda Ardern as prime minister in a landslide. By employing New Zealand's media coverage of the pandemic and its political articulations, our authors make visible how different New Zealand actors – in support or contestation of pandemic distancing policies – have been able to influence and transform the country into a model for pandemic mitigation in spite of its challenging social distancing mandates. Moreover, by refraining from an empirical use of theory, Armoudian and Duncan's chapter is able to unfurl an energized journalistic reporting of New Zealand's sociopolitical dynamics in ways that allow it to segue to our volume's conclusion, one that similarly highlights continuing frenetic energies of pandemic discontents, transposed to examine the globe.

Without further delay, we thus begin our rendering of the human and social condition under pandemic distancing as seen across many parts of the world. Through the voices of our contributors, who, in solidarity with one another, soldiered onward under duress with fellow citizens in spite of social and physical distancing mandates, we hope to make visible the struggles of actors, citizens, public authorities, the state, and ourselves, as we attempt to reconceptualize and revitalize what it means to be social thinkers at a time when our communities and publics were prevented from operating effectively.

NOTES

1 As we are dealing with social changes to our communities in the current pandemic, I hope to employ terms such as "systems in crisis" or "systemic crises" where appropriate to distinguish our work from previous scholarship on communities experiencing acute emergencies. The rationale is simple: systems shape social lives across the present and across time, and many of the world's social systems pummelled by COVID-19 have experienced operational issues. My gesture is meant to exhibit no dismissiveness towards the historically impressive efforts made by those conducting research at the Disaster Research Center or scholars that choose to employ the term "disaster" or "catastrophe."
2 The Disaster Research Center (DRC), founded in 1963 by sociologists Drs. E.L. Quarantelli, Russell Dynes, and J. Eugene Haas, has expanded its panorama to examine all forms of disasters, and conducts research projects on "group, organizational, and community preparation for, responses to, and recovery from natural and technological disasters and other community-wide crises" (Disaster Research Center website at drc.udel.edu). At the time of this writing, the DRC is actively engaged in addressing COVID-19, especially its impact on the state of Delaware. Its dynamic approach towards studying disasters has seen the centre funded by the National

Science Foundation, the US Department of Homeland Security, the US Department of Health and Human Services, the US Department of Defense, the Center for Disease Control and Prevention, and the Federal Emergency Management Agency.

3 Even as far back as the mid-1990s, Quarantelli noted how a geographical approach to natural hazards "would be definitively worthwhile" (1994, p. 31).

4 Sociologists R. Felix Geyer and David R. Schweitzer might be less enthusiastic about including our associates from psychology, a discipline envisioned as operating "on the methodological assumption that the aggregation of individual data … can lead to the discovery of social structural processes" (1976, p. xxi). What Geyer and Schweitzer discerned is not new but an important revisit of a precedent-setting 1903 meeting between Emile Durkheim and his intellectual detractor, Gabriel Tarde. Although Tarde was a fellow sociologist, he was operating from the paradigm that is the bane of Geyer and Schweitzer's approach, namely that society is an aggregate of individuals. For Durkheim, society – as a collective of institutions and groups – could not be reduced to the individual (Damle & Candea, 2008, p. 767). In 1903, the matter was brought to the fore in a debate between the two figures at Paris's École des Hautes Études Sociales. Durkheim proclaimed a sociology where societies could "be measured, their relative sizes compared," a process that would reveal "a world hitherto unknown, different from those explored by the other sciences. This world is nothing if not a system of realities" (Damle & Candea, 2008, p. 764). For philosopher Daniel Little, "Durkheim essentially won the field … and Tarde's reputation diminished for a century" (2010).

5 Not fully appreciated is Moore's chapter that methodically documented how almost every state of the US, as well as countries such as the Dominican Republic and India, donated to survivors amounts that ultimately totalled between US$1 million and US$1.2 million dollars, or over US$12.3 million dollars adjusted for inflation in 2021, all of which stemmed from primarily "voluntary and largely unsolicited contribution" at the time (1953, p. 155). The American Red Cross itself donated over US$351,923.35 (or over US$3.6 million adjusted for inflation) (1953, p. 155). Such activities would later be echoed around the world during the pandemic as food banks, more fortunate businesses, textile companies and their offering of masks, and good neighbours set into motion their efforts to promote a shared humanity in their reconfigured communities (intermittent politicized protests in some countries against government mandates notwithstanding).

6 Interestingly, Prince preferred the use of the term "catastrophe" over "disaster" (thus, the latter's mention in only the subtitle). Indeed, in the 10 lengthy chapters of the monograph, the word "catastrophe" appeared as the first word in nine chapters (e.g., "Catastrophe and Social Organization," "Catastrophe and Social Legislation," and "Catastrophe and Social Change," to name but a few).

7 The score comes from the Global COVID-19 Index that bases 70% of its calculation on big data and daily analysis from 184 countries, and 30% from the Global Health Security Index, an assessment of global health security in 195 countries prepared by the Johns Hopkins Centre for Health Security (*The Nation*, 30 March 2020). See their

methodology at https://covid19.pemandu.org/?fbclid=IwAR2k0KnoqzozyQFFh6NoiYCQDc9vvcn3PkOqQ03Curn7xkZKLbfBnJaY8vE

8 The Spanish Flu did not definitively emanate from Spain. Reporters covering the regions of Europe experiencing the conflagration of the First World War had few relatively safe countries to report from about the afflictions where this variant of the flu proliferated, except for Spain. Moreover, there is much evidence that the "Spanish" Flu was first documented in 1918, in Kansas, USA. Such an unfortunate association prompted science journalist and podcast producer Johanna Mayer to lament how the appellation became a misnomer that lasted for a century, even prompting some Spaniards at the time to defensively refer to the virus as "the French Flu" (2019, para. 10). Our work thus refers to it as the 1918 Influenza Pandemic.

REFERENCES FOR PREFACE AND INTRODUCTION

Andrews, T.M., & Paquette, D. (2020, July 29). Trump retweeted a video with false COVID-19 claims. One doctor in it has said demons cause illnesses. *Washington Post*. https://www.washingtonpost.com/technology/2020/07/28/stella-immanuel-hydroxychloroquine-video-trump-americas-frontline-doctors/.

Baum, A., & Fleming, I. (1993). Implications of psychological research on stress and technological accidents. *American Psychologist*, *48*(6), 665–72. https://doi.org/10.1037/0003-066X.48.6.665.

BBC. (2020, December 8). COVID-19 vaccine: First person receives Pfizer jab in UK. https://www.bbc.com/news/uk-55227325.

Block, D., Silber, E., & Perry, S. (1956). Some factors in the emotional reaction of children to disaster. *American Journal of Psychiatry*, *113*, 416–22. https://doi.org/10.1176/ajp.113.5.416.

Burton, I., Kates, R., & White, G. (1978). *The environment as hazard*. Oxford University Press.

Borges, J.L. (1998). *Tlön, Uqbar, Orbis Tertius*. In A. Hurley (Trans.), *Collected Fictions* (pp. 68–81). Penguin Books. (Original work published in 1940).

Butler, J. (2012). Precarious life, vulnerability, and the ethics of cohabitation. *Journal of Speculative Philosophy*, *26*(2), 134–51. https://doi.org/10.5325/jspecphil.26.2.0134.

Chandessais, C. (1966). *La catastrophe de Feyzin* [The Feyzin disaster]. Centre D'Études Psychosociologiques des Sinistres et de Leur Prevention.

Comfort, L.K., Siciliano, M., & Okada, A. (2011). Resilience, entropy, and efficiency in crisis management: The January 12, 2010, Haiti earthquake. *Risk Hazards & Crisis in Public Policy*, *2*(3, October). https://doi.org/10.2202/1944-4079.1089.

Crispin, S. (2020, May 26). Thailand's unsung COVID-19 success story. *Asia Times*. https://asiatimes.com/2020/05/thailands-unsung-covid-19-success-story/.

Daily Sabah. (2020, November 25). "I had nowhere to go": Pandemic inflames violence against women. https://www.dailysabah.com/life/i-had-nowhere-to-go-pandemic-inflames-violence-against-women/news.

Damle, A., & Candea, M. (2008). The debate between Tarde and Durkheim. *Environmental and Planning Development: Society and Space, 26,* 761–77. https://doi.org/10.1068/d2606td.

Dynes, R.R., & Drabek, T.E. (1994). The structure of disaster research: Its policy and disciplinary implications. *International Journal of Mass Emergencies and Disasters, 12,* 5–23. https://doi.org/10.1177/028072709401200101.

Fanon, F. (2005). *The wretched of the earth.* Grove Press.

Freire, P. (1970). *Pedagogy of the oppressed.* Herder and Herder.

Fritz, C.E. (1961). Disaster. In Merton, R.K., & Nisbet, R.A. (Eds.), *Contemporary social problems: An introduction to the sociology of deviant behavior and social disorganization* (pp. 651–94). Harcourt, Brace & World.

Fritz, C.E. (1996). Disasters and mental health: Therapeutic principles drawn from disaster studies. Historical and comparative disaster series #10. University of Delaware, Disaster Research Center.

Gaillard, J.C. (2019). Disaster studies inside out. *Disasters, 43*(S1), S7–S17. https://doi.org/10.1111/disa.12323.

Geyer, R.F., & Schweitzer, D.R. (1976). *Theories of alienation: Critical perspectives in philosophy and the social sciences.* Springer Publishing.

Guiberson, B.Z. (2010). *Disasters: Natural and man-made catastrophes through the centuries.* Henry Holt and Company.

Guterres, A. (2020). Make the prevention and redress of violence against women a key part of national response plans for COVID-19. United Nations COVID-19 response. https://www.un.org/en/un-coronavirus-communications-team/make-prevention-and-redress-violence-against-women-key-part.

Hewitt, K. (1983). *Interpretations of calamity from the viewpoint of human ecology.* Allen and Unwin.

Higginbotham, A. (2020). *Midnight in Chernobyl: The untold story of the world's greatest nuclear disaster.* Simon & Schuster.

Human Rights Watch. (2020, April 22). Iraq: Urgent Need for Domestic Violence Law. https://www.hrw.org/news/2020/04/22/iraq-urgent-need-domestic-violence-law.

Kerbeck, R. (2019). *Malibu burning: The real story behind LA's most devastating wildfire.* MWC Press.

Kreps, G.A. (1984). Sociological inquiry and disaster research. *Annual Review of Sociology, 10,* 309–30. https://doi.org/10.1146/annurev.so.10.080184.001521.

Lacan, J. (1979). *The four fundamental concepts of psychoanalysis.* W.W. Norton and Company.

Lindell, M.K. (2013). Disaster studies. *Current Sociology Review, 61*(5–6), 797–825. https://doi.org/10.1177/0011392113484456.

Lindell, M.K., & Prater, C.S. (2003). Assessing community impacts of natural disasters. *Natural Hazards Review, 4*(4), 176–85. https://doi.org/10.1061/(ASCE)1527-6988(2003)4:4(176).

Little, D. (2010, July 9). Gabriele Tarde's rediscovery. *Understanding Society.* https://understandingsociety.blogspot.com/2010/07/gabriel-tardes-rediscovery.html.

html#:~:text=Gabriel%20Tarde%20was%20an%20important,reputation%20 diminished%20for%20a%20century.
Lochbaum, D., Lyman, E., Stranahan, S.Q., & Union of Concerned Scientists. (2014). *Fukushima: The story of a nuclear disaster*. New Press.
Lofland, L.H. (1973). *A world of strangers: Order and action in urban public space*. Basic Books.
Lofland, L.H. (1998). *The public realm: Exploring the city's quintessential social territory*. Aldine de Gruyter.
Luhmann, N. (1982). *The differentiation of society*. Columbia University Press.
Luhmann, N. (1989). *Ecological communication*. University of Chicago Press.
Luhmann, N. (1993). *Risk: A sociological theory*. Aldine de Gruyter.
Mayer, J. (2019, January 29). The origin of the name "Spanish Flu": It's a misnomer that endured for a century. *Science Friday*. https://www.sciencefriday.com/articles/the-origin-of-the-spanish-flu/.
Mitchell, J.K. (1990). Human dimensions of environmental hazards: Complexity, disparity and the search for guidance. In Kirby, A. (Ed.), *Nothing to fear: Risks and hazards in American society* (pp. 131–75). University of Arizona Press.
Moore, H.E. (1958). *Tornadoes over Texas: A study of Waco and San Angelo in disaster*. University of Texas Press.
Okabe, K., & Hirose, H. (1985). The general trend of sociobehavioral studies in Japan. *International Journal of Mass Emergencies and Disasters*, *3*, 7–19. https://doi.org/10.1177/028072708500300102.
Oldenburg, R. (1999). *The great good place: Cafés, coffee shops, bookstores, bars, hair salons, and other hangouts at the heart of a community*. Marlowe & Company.
Phillips, B. (1997). Qualitative methods and disaster research. *International Journal of Mass Emergencies and Disasters*, *15*(1), 179–96. https://doi.org/10.1177/028072709701500110.
Prince, S.H. (2011 [1920]). *Catastrophe and social change: Based upon a sociological study of the Halifax disaster*. Nabu Press.
Quarantelli, E.L. (1987). Disaster studies: An analysis of the social historical factors affecting the development of research in the area. *International Journal of Mass Emergencies and Disasters*, *5*(3), 285–310. https://doi.org/10.1177/028072708700500306.
Quarantelli, E.L. (1994). Disaster studies: The consequences of the historical use of a sociological approach in the development of research. *International Journal of Mass Emergencies and Disasters*, *12*(1), 25–49. https://doi.org/10.1177/028072709401200102.
Said, E. (1978). *Orientalism*. London.
Slack, P. (2021). *Plague: A very short introduction*. Oxford University Press.
Stallings, R.A. (1997, April 17). Sociological theories and disaster studies. Article presented at the Disaster Research Center, Department of Sociology and Criminal Justice, University of Delaware, Newark.
Stallings, Robert A. (2002). Weberian political sociology and sociological disaster studies. *Sociological Forum*, *17*(2), 281–305. https://doi.org/10.1023/A:1016041314043.
Taleb, N.N. (2010). *The black swan*. Random House.

Taleb, N.N. (2012). *Antifragile: Things that gain from disorder*. Random House.
Thornburg, P.A., Knottnerus, J.D., & Webb, G.R. (2007). Disaster and deritualization: A re-interpretation of findings from early disaster research. *Social Science Journal*, *44*, 161–6. https://doi.org/10.1016/j.soscij.2006.12.012.
Turner, S. (2010). The silenced assistant: Reflections of invisible interpreters and research assistants. *Asia Pacific Viewpoint*, *51*(2), 206–19. https://doi.org/10.1111/j.1467-8373.2010.01425.x.
Tyhurst, J.S. (1950). Individual reactions to community disaster. *American Journal of Psychiatry*, *107*, 764–9. https://doi.org/10.1176/ajp.107.10.764.
Yeakey, C.C., Thompson, V.L.S, & Wells, A. (2014a). *Urban ills: Twenty-first-century complexities of urban living in global contexts*, Vol. 1. Lexington Books.
Yeakey, C.C., Thompson, V.L.S, & Wells, A. (2014b). *Urban ills: Twenty-first century complexities of urban living in global contexts*, Vol. 2. Lexington Books.

# The Far East

# 1 Reconsidering the Third Place: Social Distancing and Inequality in South Korea during the Era of Coronavirus

BY KELLY HUH AND HYEJIN YOON*

## Introduction

Under current globalization dynamics, the world has become interconnected by the development of information communication technology and the tightly woven global economy. Ironically, the outbreak and diffusion of a coronavirus, COVID-19, is an example of how one globalized phenomenon can influence many people's lives and change social norms and spatiality. In this chapter, we examine the case of South Korea and how the diffusion of COVID-19 and the Korean government's responses have changed people's lives. South Korea is one of the first few countries outside China that reported new cases of COVID-19 at the end of January 2020 (Normille, 2020). Within a few days, many people expected that South Korea's outbreak would keep growing exponentially, owing to its geographical proximity to China (Cohen & Kupferschmidt, 2020). However, the country's Korea Disease Control and Prevention Agency (KDCA), the South Korean government, and its citizens' responses to lower the spread of COVID-19 and reduce the death rate have been more effective than measures in other neighbouring countries (You, 2020).

We would like to answer the following questions: First, how did the KDCA, South Korean government, and South Korean citizens respond to the spread of COVID-19 differently from the previous outbreak of Middle East respiratory syndrome coronavirus (MERS-CoV) in 2015? Second, how did COVID-19 change the social and cultural norms of the Korean people by instilling in them the adoption and practise of social distancing? In the case of MERS-CoV in 2015, 186 people were infected from the outbreak in two months and 36 died (Normille, 2015). Moreover, nearly 17,000 people were further quarantined from MERS-CoV. Even earlier, in 2009, an

---

* Both authors contributed equally to the manuscript and are listed alphabetically.

outbreak of the Influenza A virus subtype (H1N1) made the government raise its four-tier alert to its highest red level to rein in the spread of the flu virus. What led to such different results from the outbreak of COVID-19 compared to MERS-CoV? Lim and Sziarto (2020) argued that, first, the neoliberal government of South Korea had privatized public health, and it was this aspect of its healthcare system that attempted to control outbreaks such as MERS in 2015. Second, the conservative political party did not provide clear information about the MERS outbreak to the general public. Lastly, the combined neoliberal and illiberal governance resulted in the absence of the state, and no public officials took responsibility for the outbreak of MERS-CoV in 2015.

Compared to the MERS-Cov outbreak, the following two reasons can partially explain the immediate government response in the case of the COVID-19 outbreak. First, the liberal government of President Moon Jae-in provided prompt control of the disease by promoting and mandating social distancing and tracking epidemiology utilizing global positioning systems (GPS) and smart city technologies (Sonn, 2020). Second, Korean cultural norms such as "respect for authority" and collectivism may be key motives for observing social distancing and hygiene practices guided by the government (Kasdan & Campbell, 2020). In addition, various groups have reacted differently to government-mandated disease management; for example, religious groups, particularly Protestant churches, have politicized and cooperated with conservative political groups. Also, we will explain how the neoliberal economy excluded labour and marginalized people since the late 1990s, followed by the account of COVID-19 and how it changed social distancing practice policies driven by the government.

## Territorializing Politics and Politicizing the Protestantism of South Korea

In this section, we focus on how South Korea's conservative political groups and Korean Protestantism have cooperated to frame issues related to social distancing based on political and religious differences among people. Two major attitudes that have affected South Korean society since the ceasefire agreement in 1953 with North Korea are anti-communism and an anti-North Korea stance. In particular, before the late 1980s, the framing of pro-North Korea followers as the enemy was an effective tactic for the military authoritarian governments to legitimize their regimes (Sung, 2017; Hwang, 2016; Park, 2003). "Othering" groups of South Koreans based on these frames is still a common occurrence (Sung, 2017; Kang, 2016; Shin, 2012). Thus, regionalism and territorializing politics have been deeply rooted in the legacy of governance exhibited by the authoritarian regimes of South Korea (Park, 2003).

The politics of South Korea have been regionalized by the government and have resulted in political and economic cleavages between South Koreans. For example, under state-led economic plans of the 1970s and the 1980s, the southeast region enjoyed more advantages because of investments in heavy chemical industrialization projects, which deepened the economic gaps between the different regions in South Korea (Park, 2003). Since then, the southeast provinces have, until recently, been more supportive of the military regimes of Park Jung-hee and Chun Doo-Hwan and the conservative political party (Kang, 2016). Interestingly, "nostalgia for Park's regime" or advocating for authoritarian regimes is an expression of disappointment in the widening inequality among people under neoliberal economic restructuring after the 1997 Asian economic crisis (Kang, 2016). Koreans' nostalgia for authoritarian regimes envisions the dictator as a strong leader who brings economic growth; they do not recognize democratization of the state as a huge achievement of the citizens (Kang, 2016). In addition, South Korean Protestantism has participated in various political activities with conservative civic organizations to contest public policies espoused by the country's liberal regimes (Shin, 2012).

How did South Korean Protestantism and conservative political civic groups unite, and when did this begin? South Korea has experienced rapid changes in the economy and society since the 1950s. After the Korean War, economic development and industrial modernization was the foremost goal of the state. During Japanese colonialism, the Japanese empire had benefited from taking natural resources and agricultural products from the Korean peninsula, and the economic system of Korea had been exploited as a colony of the Japanese empire (Hong, 2009). To overcome such difficulties, Korea built a centrist populace involving political cohesion and a model religion to facilitate state-led development (Cho, 2014). Education and medicine from American Protestant missionaries had enabled the general public to recognize the Protestantism that would facilitate modernization of the nation (Kang, 2016). In this regard, two major goals of South Korean Protestantism, anti-communism and pro-Americanism, link them with South Korean conservative political parties (Hong, 2009). Initially, Korean Protestantism had been pro-America because of the influence of American missionaries who also ran educational institutions to deliver pro-American and anti-communist messages to the Koreans. Before the liberation of Korea from Japan in 1945, however, the stance of the majority of Protestant churches in Korea had been apolitical about social movements and local politics. During Syngman Rhee's regime (1948–60), Christian refugees from North Korea and their Protestant churches eventually supported the president, who was a Methodist (Hong, 2009).

Since the Korean War, South Korean Protestant churches have created a unique culture and organization by combining Korean Shamanic and Confucian traditions (Cho, 2014). During the Park Jung-hee regime (1961–79), rapid industrialization and urbanization continued. Both Korean Protestant and Catholic churches supported the democratization of South Korea and the populism of the general public and especially helped urban low-income groups (Lee, 2020). Owing to rapid industrialization and urbanization during state-led economic growth, huge numbers of people moved from rural areas to urban areas and thus constituted the working class and urban poor during the 1970s and 1980s. Therefore, South Korean churches – Catholic and Protestant – paid attention to emergent urban working-class groups as niche markets to expand their influences, and this resulted in quick growth for the churches in Korea (Im, 2004). In addition to expanding the niche markets targeting the urban poor, Korean Protestant churches engaged tightly with the South Korean government up to the 1980s, including the military regimes of Park Jung-hee and Chun Doo-hwan (Lee, 2020; Cho, 2014). The military regimes supported Korean Protestantism as a model religion from developed countries, in particular the United States, where the people's incomes were higher. In this regard, many Protestant churches advocated for the legitimacy of military governments and their austerity policies (Lee, 2020).

Another interesting aspect of Korean Protestantism is its integration of Christian and Confucian traditions. Traditional Confucian familial relations and the patriarchal system that have integrated into Korean Protestantism have created a unique culture: as a reward for their beliefs in Korea, families that attend the churches are blessed to live healthier and richer lives than those who do not believe in God (Chong, 2006; Jones, 2006). Therefore, South Korean churches have been rigidly hierarchical, and women have been excluded from the decision-making processes within them (Chong, 2006). The clergyman in each church is called the spiritual father and has a strong authoritarian power over the laity (Hong, 2009). These unique characteristics of Korean Protestantism empower the evangelical mega-churches and other cult churches such as *Shincheonji*, a religious movement based on South Korean Protestantism (Kwon, 2020).

During the Kim Dae-Jung government (1998–2003), South Korean society changed in many ways. First, it adopted neoliberal policies to adjust economic structuring when the Asian economic crisis affected South Korea in the late 1990s (Shin, 2012). Second, overall, South Korean society became more liberalized in politics, and this resulted in political and ideological polarization (Han & Shim, 2018; Cho, 2014). Third, socio-economic polarization emerged because of the forces of economic globalization (Han & Shim, 2018). The liberal government regimes of both Kim Dae-Jung and Roh Moo-hyun (2003–8) placed enormous pressure on the Protestant churches and their ideological values of pro-Americanism and anti-communism. First, the liberal governments adopted a different approach to deal with South and North Korean relations (i.e., the Sunshine Policy). They treated North Korea as a "brother country"

(Cho, 2014), and increased people's attention to socio-economic diversity and international relations. The changes in geopolitics between North Korea and South Korea after the end of the Cold War ushered in values that the South Korean Protestant churches even supported for many years. Second, the liberal government also promoted cultural diversity, a position it viewed as needed because of changes in demographics and industrialization trends in the country. Its embrace of cultural diversity, however, resulted in a "religious pluralism" that tended to contest Protestantism, further exacerbated by episodes of internal corruption and embezzlement by the clergy in the big churches that left negative impressions on the general public (Cho, 2014).

To better respond to such social changes, Protestant churches have attached themselves to conservative political groups that share two similar orientations, anti-communism and pro-Americanism, and have politicized themselves since the mid-2000s (Cho, 2014). They have actively encouraged the laity to vote for conservative presidential candidates such as Lee Myung-Bak, a former businessman of Hyundai, a large Korean *chaebol* – a large family-operated corporation or conglomerate with strong ties to the state – and Park Guen-hye, a daughter of the long-time military dictator Park Jung-hee. Moreover, Protestants have founded their own political party even though it has failed to become popular. In short, the Protestant churches in South Korea still cannot move from the legacy of the Cold War towards globalization and its twenty-first-century geopolitics to become more politicized through their consolidation of South Korean nationalism.

**Urban Space and Exclusion in South Korea in the Neoliberal Economy**

In the neoliberal economy, widened socio-economic inequalities have disproportionately affected young people (Lee-Caldararo, 2020; Jung & Lee, 2011). Distinctions of personal and private spaces are not easy in South Korea, especially for young urban dwellers who do not own their homes and live in poor-quality, short-term, contract-based housing, such as a *goshiwon*, a low-expense boarding house facility (Yonhap, 2018). Owing to urbanization and industrialization, many young people resided in the Seoul metropolitan area to pursue their degrees in higher education and to seek job opportunities. However, the limited space resulted in skyrocketing rents and an overheated housing market. Because the South Korean economy has evolved since the 1997 Asian economic crisis, real estate has become an important way of accumulating wealth through deregulation and privatization of housing policies by the government (Jung, 2017). The neoliberal economic reform that was forced upon South Korea by the International Monetary Fund and the state government led to deregulation of the labour market and resulted in a flexible labour force. For example, irregular work with short-term contracts instead of lifetime employment became more common to many young South Koreans. Therefore, many young South Koreans have since faced uncertainty and insecurity in employment at many points in their life cycle.

Overall, the wide range of transformations in South Korean society brought enormous changes to young people's lifestyle dynamics in jobs, family, and housing. Social and economic inequality has increased (Jung, 2017; Kang, 2016), and many young people have delayed their marriages and changed their life-courses because of insecure employment (Lee-Caldararo, 2020; Jung & Lee, 2011). Meanwhile, commodification of land and housing has accelerated and intensified as a means of reproducing capital, a trend that strengthened the status of the upper class (La Grange & Jung, 2004). Therefore, since the 2000s, the majority of young South Koreans are unable to purchase their own homes (Ronald, 2017). The unaffordability of housing brought about intense reliance on alternative spaces such as cafés, utilized for studying (Lee-Caldararo, 2020), and PC *bangs* – an informal term for South Korea's internet cafés – utilized both for leisure, as in playing Internet games, and for business purposes (Choi et al., 2009). Thus, informal spaces have been widespread, but the characteristics of the third place and its informal public life in South Korea are not the same as in North America.

In North America, "a two-stop daily routine" between home and the workplace is common, and people rarely use the third place, such as a neighbourhood café, for long durations on days of work (Oldenburg, 1999, p. 9). Oldenburg examined the major characteristics of the third place as exhibiting a low profile and a playful mood, attributes that are further enhanced by its offering to patrons a sense of community and home. Discussion on the importance of third places in the US is based on the suburbanization of US cities after the 1960s and the lack of informal public life (Oldenburg, 1999). With the construction of the US highway system, urban areas have deteriorated and migration from older urban centres to newer suburban residential areas resulted in the lack of an informal public life for many Americans. In contrast, third places in South Korea are perceived as exhibiting more security for young people to work, or as offering alternative environments that are difficult to replicate elsewhere because of limited residential spaces and unaffordable housing (Jung & Lee, 2011). The first place, the home, in urban South Korea rarely provides a pleasant environment for young people to live their everyday lives (Jung, 2017). In the extremely small space of the first place for young South Korean urbanites who live separately from their parents, the lack of sunlight and creature comforts often forces people to migrate from first places towards third places. Therefore, the territories of the first and third place or the boundaries between private and public spaces are blurred in urban South Korea (Lee, 2011). People thus territorialize their blurred third places by personalizing public spaces.

Ironically, many female urban residents feel safer when they privatize the third place such as a café than when they stay at their *goshiwon* (Jung & Lee,

2011; Lee, 2011). To some of them, using the bright and open space of a café as their studying room gives them a feeling of escape from the dark and small micro-housing environments. Recently, female urban dwellers have been exposed to unsafe and insecure environments because of the greater number of crimes that target single female urban residents (Yu, 2020). Moreover, the territorializing of South Korean third places by young people is a coping strategy not to be excluded from the corporate neoliberal society that enhances competition among young people to enter into society (Lee-Caldararo, 2020). Lee-Caldararo (2020) argues that young people spending nights at cafés (the third place) to prepare for various examinations to get certificates for their careers are not merely driven by their reactions to neoliberal corporate society. Rather, choosing this strategy is understood as a means by which young people can refuse to be "a neoliberal subject" by empowering themselves through third places and surviving within them. In sum, the third place of South Korea has been constructed as a web of social, economic, and political relations for young people (Lee-Caldararo, 2020; Jung, 2017). The third place as informal public space implies the use of coping strategies in order to survive under the neoliberal state, a process that reflects inequalities seen in many spaces of South Korea.

**Outbreak and Diffusion of COVID-19 in South Korea**

The spread of COVID-19 in South Korea has exposed widening political and economic gaps in Korean society. South Korea was the third country outside China to report a case of COVID-19 just before the lunar new year on 20 January 2020 (Normille, 2020). Moreover, the case numbers of confirmed COVID-19 patients abruptly increased in February 2020 because of the country's geographical proximity to China (Cohen & Kupferschmidt, 2020). This hyperconnected world seemed to make the global spread of the virus easier, but at the time the transmission of COVID-19 was not fully studied compared with other more conventional influenza strains.

Interestingly, the majority of new afflicted patients recorded in February and March were related to two unique characteristics of Korean society: religion and the poor working conditions of labour. First, more than five thousand cases, about 60% of the nation's total at the time, were linked to the *Shincheonji* Church of Jesus, a secretive and messianic megachurch based in the city of Daegu (Choe, 2020). Many people outside of Daegu were less sympathetic to the large number of cases from Daegu. This mainly results from territorial politics in the South Korean context. Many conservatives harshly critiqued the massive outbreak in Daegu. Second, another group of new patients emerged from workers employed in call centres in Seoul, South Korea's capital city. Generally, working conditions in the country's call centres are poor. Moreover, most of the workers are women employed in intense working environments with unstable contracts and low incomes (H. Kim, 2020). Furthermore, the call centres are mainly outsourced companies that hire many irregular workers who work in small offices without wearing masks. Therefore, social distancing practices were overlooked in such poor working environments. Although new case numbers declined by mid-March, aided by the majority of South Koreans complying with social distancing measures, new clusters emerged in a few smaller church groups and in other contexts, including a call centre in Seoul in March, and through actions directed by Protestant churches against the country's liberal policies (Choe, 2020). In sum, the outbreak of COVID-19 has shown the socio-economic and geographic unevenness exhibited by the people of South Korea.

**Social Distancing and New Changes in South Korea**

Social distancing has influenced the country in various ways. Under the highest-level alert in the fourth weekend of February, issued when the virus started to spread nationwide, the South Korean government had already set up the Central Disaster and Safety Countermeasures Headquarters, headed by Prime Minister Chung Sye-kyun. The government issued a mandate to restrict the operation of public transport, reduce the number of international flights, ban large-scale events, and allow employees to telework from home (Ock, 2020a). Many South Koreans were shocked by both the megachurch-based clusters in February and the number of new cases that had arisen despite the nation's mandated social distancing measures. Those conceding to state mandates preferred participation in online religious services to actual church attendance. Moreover, for this same group, who learned from the country's previous experience with preventing contagious diseases such as MERS and H1N1, wearing masks and avoiding crowds, meetings, and non-essential work and, if possible, working remotely from home became more sensible options.

*Workplace*

The workspace landscape in South Korea has drastically changed since the coronavirus outbreak (Shin, 2020). In pre-pandemic South Korea, working overtime in the office was common practice. Similarly, many companies neglected the importance of their employees' personal matters, such as nurturing children. However, during the country's pandemic period, important South Korean companies and corporations actively promoted social distancing in various ways: allowing work from home except for a few essential personnel, offering flexible working hours, suspending collective training, events, and, after the government raised its pandemic alert to its highest level in February 2020, complying with a cessation of gatherings until the end of March 2020. South Korean *chaebols* such as Samsung and LG Corporation permitted pregnant employees vulnerable to the disease to telecommute on a rotating basis from home. Also, many e-commerce firms and internet firms ordered all employees to work from home. A few of these major companies allowed employees to work at home if those who had young children needed more hours to attend to them because of systemic delays in the opening of kindergartens and elementary schools.

## Schooling

The South Korean government seriously considered temporarily postponing students' return to academic instruction. The quality of schooling and education has been the nation's foremost concern for many decades, so such a decision reflects how the country's government took a cautious approach to prevent the spread of the disease. For example, the Ministry of Education announced that the start of the new academic year for K–12 and special schools would be delayed by one week to 8 March 2020 (Ock, 2020a). The academic year in South Korea begins in early March. However, the opening of schools was postponed again by two more weeks to 23 March, and once more to early April to prevent the spread of infection among school children as the country continued to report new clusters of cases in metropolitan areas.

The Ministry of Education had already ordered all kindergartens and elementary schools in the country to run an emergency childcare service with a maximum of 10 students per class (Cho, 2020). However, in spite of having access to such a service, anxious parents did not send their children to daycare and many parents of toddlers cancelled registration for the entire month of March because of the fear of infection. The government judged that it was still too early to relax mandates and added two to three more weeks to rule out the possibility of cluster infections in schools. Many parents struggled to realize their backup plans to take care of their children and to prepare for the possibility of remote education at home. A nationwide social distancing campaign was enacted on 22 March 2020, with government appeals for public compliance with social distancing guidelines that required staying indoors as much as possible, while it tightened restrictions on potential hotspots for the virus, including indoor sports facilities, entertainment establishments, and religious gatherings.

Changes in school schedules have reflected the changes in social distancing measures by the government since the spring of 2020. The South Korean Education Ministry announced the first Monday in May 2020 that schools would open in phases, starting with senior-year high school students on 13 May followed by senior-year middle school students and other grades in high schools in the subsequent week, and the gradual opening for the rest of K–12 grades by 8 June. The minister urged all students to routinize distancing lifestyles by complying with social distancing guidelines set by their schools, such as the wearing of face masks at all times in and out of classrooms settings.

*Government Regulation of Face Masks*

Because of the enforcement of social distancing measures by the state government, crowds of shoppers abruptly disappeared from city streets, apart from queues to buy new disposable face masks. At the time, South Korea grappled with a temporary shortage of masks. It soon experienced a mask crisis because of increased COVID-19 cases. Amid public outcry over a lack of face masks, its government first informed people to wear their masks but later changed its position, owing to a lack of supply. The government then enacted new measures to have enough face masks acquired domestically. It announced a series of plans to ensure control of the mask supply and implemented a five-day rotation system for public mask distribution, which seemed to be an amicable solution.

The mask ration system stipulated that people could purchase only two face masks per week from pharmacies on designated days of the week depending on their year of birth. Also, the Korea Customs Service (KCS) only allowed a limited number of masks to be sent overseas and only to immediate family members. Once the mask ration system ended in May 2020 because of increased domestic production and imported quantities from China (Yonhap, 2020b), South Koreans purchased up to five masks per week, a pattern that coincided with the KCS allowance of an increased number of masks to be sent to families overseas. By July 2020, the government relaxed all regulation, enabling people to purchase face masks without any restrictions. However, it was expected that South Koreans travelling abroad would report the number of face masks they carried to the government officers while at the airport. In short, the KDCA played an important role in preventing COVID-19 by adopting various social distancing practices, allowing the central government to actively manage the pandemic. Furthermore, new sociocultural practices, such as observing sick leaves at workplaces and schools as well as engaging in behaviours to prevent COVID-19 transmission to the general public were vigorously adopted. However, socio-economically marginalized groups were overlooked, and the gaps between various groups of people began to deepen.

## Mechanics of Contact Tracing in South Korea

Two major features of contact tracing in South Korea are speed and transparency of information based on techniques and technology for collecting data in order to manage smart cities like Seoul. Contraptions such as closed-circuit TV and tracking credit card usage and mobile phone usage were harnessed for such actions (Sonn & Lee, 2020). In addition, the KDCA actively employed rapid COVID-19 testing, implemented drive-through clinics for mass testing, maintained a well-established digital communications network, and employed an effective information and communications technology (ICT) app called the Special Immigration

Procedure, one that was provided to new airport arrivals (Campbell, 2020). Park Neung-hoo, the minister of the Ministry of Health and Welfare (MOHW), noted that the ICT can accurately identify people and locate their contacts swiftly by tracing them through credit card usage, mobile phone tracking, and surveillance cameras (Campbell, 2020). Once any confirmed patients were identified, the ICT app helped to quickly identify a patient's travel time, route, and location with their close contacts. Such persons who were near the new patient's residential area and movements then received a mass text message with anonymous information about the geographic flow of new patients in the area.

The utilization of such technologies is made possible through South Korea's smart-city infrastructure. This digital infrastructure has been applied to public health surveillance for tracking confirmed patients and monitoring them for the 14 days of the self-quarantine period (Yoon, 2020). The fast implementation of public health surveillance has been possible because the country had already developed a cloud-based data hub through the Smart-City Innovation and Development Project of 2018 employed by the Korean Ministry of Land, Infrastructure, and Transport. The data hub is a large-scale city data tool to exchange and analyse 28 related organizational data in real time about cities such as transportation patterns and safety practices. The tracing of credit card usage to track virus spread became legal amid the MERS outbreak in 2016 (Choi, 2020), prompting the KDCA during spring 2020 to employ its public surveillance system for engaging in fast contact tracing based on big data in its governance of the smart city of Seoul.

In the midst of the danger of leaking personal information, the KDCA announced plans to apply security systems to prevent an invasion of privacy and loss of personal information from any misuse of the public surveillance system. They also employed a virtual private network to prevent external hacking by making data accessible only to authorized personnel in the organization, collecting information on a legal basis, and implementing data disposal procedures in accordance with the law. At the citizen level, the public voluntarily developed multiple free GPS-based, self-quarantining, safety protection smartphone apps with government agencies, which contributed to the detection of movements of confirmed patients and estimations of both regionally and locally proliferating cases (M. Kim, 2020). Anyone who came within six feet of a confirmed patient or was in the same space either in public or private was subject to a mandatory two-week self-quarantine (Image 1.1). Once under self-quarantine, subjects received instructions from local medical authorities and were legally prohibited from leaving their homes or quarantine areas. Citizens could use the app to report any developing symptoms and provide their status to officials who monitored them through the apps. If any self-quarantine subjects left their quarantine area, the local case officer would be alerted. Such practices became a new normal during the height of the pandemic in South Korea. However, officials also monitored

Image 1.1. The mandatory self-quarantine app (left) and public notice of the mandatory 14-day quarantine at Incheon International Airport (screen capture by Minji Kim (left) and photograph by Gainbi Park (right), 16 June 2020).

self-quarantining subjects through traditional telephone calls and took a flexible approach to GPS tracking because the apps were not mandatory for those who had difficulties downloading or using them. People also tried to retrace their own steps over the past several days, hoping that their paths did not cross any active cases. Locally confirmed cases were posted on each district's website to follow the patients' movements, and multiple alerts were sent to mobile devices about new confirmed cases and the number of daily deaths (Strother, 2020). In South Korea, privacy concerns are largely bypassed for the collective, public good.

*The Surge of COVID-19 Based on the Guidance of Social Distancing Measures Enacted by the Central Government*

COVID-19 appears to be under control in South Korea at the time of this writing during late fall 2020. The rapid and thorough mass testing facilitated by a well-established digital communications network has received worldwide attention and prompted several countries to emulate the model. COVID-19 has not disappeared completely from South Korea. Nonetheless, the KDCA has reported a significant drop in fatalities, maintaining the fatality rate at 1.99% since May 2020, thus instilling in many a sense of confidence that lives can be saved if people are informed about newly diagnosed cases that have been under quarantine and away from the public (Ock, 2020b). With such compliance, the social distancing campaign of South Korea as a whole was eased to level 1 in late April 2020. This level allowed citizens to carry on a semblance of familiar social and economic activities under quarantine rules. According to daily briefings of MOHW on 2 July (2020a) and 12 July (2020b), gradual increases in both public transportation in metropolitan areas and cell phone–based movement over the entire country occurred immediately after the easing of social distancing regulations effective on 20 April 2020 (see figures 1.1 and 1.2). For example, weekend public transportation usage patterns slightly increased by early July 2020 for both metropolitan and non-metropolitan areas (see Figure 1.1).

In addition to public transport usage data framing weekend dynamics, other data about cell phone users who visited outside their residence areas for at least 30 minutes were used to investigate how people complied with social distancing guidelines promoted by the government (Figure 1.2). It should be noted, however, that the Korea Ministry of Public Administration and Security (KM-PAS) cautioned that the data explain only approximately 15% of the total population of South Korea, a country of approximately 52 million citizens, during September 2020. Nonetheless, cell phone–based movement data on weekends show a slight decline in non-metropolitan areas after routine distancing in daily life effective from 6 May 2020 onward, a pattern that is opposite to the social dynamics seen in metropolitan areas, where figures show community activities exhibiting a slight increase in cell phone use (see Figure 1.2).

The trend in economic activity estimated from credit card spending data between 18 April and 11 October 2020, by a major credit card company in South Korea displays a sharp rise after distancing in daily life started on 6 May 2020, near the long weekend holiday of Children's Day, and an insignificant decrease about one week before enhanced social distancing measures were enacted, effective 28 May 2020. The data reveal a steady pattern through midsummer (see Figure 1.3).

The South Korean government started to implement a three-level social distancing plan with countermeasures on 2 July (see Table 1.1) and set social distancing at level 1 as the country's daily new confirmed cases rebounded without signs of a slowdown. The government, however, reassured the public that cases were manageable by the country's medical system, a certainty conveyed at the end of June 2020 (A. Kim, 2020; Yonhap, 2020a). Many Koreans felt relieved and resumed their daily routines, since these offered a semblance of pre-pandemic normalcy, and also because most counts of new daily infections were

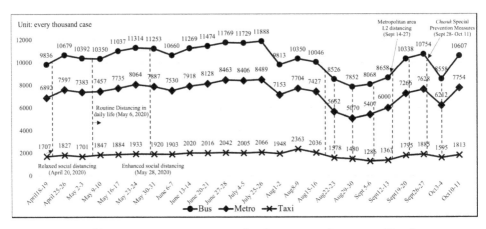

Figure 1.1. Public transportation usage on weekends in metropolitan areas of South Korea between 18 April and 11 October 2020. (Data for two weekends, 11–12 July and 18–19 July 2020, were not available.). Source: This figure is reproduced by the authors based on the regular briefings of MOHW on 2 July (2020a), 12 July (2020b), 18 September (2020c), and 16 October (2020d).

lower than 50, with the vast majority of new cases, between 70% and 80%, being attributed to people arriving from abroad.

In mid-August, however, South Korea's infections swelled to a five-month high because of a surge in church-related cases based on massive anti-government and anti-North Korea protests organized by Protestant churches and conservative political organizations in Seoul (Ock, 2020b). Protesters did not entirely follow social distancing guidelines established by the KDCA and generated a surge in outbreaks again after mid-August. Cluster outbreaks at churches had put a strain on the health authorities' virus fight since the explosion of infections in a megachurch in South Korea during February 2020. Health experts warned that the country might be facing another wave of infections, and the situation could be worse than the clustered nightclub- and warehouse-connected cases in specific locations of Seoul from May to June (Yonhap, 2020c). The resurgence of new cases, reaching triple digits for three straight days, alarmed the government and South Koreans. Thus, on 19 August 2020, the Korean Central Disaster and Safety Countermeasures Headquarters determined that they had to immediately raise social distancing measures to level 2 in Seoul and Gyeonggi Province and its neighbouring areas for two weeks (KBS World, 2020). Under stricter guidelines of level 2 social distancing, all public facilities, including libraries, museums, and galleries, were to be closed, while all private and public indoor meetings of 50 or more and outdoor meetings of over 100 were banned.

Social Distancing and Inequality in South Korea    47

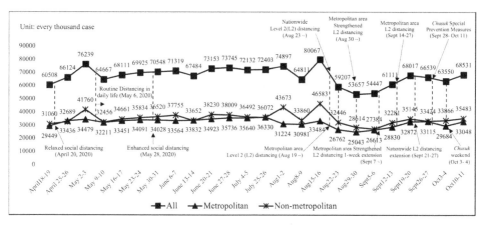

Figure 1.2. Cell phone–based movement on weekends between 18 April and 11 October 2020 in South Korea. (Data for two weekends, 11–12 July and 18–19 July 2020, were not available.). Source: This figure is reproduced by the authors based on the regular briefings of MOHW on 2 July (2020a), 12 July (2020b), 18 September (2020c), and 16 October (2020d).

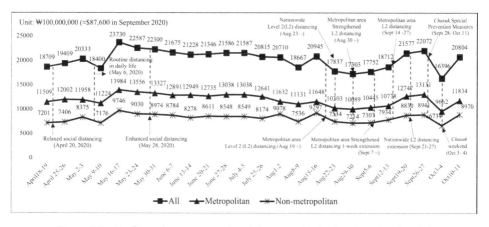

Figure 1.3. Credit card usage on weekends between 18 April and 11 October 2020 in South Korea. (Data for two weekends, 11–12 July and 18–19 July 2020, were not available.) Source: This figure is reproduced by the authors based on the regular briefings of MOHW on 2 July (2020a), 12 July (2020b), 18 September (2020c), and 16 October (2020d). Data do not include paying insurance, communication, home shopping, and other online businesses. The entire credit card spending trend in South Korean currency (₩) is estimated from calculated credit card spending data for the country.

Table 1.1. Three-level social distancing in South Korea: Released on 2 July 2020, by the Central Accident Control Headquarters of the Ministry of Unification Health and Welfare. Source: Korea Ministry of Public Administration and Security (KMPAS) (22 August 2020).

| Occasion | Level 1 | Level 2 | Level 3 |
|---|---|---|---|
| Daily new confirmed cases in the community within 2 weeks | Less than 50 people | 50-100 people | 100-200 people, or if daily new confirmed cases double twice in a week |
| Gathering/meetings | Allowed when upholding COVID-19 prevention guidelines | (Indoor) 50 people or more (Outdoor) 100 people or more are prohibited | More than 10 people are prohibited |
| Professional sports | Sports events with limited audience | Sports events without audience | No sporting events |
| Public multi-facility | Permitted if needed, with limited audience | Closed temporarily | Closed |
| Private multi-facility | Permitted, limited operation on High-risk facility | Suspension of high-risk facility operations. | Suspension of high-risk facility operations; no operations of private multi-facility located at ground-level; operations of other high risk facility above ground level will be suspended after 9:00 p.m. |
| K-12 schools | In-person and distance learning | Schools in districts or cities affected by ongoing cluster outbreaks are required to switch to distance learning. Schools in other areas to reduce in-person class size to 1/3 of current level (2/3 for high schools) and distance learning | Remote learning or Closed temporarily |
| Social welfare facilities and preschool/daycare facilities | Allowed when keeping COVID-19 prevention guidelines | Recommended to close temporarily | Closed temporarily |
| Public workplaces | Reduce on-site personnel (1/3 of workers) through flexible work hours or work-from-home options | Reduce on-site personnel (1/2 of workers) through flexible work hours or work-from-home options | Work-from-home only except required on-site personnel only |
| Private workplaces | Private employers are recommended to implement measures | Private employers are recommended to implement measures to reduce on-site personnel to similar level | Work-from-home only except required on-site personnel only |

As a consequence, students attended schools both remotely and in person (with precautions undertaken), while sporting events reverted to crowd-less matches. Small to regular-sized cafés and restaurants were allowed to open, but customers had to don their masks when not dining and business proprietors were required to record the identities of their customers digitally by using quick response (QR) codes. Buffet restaurants were subject to closing during the enforcement of level 2 social distancing. High-risk facilities, including bars, indoor gyms, concert halls, and cram schools with more than 300 people were also subject to closures. Such measures were balanced by both public and private companies enacting flexible working options to reduce the density of their employees at workplaces, even though many were critical of certain citizens who could not access such health-related technologies, such as homeless people and those who did not own cell phones. Again, practising social distancing to lower new cases of the disease was enforced by restrictions on informal public life and limited use of third places. As a result of the strengthened social distancing of level 2, weekend public transportation usage (Figure 1.1), cell phone–based movement (Figure 1.2), and credit card spending (Figure 1.3) between 19 August and 30 August show a decrease within the reinforced social distancing periods from 19 August to 13 September 2020. However, all the data show gradual increases towards the end of September, since the strengthened social distancing period dates back to regular level 2 social distancing regulations from 14 September. Such an increase was likely due to the official start of the *Chusuk* season, or Korean Thanksgiving (30 September–4 October 2020), when many people began their vacations during the previous weekend, thus exhibiting decreased patterns for social distancing.

Sporadic infections were again reported in August 2020 stemming from gatherings of employees at fast-food franchises, coffee shops, offices, and schools. However, the KDCA and health authorities in the country had forewarned about locally transmitted cases, thus preventing additional fatalities at the time of this writing. For example, when local governments have traced sporadic cluster infections to churches in metropolitan areas, such facilities have been placed on high alert, with restrictions placed on gatherings and religious activities to further contain the spread of the virus. In short, the South Korean government did not exclusively adopt a full lockdown approach to lower the outbreak of COVID-19. Instead, the KDCA managed social distancing on a continuous basis based on the number of outbreaks as well as on when the outbreaks surfaced, as during national holidays.

## Conclusion

Our chapter examined how advanced ICT and its efficient utilization have enhanced social distancing practices to lower COVID-19 rates in South Korea. The fast, reliable, and aggressive responses to COVID-19 in South Korea became a model for other countries to follow. For example, there are daily briefings by the first female chief of the KDCA, Commissioner Jun Eunkyeong, designed to maintain transparency as well as update the public on further COVID-19 cases, including the number of confirmed cases, the origins of their infections, and the status of tests, quarantine, and treatment. As a consequence, most Koreans have been following COVID-19 guidelines and maintaining social distancing protocols since the first case was confirmed in the country (Moon, 2020). Furthermore, based on South Korea's continuing subscription to Confucian culture, hierarchical collectivism has effectively managed COVID-19 from the state to the individual level. Critics argue that such efficiency has resulted in a trade-off where there is also a loss of privacy (You, 2020; Sonn & Lee, 2020). Thus far, tracking and sharing new patients' geographical locations and their movements have been important factors in enhancing social distancing practices. However, privacy issues remain an unavoidable topic in the process of COVID-19 mitigation. Moreover, backlash from conservative political groups and politicized Korean Protestant churches against South Korea's liberal government policies is another issue that hampers the central government's management of the disease. In particular, the regional outbreak cluster of Daegu in February 2020 clearly showed territorialized politics and politicized religious groups of South Korea engaged in shaping the narrative of COVID-19.

While the central government strictly enforces social distancing practices, the othering of marginalized groups in the third place became a critical issue. Accessibility to high-tech ICT infrastructure is a mechanism for maintaining high levels of social distancing as well as lowering the death toll and diffusion of the disease, although the process has also become a way of excluding certain disadvantaged socio-economic groups – that is, simply stated, if people do not have their own cell phones they will have difficulties acquiring full information about social distancing and related information. Moreover, these socio-economically disadvantaged groups have not been allowed to enter and dine at restaurants because they cannot use QR codes, for example. Ironically, keeping and following social distancing guidelines by the central government turned out to be effective in preventing the spread of the infection, a process that has ushered in a new normal where quality of life of individuals is emphasized, such as allowing the employed to use sick leave to remove themselves from social settings that enable the spread of the disease. This is an important transformation in South Korean society, where collectivism has long been a virtue for economic development. However, such a new norm functions as another means to deepen socio-economic and political cleavages between Koreans. As we noted, finding refuge in the third place became a strategy for many South Koreans to survive skyrocketing rents and housing issues, since it is primarily the country's youth who study and spend their time in spacious cafés and public libraries where they enjoy an informal public life, thus shunning lifestyles that trap them in their extremely small *goshiwon*.

In this chapter, we addressed some critical issues that South Koreans have faced while they practise social distancing to respond to the spread of COVID-19, and how this process has othered socio-economically poor groups of people. However, some unanswered questions still remain. For example, how has this strict approach to disease control, combined with the enforcement of social distancing, influenced the lives of Korean people and their space? How have people adapted to the strict enforcement of social distancing that is tightly engaged with current socio-economic disparities under the neoliberal and technology-oriented economy of South Korea? Further studies on how people negotiate their individual needs in order to maintain balance with state-led disease control protocols will need to be undertaken.

REFERENCES

Arin, K. (2020, June 24). Korea hosting increasingly more foreign patients of COVID-19. *Korea Herald*. http://www.koreaherald.com/view.php?ud=20200624000945.

Campbell, C. (2020, April 30). South Korea's health minister on how his country is beating coronavirus without lockdown. *Time*. https://time.com/5830594/south-korea-covid19-coronavirus/.

Cho, J. (2020, March 4). South Korea's drastic measures against the coronavirus offers a glimpse of what the US may need to do. ABC News. https://abcnews.go.com/International/south-koreas-drastic-measures-coronavirus-offers-glimpse-us/story?id=69383034.

Cho, K. (2014). Another Christian right? The politicization of Korean Protestantism in contemporary global society. *Social Compass*, *61*(3), 310–27. https://doi.org/10.1177/0037768614535699.

Choe, S.H. (2020, February 21). Shadowy church is at center of coronavirus outbreak in South Korea. *New York Times*. https://www.nytimes.com/2020/02/21/world/asia/south-korea-coronavirus-shincheonji.html.

Choi, J.H. (2020, April 8). South Korea's best method of tracking COVID-19 spread – credit-card transactions. *Korea Herald*. http://www.koreaherald.com/view.php?ud=20200408000918&ACE_SEARCH=1.

Choi, J.H., Foth, M., & Hearn G. (2009). Site-specific mobility and connection in Korea: Bangs (rooms) between public and private spaces. *Technology in Society*, *31*(2), 133–8. https://doi.org/10.1016/j.techsoc.2009.03.004.

Chong, K.H. (2006). Negotiating patriarchy: South Korean evangelical women and the politics of gender. *Gender & Society*, *20*(6), 697–724. https://doi.org/10.1177/0891243206291111.

Cohen, J.C., & Kupferschmidt, K. (2020). Strategies shift as coronavirus pandemic looms. *Science, 367*(6481), 962–3. https://doi.org/10.1126/science.367.6481.962. https://science.sciencemag.org/content/367/6481/962.full.

Han, S.J. & Shim, Y.H. (2018). The global economic crisis, dual polarization, and liberal democracy in South Korea. *Historical Social Research/Historische Sozialforschung, 43*(4), 274–99. https://doi: 10.12759/hsr.43.2018.4.274-299.

Hong, J.Y. (2009). Evangelicals and the democratization of South Korea since 1987. In Halloran, D. (Ed.), *Evangelical Christianity and democracy in Asia* (pp. 185–234). Oxford University Press.

Hwang, J.T. (2016). The Chun Doo-Hwan authoritarian regime's securitization of water: The case of the peace dam, South Korea. *Scottish Geographical Journal, 132*(3–4), 234–45. https://doi.org/10.1080/14702541.2016.1178799.

Im, H.B. (2004). Christian churches and democratization in South Korea. In Chang, T., & Brown, D. (Eds.), *Religious organizations and democracy in contemporary Asia* (pp. 95–115). M.E. Sharpe.

Jones, N. (2006). *Gender and the political opportunities of democratization in South Korea*. Palgrave Macmillan.

Jung, M. (2017). Precarious Seoul: Urban inequality and belonging of young adults in South Korea. *Positions: Asia Critique, 25*(4), 745–67. https://doi.org/10.1215/10679847-4188398.

Jung, M.W., & Lee, N.Y. (2011). 청년세대, '집'의 의미를 묻다: 고시원 주거 경험을 중심으로 [Questioning the meaning of normative "home": Youth experience living in gosiwon]. *Korea Journal of Sociology, 45*(2), 130–75.

Kang, W.J. (2016). Democratic performance and Park Chung-hee nostalgia in Korean democracy. *Asian Perspective, 40*, 51–78. https://doi.org/10.1353/apr.2016.0002.

Kasdan, D.O., & Campbell, J. (2020). Dataveillant collectivism and the coronavirus in Korea: Values, biases, and socio-cultural foundations of containment efforts. *Administrative Theory and Praxis, 42*(4), 604–13. https://doi.org/10.1080/10841806.2020.1805272.

KBS World. (2020, August 22). S. Korea to expand level 2 social distancing nationwide from Sunday. https://world.kbs.co.kr/service/news_view.htm?lang=e&Seq_Code=155744.

Kim, H.K. (2020, March 13). 코로나19: 집단감염 사태로 드러난 콜센터의 민낯 [COVID-19: The truth of the call centre exposed by mass infection]. BBC Korea. https://www.bbc.com/korean/features-51867202.

Kim, M.S. (2020, March 6). South Korea is watching quarantined citizens with a smartphone app. *MIT Technology Review*. https://www.technologyreview.com/2020/03/06/905459/coronavirus-south-korea-smartphone-app-quarantine/.

Korea Ministry of Public Administration and Security (KMPAS). (2020, August 22. Elevating social distancing to step 2: Let's learn step-by-step instructions for social distancing? https://blog.naver.com/mohw2016/222067660451.

Korea Ministry of Public Administration and Security (KMPAS). 2020. Resident registration demographics in Korea. https://jumin.mois.go.kr/.

Kwon, G.S. (2020, March 2). Shincheonji foundation is the Korean mega churches. https://www.hankookilbo.com/News/Read/202003011687098443.

La Grange, A., & Jung, H.N. (2004). The commodification of land and housing: The case of South Korea. *Housing Studies*, *19*(4), 557–80. https://doi.org/10.1080/0267303042000221963.

Lee, K.Y. (2020). *Religious experience in trauma: Koreans' collective complex of inferiority and the Korean Protestant church*. Palgrave Macmillan.

Lee, S.A. (2011). 혼종공간으로서 카페와 유목민의 문화풍경 [Cafés as hybrid spaces and cultural scapes of nomadic subjects]. *Culture and Society*, *11*(10), 34–66. https://doi.org/10.17328/kjcs.2011.10.1.002.

Lee-Caldararo, J. (2020, forthcoming). Sleepless in Seoul: Understanding of sleepless youth and their practices at 24-hour cafés through neoliberal governmentality. In Horton, J., Pimlott-Wilson, H., & Hall, S.M. (Eds.), *Growing up and getting by? Poverty, precarity and the changing nature of childhood and youth*. Bristol University Press.

Lim, S.H., & Sziarto, K. (2020). When the illiberal and the neoliberal meet around infectious diseases: An examination of the MERS response in South Korea. *Territory, Politics, Governance*, *8*(1), 60–76. https://doi.org/10.1080/21622671.2019.1700825.

MOHW (Ministry of Health and Welfare of the Republic of Korea). (2020a, July 2). COVID-19 Central Disaster and Safety Countermeasure Headquarters regular briefing on June 28, 2020. https://www.mohw.go.kr/eng/nw/nw0101vw.jsp?PAR_MENU_ID=1007&MENU_ID=100701&page=1&CONT_SEQ=355256_.

MOHW (Ministry of Health and Welfare of the Republic of Korea). (2020b, July 12). COVID-19 Central Disaster and Safety Countermeasure Headquarters regular briefing on July 12, 2020. http://ncov.mohw.go.kr/tcmBoardView.do?brdId=&brdGubun=&dataGubun=&ncvContSeq=355376&contSeq=355376&board_id=&gubun=ALL.

MOHW (Ministry of Health and Welfare of the Republic of Korea). (2020c, September 18). COVID-19 Central Disaster and Safety Countermeasure Headquarters regular briefing on September 18, 2020. http://ncov.mohw.go.kr/tcmBoardView.do?brdId=&brdGubun=&dataGubun=&ncvContSeq=359943&contSeq=359943&board_id=&gubun=ALL.

MOHW (Ministry of Health and Welfare of the Republic of Korea). (2020d, October 16). COVID-19 Central Disaster and Safety Countermeasure Headquarters regular briefing on October 16, 2020. http://ncov.mohw.go.kr/tcmBoardView.do?brdId=&brdGubun=&dataGubun=&ncvContSeq=360312&contSeq=360312&board_id=&gubun=ALL.

Moon, J.A. (2020, September 22). The 100 most influential people of 2020 – Jung Eun-kyeong. *Time*. https://time.com/collection/100-most-influential-people-2020/5888333/jung-eun-kyeong/.

Normille, D. (2015, December 23). South Korea finally MERS-free. *Science*. https://www.sciencemag.org/news/2015/12/south-korea-finally-mers-free.

Normille, D. (2020, January 20). China reports more than 200 infections with new coronavirus from Wuhan. *Science*. https://www.sciencemag.org/news/2020/01/china-reports-more-200-infections-new-coronavirus-wuhan.

Ock, H.J. (2020a, February 23). Seoul raises virus alert level to "highest." *Korea Herald*. http://www.koreaherald.com/view.php?ud=20200223000229.

Ock, H.J. (2020b, August 19). COVID-19 infections among Aug. 15 rally participants sound alarm. *Korea Herald*. http://www.koreaherald.com/view.php?ud=20200819000196.

Oldenburg, R. (1999). *The great good place: Cafés, coffee shops, bookstores, bars, hair salons, and other hangouts at the heart of a community*. Marlowe & Company.

Park, B. (2003). Territorialized party politics and the politics of local economic development: State-led industrialization and political regionalism in South Korea. *Political Geography, 22*(8), 811–39. https://doi.org/10.1016/S0962-6298(03)00102-1.

Park, B.G. (2011). Territorial politics and the rise of a construction-oriented state in South Korea. *Korean Social Science Review, 1*(1), 185–220.

Ronald, R. (2017). The remarkable rise and particular context of younger one-person households in Seoul and Tokyo. *City & Community, 16*(1), 25–46. https://doi.org/10.1111/cico.12221.

Shin, J.H. (2020, February 25). Coronavirus changes the South Korean workplace landscape. *Korea Herald*. http://www.koreaherald.com/view.php?ud=20200225000733.

Shin, K.Y. (2012). The dilemmas of Korea's new democracy in an age of neoliberal globalisation. *Third World Quarterly, 33*(2), 293–309. https://doi.org/10.1080/01436597.2012.666013.

Sonn, J.W. (2020, March 19). Coronavirus: South Korea's success in controlling disease is due to its acceptance of surveillance. *The Conversation*. https://theconversation.com/coronavirus-south-koreas-success-in-controlling-disease-is-due-to-its-acceptance-of-surveillance-134068.

Sonn, J.W., & Lee, J.K. (2020) The smart city as time-space cartographer in COVID-19 control: The South Korean strategy and democratic control of surveillance technology. *Eurasian Geography and Economics, 61*(4–5), 482–92. https://doi.org/10.1080/15387216.2020.1768423.

Strother, J. (2020, March 3). Is South Korea's approach to containing coronavirus a model for the rest of the world? *The World*. https://www.pri.org/stories/2020-03-03/south-korea-s-approach-containing-coronavirus-model-rest-world

Sung, M. (2017). Surveillance and anti-communist authoritarianism in South Korea. *Surveillance & Society, 15*(3/4), 486–90. https://doi.org/10.24908/ss.v15i3/4.6592.

Yonhap. (2018, November 23). "Goshiwon," housing alternative for low-income urban tenants. *Korea Times*. https://www.koreatimes.co.kr/www/nation/2018/11/119_259187.html.

Yonhap. (2020a, June 28). Seoul adopts three-level social distancing as virus cases rise. *Korea Herald*. http://www.koreaherald.com/view.php?ud=20200628000046.

Yonhap. (2020b, July 12). Mask ration system ends. Yonhap News. https://en.yna.co.kr/view/PYH20200712063100325.

Yonhap. (2020c, August 16). New virus cases hit 5-month high of 279. *Korea Herald*. http://www.koreaherald.com/view.php?ud=20200816000040.

Yoon, J.S. (2020, April 10). Tracking the movement of confirmed patients within 10 minutes … Korea's epidemiological investigation that 50 foreign media are curious about (comprehensive). Yonhap News Agency. https://www.yna.co.kr/view/AKR20200410149451003?input=tw.

You, J. (2020). Lessons from South Korea's COVID-19 policy response. *American Review of Public Administration*, *50*(6–7), 801–8. https://doi.org/10.1177/0275074020943708.

Yu, H. (2020, February 10). 여성청년 1인 가구, 범죄 피해 가능성 더 높아 [The one-person female household has higher likelihood of being victims of crime]. *Redian*. http://www.redian.org/archive/140748.

## 2 The Anthropocene, Zoonotic Diseases, and the State – Japan's Response to the COVID-19 Pandemic: Criminal Negligence or Crimes against Humanity?

BY HIROSHI FUKURAI

### Introduction

On the morning of 24 March 2020, Japanese Prime Minister Shinzo Abe and International Olympic Committee (IOC) President Thomas Bach announced that the 2020 Summer Olympic Games would be postponed until 2021. Both indicated that their shared concern about the coronavirus pandemic had played a principal role in their decision to postpone (Young, 2020). Two weeks later, on 7 April, Prime Minster Abe declared a state of emergency for the first time in Japan, although this governmental warning was not issued for the entire national population, but only for seven populous prefectures (the equivalent to "states" in the US). These were all in metropolitan regions, including Tokyo and its three surrounding prefectures in eastern Japan; Osaka and Hyogo in southwest Japan; and Fukuoka, the most populous prefecture in the Island of Kyushu in southern Japan. As the infection began to travel beyond prefectural barriers and borders, the government finally issued a nationwide state of emergency on 16 April. By that time, most Asian governments had already implemented numerous protocols to deal with the growing pandemic. For example, Taiwan's Centers for Disease Control (CDC) had decided to implement an inspection measure on 31 December 2019, including "onboard quarantine inspections" of all direct flights from Wuhan, China, in response to reports of an unidentified virus outbreak (Chang et al., 2020). On 5 January 2020, the government organized a coronavirus advisory committee to deal with the possible outbreak. By then, Taiwan's CDC had already inspected a total of 1,317 passengers and cabin crews on 14 flights, including all individuals who had also travelled to Wuhan within the previous 14 days, to detect any exhibited symptoms of infections (Wang et al., 2020). By 26 January 2020, the Taiwanese government suspended all air travel to and from China to prevent the potential spread of the virus (Wang et al., 2020). These key government decisions were all established and implemented in the midst of the Taiwanese presidential and vice-presidential election in January 2020. During the same period, the Japanese government failed to issue any suspension of flights between China and Japan, leaving such a crucial decision to the discretion of each airline. For example, Japan Airlines (JAL) continued with "'a business as usual' mentality," although lack of demand eventually led to JAL's suspension of flights between Narita International Airport and Beijing from 17 February to 28 March (Singh, 2020).

By the close of 2020, Taiwan had suffered limited consequences from COVID-19, with 7 deaths overall and 700 nationwide infections. Meanwhile, Japan had a death toll of well over 3,000, and 200,000 infections. By the end of November 2020, the numbers of infections and deaths in Japan began to increase exponentially, and the government reluctantly admitted that those numbers would continue to rise, at least until the end of the winter season in 2021 (Steen, 2020).

The analysis in this paper is structured as follows. As the Japanese government's response to COVID-19 has significantly affected the public health and safety of the general population, the first section interrogates the reasons for the delayed response and the failure to implement effective measures to deal with the pandemic, examining whether the concern over Japan's economy, financial investments, and the planned Olympic games superseded public health concerns and welfare measures. Despite strong public demand, grassroots efforts, and media pressure for prompt, effective governmental responses, the Abe administration's announcement of the state of emergency came only after the reluctant decision in late March to postpone the 2020 Tokyo Summer Olympic Games. Furthermore, this initial emergency warning and delayed governmental response had negative impacts on the national containment of the pandemic, as did the subsequent emergency declaration and stop-gap measures. The second section offers cross-national comparisons among multiple Asian countries' intrepid responses to the outbreak, including Taiwan and Vietnam among others. Also examined is the significant contribution of Cuban doctors and healthcare professionals in assisting the governments of Vietnam, China, and other Asian countries and regions, as well as those in Africa, Latin America, Europe, and the Caribbean, in response to the pandemic threat. This section also probes why many economically poor, "socialist-oriented" countries in the Global South were able to subdue the pandemic more effectively and successfully than did many so-called economically prosperous, "highly advanced capitalist economies," such as Japan, the US, the UK, and other European and North Atlantic countries in the Global North.

The third section examines the possible allegation of a "state crime (or government crime)" committed by the Abe and subsequent Japanese administrations in their failure to protect the Japanese public from the deadly virus. An overt neglect of governmental responsibility that leads to innocent civilians' unnecessary deaths and injuries has been condemned as a "crime against humanity" under international law, and is viewed as an indictable offence adjudicated by the

International Criminal Court (ICC). Such governmental criminal negligence has also been classified as a "white-collar crime (WCC)," i.e., a serious criminal offence that is also adjudicated in domestic courts. For example, in July 2020, Brazil's trade and labour unions filed a lawsuit at the International Criminal Court, charging that Brazilian President Jair Bolsonaro's lacklustre response to the COVID-19 crisis constituted a crime against humanity, endangering many lives, including Indigenous peoples, aboriginal communities in the Amazon, economically impoverished sectors of urban regions, and other marginalized segments of the overall Brazilian society. The final section examines the coming crisis of the Anthropogenic threat, including emerging trans-species, zoonotic virus pandemics and their impacts in Japan, Asia, and beyond. This section also provides suggestions on how humanity can more effectively deal with emerging pandemic crises in the coming years and decades of the Anthropocene.

## I. The Japanese Government's Delayed Responses and Deadly Consequences

Japan's delayed decisions and responses to the COVID-19 pandemic have resulted in serious economic setbacks and have had deadly consequences. The significant delay in the government's response to the threat of COVID-19 has been seen as largely due to the prioritization of the Olympic Games and Japan's economy over the public health and welfare of the general population (Excite News, 2020). It has been suggested that the Japanese government's admission of the pandemic and subsequent decision to delay, postpone, or even cancel the 2020 Tokyo Summer Olympic Games scheduled in July and August 2020 would have significantly harmed the Japanese economy by preventing potential international investments in new sectors of the economy, the creation of new employment and businesses, and the building of socio-economic infrastructures necessary to revitalize Japan's stagnated economy. According to the government estimate, the delay or significant postponement of the Olympic Games was projected to cost between US$2.7 billion and US$5.8 billion, while the complete cancellation was estimated to result in a potential loss of at least US$45 billion (Macnaughtan, 2020).

The 2020 Tokyo Summer Olympics was promoted by the Japanese government as a major economic stimulus to an already stagnating Japanese economy that was due in large part to a series of serious economic setbacks and crises during the so-called lost three decades of the Heisei Era, from 1989 to 2019. The Heisei Period witnessed the bursting of Japan's economic bubble in the 1990s, leading to serious economic recession, the 2008 Great Global Recession that depressed Japan's main export industries and economies, and more than US$360 billion in economic losses because of the triple catastrophes of earthquake, tsunami, and nuclear disaster in the northeastern regions of Japan in March

2011. The tsunami alone killed nearly 20,000 people, and the nuclear reactor explosions at the Fukushima Nuclear Power Plant not only led to the forced relocation of more than one million Fukushima residents but also to the significant financial compensation and reparation paid to nuclear fallout victims, thus paralysing Japan's economy, creating energy crises in commercial and manufacturing sectors, and further destabilizing socio-economic infrastructures.

In September 2011, not long after the triple disasters in Northern Japan, both the Japanese government and the Japanese Olympic Committee (JOC) created the Tokyo 2020 Bid Committee to help bring the Olympic Games to Tokyo, promoting it as a major stimulus to revitalizing the economy. Its advisory board comprised 64 corporate CEOs and state political elites, including the JOC president, chairpersons of both the Japanese Business Federation and the Japanese Chamber of Commerce and Industry, and then-Prime Minister Yoshihito Noda, who served as the top advisor. At the 2013 IOC Session in Buenos Aires, Argentina, it was announced that Tokyo had won their bid to host the 2020 Summer Olympics (CBS News, 2013). Soon after the IOC decision, in 2014, the Tokyo 2020 Organizing Committee strategized the logistics of the 2020 Games, with the committee's key positions, once again, filled by state and corporate elites, including former Prime Minister Yoshio Mori as chair, plus many members from such organizations as the JOC, the Japanese Paralympics Committee, the Tokyo Metropolitan Government, and multiple national government agencies. The recovery of national economic strength had become the number one priority of Japan's ruling elites, who had witnessed Japan's significant decline in economic standing in the international community since the early 1990s. Today, according to the World Bank (WB) report, Japan's GDP per capita ranks 27th in the world, one of the lowest among OECD countries (World Bank, 2019). This clearly contrasts with Japan's GDP per capita in the 1990s, when it had been ranked third in the world. During the Heisei Era, the People's Republic of China (PRC) had surpassed Japan to become the world's second largest economy after the US. It is thus not surprising that the Japanese government and corporate elites viewed the 2020 Tokyo Summer Olympics as a major economic stimulus program that could provide a turning point to revitalize, if not restore and even cause to flourish, the once formidable Japanese economy and its former esteemed standing in the international community.

*The Prioritizing of the Olympic Games and the Economy over Public Health and Public Welfare*

After the G7 Summit video teleconference meeting on 16 March 2020, Prime Minister Shinzo Abe stated, "I want to hold the Olympics and Paralympics perfectly, as proof that the human race will conquer coronavirus" (Ingle, 2020). As stated earlier, Japan's ruling elites had long predicted that such mega sports events as the 2020 Olympic Games, and even the 2019 World Rugby Cup, would encourage international investment in new sectors of Japan's economy, including tourism, travel, education, culture, health care, entertainment, amateur sports, and a gambling market including the liberalization of casinos, the development of land-based casino resorts, and the boosting of international gambling tourism (O'Connor, 2020). Corporate and government sectors had also projected that the 2020 Summer Olympics in July and August would bring in additional tens of billions of dollars, through hosting nearly 11,000 athletes from 206 countries and independent regions, participating in 339 games in 33 different sports. The 2019 Rugby World Cup in Japan, with only 20 countries participating, for example, delivered Japan's economy more than US$6 billion in economic benefits (Kyodo News, 2020e). During the 2020 Summer Olympics, nearly 10 million visitors were expected to travel to Japan, and the Japanese public had already bought 4.5 million out of a total 7.8 million tickets. The Tokyo Olympics had also generated record-breaking domestic corporate sponsorship revenues of more than US$3 billion, as well as additional revenues and income from overseas sponsors such as NBC Universal, Discovery Communications, and other international telecommunication corporations (Macnaughtan, 2020).

The government and corporate sectors had also made significant investments of between US$12 billion and US$25 billion, including the construction of the 5,500-unit housing and apartment complex in the Olympic Athletes' Village. A delay or postponement would have affected the projected sales of these multiple units to public owners and potential private investors for use after the Olympic Games (Ingle, 2020). Since the Japanese government's annual budget in 2021 was estimated to be approximately US$1 trillion, a decision to delay, postpone, or cancel the Olympic Games could be expected to cause significant economic backlash and irreparable financial damage to an already stagnated Japanese economy (Asahi Shimbun, 2020a; Kyodo News, 2020f).

When the Japanese government's eventual decision to postpone came in late March, it was amid growing public and media speculation that the Abe administration had purposefully downplayed the pandemic threat in a desperate attempt to save the Summer Olympics, the opening ceremony of which was scheduled to begin in only a few months. When the government finally declared a nationwide pandemic emergency on 16 April, it was only designed to last until 6 May. But because of the significant increase in infections and the continuing spread of COVID-19, Abe announced on 4 May that the nationwide state of emergency would extend until the end of May. However, on 25 May, the Abe administration abruptly terminated the state of national emergency, declaring that the serious coronavirus crisis was finally over in Japan (Kyodo News, 2020d).

In July 2020, the Japanese government's insistence on prioritizing the economy and protecting business investment and corporate profits over public health resurfaced with the proposal of the so-called Go To Campaign (GTC) programs. This government-corporate joint proposal was intended to stimulate the Japanese economy through promotion of tourism, despite fears that promoting domestic travel during an ongoing pandemic would only exacerbate the situation (Takahashi, 2020a). After the Olympic postponement was announced, the Japanese government shifted its priority by allocating heavy government spending aimed at stimulating the already lagging Japanese tourism industries. These GTC programs were started in July 2020, and had five distinct pro-business components: (1) "Go To Travel," with special 50% discount packages available through tourism booking companies, including reduced accommodation fees, prefectural goods coupons, and other travel benefits; (2) "Go To Eat," which offered discount restaurant coupons, special point systems for further food discounts, and designated premium food coupons; (3) "Go To Event," providing up to 20% discounts for events and entertainment; (4) "Go To *Shotengai* (Shopping Arcades)," with companies receiving extra funding for organizing local commercial events; and (5) "Go To Promotion," providing funding for promoting GTC programs (Perez, 2020). The Abe administration intended these commercial stimulus packages to promote hotel, restaurant, entertainment, attractions, and domestic tourism sectors that had been hardest hit by the pandemic.

The government was forced to review the societal effects of the GTC programs, as the greater mobility of people in the midst of the pandemic had begun to contribute to a significant spike in infections and deaths in many metropolitan regions, including Tokyo, Osaka, Hokkaido, and others (NHK, 2020). Nevertheless, the government insisted on the GTC's significant economic contributions to Japan's tourism industries and overall economy and decided to extend the GTC programs until March 2021. With coronavirus infection and death tolls beginning to show signs of a significant spike in December 2020, the media and civic groups demanded that the government stop GTC altogether. On 13 December, new Prime Minister Yoshihide Suga finally announced the suspension of GTC programs, but only from 28 December 2020 to 11 January 2021 (Takahashi, 2020b). These pro-business governmental decisions clearly indicated that the Japanese government was placing greater priority on the health of Japan's economy, industries, and corporate profits than on the public health, safety, and welfare of the general population as a whole.

*The Japanese Replication of US Decisions to Decentralize Responsibility for Pandemic Responses*

As a result of the Japanese government's delayed response and failure to install effective nationwide health policies and preventive programs in response to the COVID-19 pandemic, local governments and organizations were forced to act independently in efforts to control the spread in their own jurisdictions. In late February 2020, well before the Japanese government's declaration of a national emergency, Hokkaido Governor Naomichi Suzuki decided to issue his own Declaration of a New Coronavirus Emergency, asking Hokkaido residents to remain indoors and refrain from unnecessary outdoor activities (Kyodo News, 2020b). In addition, the Education Board of Hokkaido called for the temporary closure of 1,600 public and private elementary and junior high schools. On 19 March, both the Osaka and Hyogo governors in southwestern Japan also asked their residents to avoid non-essential travel in order to contain and manage the unnecessary spread of virus between the two populous and neighbouring prefectures. And on 23 March 2020, one day before the announcement of the Olympic Games postponement, Tokyo Governor Yuriko Koike issued her first warning to Tokyo residents that a lockdown might be forthcoming as an important measure to stop the spread of COVID-19 (Park & Takenaka, 2020).

The Abe administration was not completely dormant in their fight against the pandemic, having established the Japan Anti-Coronavirus National Task Force on 30 January 2020, in order to oversee the government's response to the pandemic. On 27 February, the Abe administration requested the temporary closure of all Japanese elementary, junior high, and high schools until early April (Kyodo News, 2020a). The request came days after Hokkaido's education board had already called for the closure of its own school system to stop the virus spread there. The Abe government's request, however, failed to address concomitant issues related to the school closure, including the failure to consider possible uses of online education, reconsider school closures in areas where the health crisis was absent, deal with childcare problems at home and work, and rectify the limited availability of internet access, among many other crucial issues and problems. In early April, the Abe administration decided to distribute two cloth reusable masks to every household in the country in order to solve mask shortages caused by the pandemic. Soon after they had been distributed to 143 municipalities, the Ministry of Health, Labour and Welfare was flooded with complaints about tainted masks, some of which were contaminated with human hairs, insects, mould, and other defects (Japan Times, 2020b). Dubbed as "Abenomasks," they were largely viewed as a symbol of the Abe administration's utter failure to meet people's demands for effective action in the suppression of the outbreak. The huge cost of US$450 million for the derided project also angered the public, and the mask shortage was eventually eliminated by initiatives of private-sector leaders (Kyodo News, 2020c; News on Japan, 2020). The policy of distributing washable gauze masks had also been criticized by experts because of their poor quality and utter ineffectiveness in preventing the coronavirus infection (Beusekom, 2020).

In their effort to keep from postponing or cancelling the Tokyo Summer Olympics, the Abe administration tried to downplay the total number of infections and/or infected pandemic patients. Masahiro Kami, a chairperson of the Medical Governance Institute in Tokyo, accused the Abe administration of imposing strict constraints on testing for the virus by Japanese health authorities, of lacking sufficient testing facilities, and of having limited procedural protocols. Kami asserted, "the Japanese Government wants to go ahead with the Olympics no matter what, so it doesn't want a large number of infected people … Not letting people take tests means it's a cover-up" (Sturmer & Asada, 2020). The government's efforts to downplay the pandemic led to a limited number of polymerase chain reaction (PCR) tests being conducted in Japan, and many patients with high fevers were forcefully turned away from clinics and checkpoints (Dooley, Motoko, & Inoue, 2020). The Abe government authorized a few public health facilities to test for the virus, but declared that the test results could only be processed by five government-approved firms. Consequently, Japan was ranked 23rd in the world for total COVID-19 tests performed per million people (80.5 per million), behind Hungary and Croatia (21st and 22nd with 87.7 and 83.4 per million respectively). In contrast, South Korea ranked 2nd (4,831.3 per million), China was 3rd (2,820.4 per million), and Taiwan was 9th (676.6 per million) in the number of tests performed per million people (Our World in Data, 2020a).

The government's decision to expand testing was finally issued on 13 April, three days before the declaration of the national emergency (Takenaka & Slodkowski, 2020). Japan's initial strategy of keeping the number of PCR tests low, thus indicating fewer infected people, also made it difficult to trace the real infection rates and total number of COVID-19 patients, as the infection clusters had already spread throughout Tokyo and other metropolitan regions, leading to multiple waves of in-hospital infections and leading the healthcare system to the brink of collapse (Takenaka & Slodkowski, 2020). The Japanese government's approach to the under-testing of people, including those with high fevers and other COVID-19 symptoms, led to the under-counting of a large number of real infections and fatalities and contributed to the low credibility of statistical data compiled by the government (Brown, 2020). Professor Kenji Shibuya, director of the Institute of Population Health at Kings College, London, stated that the actual number of infections in Japan was likely to be at least ten times higher than that reported through government channels. Hokkaido Professor Hiroshi Nishiura, a key member of the Ministry of Health, Labour and Welfare's task force on coronavirus clusters, argued that the number of infections was likely even higher than Professor Shibuya's estimate (Brown, 2020). The Abe administration's efforts to downplay the pandemic and its mishandling of the COVID-19 outbreak led Abe to invest US$22 million in public relations (PR) industries to shore up his image and repair his domestic and international reputation. The government's public relations budget had already risen sharply since Abe took over as prime minster in 2012 (Denyer, 2020). Abe's obsession with the economy, stock market, and PR campaigns displayed many similarities to the strategies used by US President Donald Trump, who was Abe's closest ally on the world stage.

## II. Other Governments' Intrepid Responses to the Pandemic in Asia and Cuba

Medical experts from Taiwan and New Zealand published an influential article in the October 2020 issue of the *Lancet*, one of the most respected journals on global health, in which they examined the success of their governmental programs in the fight against COVID-19 (Summers et al., 2020). By the end of 2020, New Zealand's approach to the pandemic resulted in 25 casualties and 2,000 infections. Taiwan also had seen limited consequences of COVID-19, with 7 deaths and 700 infections. This is despite the fact that tens of thousands had travelled between the People's Republic of China and Taiwan on a daily basis prior to the discovery of the novel strain of coronavirus in December 2019. Medical experts from these countries have identified five areas contributing to the success of their programs, and to the superiority of their approach as compared to that of Japan, the US, and European countries, including (1) the establishment of extensive public health infrastructures that enabled fast and coordinated response; (2) centralized, quickly coordinated responses, including early screening and detection; (3) effective methods of isolation and quarantine; (4) efficient uses of digital technology in identifying and tracing potential cases; and (5) mandated use of masks in public places and spaces (Summers et al., 2020). The article also noted the even greater success of Taiwanese programs as compared to those in New Zealand, for New Zealand had a COVID-19 incidence of 278.0 cases per million, versus 20.7 cases per million in Taiwan. Taiwan was never forced to declare a national economic lockdown, while New Zealand had imposed a 12-day national lockdown because of outbreaks in August (Hallingsworth, 2020). Their studies also identified a number of important characteristics distinguishing the programs instituted in Taiwan and New Zealand from those in Japan, the US, the UK, and others, including (1) science-based government actions, in which scientific specialists played a prominent role in the formulation, implementation, and facilitation of anti-coronavirus measures; (2) the public sector production of essential anti-virus materials and equipment, including masks and personal protective equipment (PPE); (3) public action to maintain the "normality" of social life, including eviction prohibition, rent freezes, lease suspension, and a ban on telephone and/or internet disconnection; and (4) the promotion of collaborative and cooperative international support to other countries in need of PPE, doctors, and/or health workers. Taiwan, for example, donated 10 million masks and PPEs to various countries in Asia, North America, and the European Union (Summers et al., 2020). Vietnam also sent five million protective medical supplies, including

450,000 protective suits, to New York City, which had experienced an early and extensive COVID-19 outbreak, and to other US cities (Johns Hopkins University, 2020).

Vietnam, with a population of nearly 100 million, had a record of 35 deaths and 1,400 infections by the beginning of 2021. The socialist government was successful in handling the COVID-19 pandemic, having prepared for the outbreak long before it occurred. In the wake of the SARS epidemic in 2003, the government had increased investments in public health infrastructure, developed a national public health emergency centre, and instituted a national public surveillance system, including national emergency operation centres in 2013 and four regional centres in 2016. The network of emergency operation centres had successfully dealt with measles, Ebola, MERS, and Zika in the past (Our World in Data, 2020b). Under strong centralized leadership, these localized emergency centres have closely collaborated with hospitals in case detection, isolation, and treatment, while the general public has been kept well informed of protective measures and the private sector has given their support to government recommendations and programs (Ivic, 2020).

Along with its own effective strategies in combating the pandemic, the Vietnamese government accepted Cuban doctors and healthcare workers to help suppress the potential spread of COVID-19. Vietnam shares nearly 1,300 kilometres of national border with the People's Republic of China (PRC) in the north and had therefore imposed lockdowns in small villages and towns. In order to help assist the Vietnamese government in their fight against the pandemic, Cuba sent a team of medical experts as well as thousands of vials of an anti-viral drug called Interferon *alfa-2b* that was designed to induce the human body to produce antibodies against the novel coronavirus. In return, the socialist government of Vietnam donated 5,000 tons of rice to the Cuban people as a token of Vietnam's solidarity with, and respect and deep appreciation for, Cuba's strong commitment to humanitarian support of the global community (Our World in Data, 2020b).

Cuban doctors and healthcare workers worked jointly with the Vietnamese government and its emergency centres, as Cuban specialists have also worked in multiple countries in Asia, including China. In February 2020, Cuban doctors worked with the Chinese government to reinforce and coordinate the proper response to the coronavirus pandemic in China (Granma, 2020; On Cuba News, 2020a). Cuba's anti-viral drug, Interferon *alfa-2b*, was used extensively by Chinese doctors in Wuhan in treating infected patients. It was easy for Chinese medical professionals to use the drug because it had been developed by a Cuban-Chinese joint biotechnology venture called Changchun Heber (ChangHeber) in Jilin Province, China (TeleSUR, 2020a). China and Cuba had a long relationship of scientific collaboration and cooperation. In January 2020, when the coronavirus pandemic was heating up in Hubei Province, Cuba declared the establishment of the new China-Cuba Biotechnology Innovation Centre in Hunan Province, Hubei's neighbouring region (Fawthrop, 2020). In December 2020, three Cuban doctors received China's prestigious national award in honour of the 60th anniversary of diplomatic relationship between Cuba and China, acknowledging Cuba's dedication in training many medical and healthcare professionals based on ethical, humanistic, and supportive principles (CubaSi, 2020). Cuba has already signed a biotechnology Memorandum of Understanding (MOU) with numerous medical companies in other Asian countries, such as Singapore, Malaysia, Japan, and Thailand, for advanced biotechnology and biosimilar production, including generic drugs and affordable medicines designed for wider circulation and consumption by "developing" countries in the Global South (CubaSi, 2020).

Cuba, the socialist country that emerged out of the 1959 Cuban Revolution, was led by prominent figures such as Fidel Castro, his brother Raul Castro, and Argentine medical doctor Che Guevara who travelled to, and participated in, the Cuban Revolution. Since 1960, the new revolutionary Cuban government has sent their doctors and healthcare workers around the globe. Cuban doctors and medical teams have saved many lives following some of Asia's worst natural disasters in recent years, including the 2004 tsunami in Sri Lanka, the 2005 earthquake in Pakistan, and the 2006 earthquake in Indonesia. In the 2015 Nepal earthquake, Cuban doctors and healthcare workers provided much-needed humanitarian assistance with medicines, medical equipment, and other relief materials for earthquake victims (Radio Havana Cuba, 2015). Since 1960, the Cuban government has sent 400,000 Cuban doctors to 164 countries, regions, and conflict areas around the globe. In 2020, more than 30,000 Cuban doctors were dispatched to 67 countries, in addition to nearly 70,000 healthcare professionals sent to 94 countries engaged in fights against the COVID-19 pandemic.

Cuba's international medical diplomacy has extended to other socialist countries around the world, including Venezuela in Latin America. In the socialist government of Venezuela, whose vision was based on the so-called Bolivarian Revolution (in honour of Simon Bolivar, who had freed Latin America from Spanish colonial rule in the nineteenth century), there has been much suffering from the unilateral economic and trade sanctions imposed by the US since 2015. Cuban doctors and healthcare specialists were dispatched to Venezuela to work in collaboration with Venezuelan doctors, local medical experts, and government officials. Cuban doctors helped the Venezuelan government to establish multiple diagnostic centres; to promote the mandated use of masks in public spaces; to quarantine multiple states and large cities, including Caracas, the capital city; and to establish hospitals dedicated primarily to treating COVID-19 patients (Human Rights Watch, 2020; On Cuba News, 2020b). As a result, Venezuela, a country with a population of 28 million, roughly the equivalent of Taiwan's, had one of the lowest numbers of COVID-19 deaths (n=944) and infections (n=106,000) among

Latin American countries. Cuban doctors brought Interferon *alfa-2b*, which has proven to be an effective drug against the viral infection (Granma, 2020). China had also sent healthcare workers, medical experts, PPE supplies, and medicines to Venezuela to help fight COVID-19 (TeleSUR, 2020b; Tiwari, 2020). In addition, Cuban doctors had been working in multiple countries in Latin America, including Mexico, Belize, Honduras, Nicaragua, and Peru, as well as Anguilla, Martinique, Grenada, Haiti, Jamaica, Trinidad and Tobago, and Barbados in the Caribbean (Padilla Torres et. al., 2020). Under former Brazilian President Dilma Rousseff, Brazil hosted 8,500 Cuban doctors to serve Brazil's most vulnerable communities, with 90% of medical cases being treated in the small public health clinics staffed by Cuban doctors and healthcare professionals (Faiola & Brown, 2020). Bolivia similarly accepted the help of Cuban doctors and medical experts. The new governments of Brazil and Bolivia had earlier decided to expel Cuban doctors, after Jair Bolsonaro became president of Brazil in 2019 and Indigenous President Evo Morales of Bolivia was ousted in the November 2019 coup (Wyss, 2019). It seems ironic that in 2020, in the midst of the COVID-19 crisis in Latin America, Brazil and Bolivia wished to bring back Cuban doctors and healthcare professionals to help deal with the pandemic (Faiola & Brown, 2020).

The Cuban contribution to peace and solidarity with people in Asia and around the globe has largely been overlooked by the Western-led corporate media. Cubans have served as the ambassadors of humanitarian assistance in countries across the globe for the last 60 years. For instance, after the devastating 2010 earthquake in the island of Hispaniola that killed more than 250,000 Haitians, Cubans arrived immediately after the earthquake struck. One close observer declared, "It is striking that there has been virtually no mention in the media of the fact that Cuba had several hundred health personnel on the ground before any other country" (Fawthrop, 2010). In recent years, Cuban doctors and healthcare workers have also been principally responsible for the suppression of numerous zoonotic viruses, including the Ebola pandemic in West Africa in 2014, despite the dominant Western-centric narrative, in which US President Obama's pandemic team was the key to addressing the West African pandemic (Green, 2014; Schumaker, 2014). A group of 256 dedicated Cuban doctors, nurses, and other health professionals provided direct care during the Ebola pandemic in multiple West African countries, including Sierra Leone, Liberia, and Equatorial Guinea, from October 2014 to April 2015, actions which were ignored by the Western corporate media (Chaple & Mercer, 2017). In its own fight against the COVID-19 pandemic on the island of Cuba, 28,000 Cuban medical students left their dormitories in May 2020, tested the entire Cuban population of 11 million, identified cases of the coronavirus, and provided door-to-door virus care all across the country to stop it from spreading throughout the island (Abiven, 2020).

Because of Cuba's significant contribution to humanity since its founding in 1960, many human rights organizations and progressive political institutions around the globe have nominated Cuban doctors and healthcare professionals for the Nobel Peace prize (Cubanobel.org, 2020; Prashad, 2020b). Cuba, an economically impoverished country, has accomplished medical, peacekeeping, and humanitarian diplomacy for the last sixty years, despite the unilateral and harsh economic and trade sanctions imposed by the US since 1960. While Asian countries such as Vietnam, Laos, China, and Taiwan have been able to successfully deal with COVID-19, more advanced and economically affluent countries such as the US, Japan, and many European states such as Italy, France, and Germany have suffered greatly from the COVID-19 pandemic. Italy, for example, became one of the worst affected countries in Europe, ranking third in number of COVID-19 deaths per one million population, followed by Belgium and Peru (Statista, 2020). After the outbreak in March 2020, the Cuban government sent a 52-member team of Cuban doctors and medical professionals to Italy to help fight against COVID-19, while Germany, France and other EU countries refused Italy's plea for such help (Herszenhorn, Paun, & Deutsch 2020). The region of Lombardy in northern Italy had over 15,000 deaths by mid-March, and people in the northern city of Crema in the province of Cremona welcomed Cuban doctors, nurses, and health workers from their balconies, windows, and streets (Phillips & Giuffrida, 2020).

The following section examines whether the great suffering and extensive fatalities due to the Japanese government's delayed pandemic response might have been avoidable, had the Abe administration placed greater priority on people's public health and safety than on economic health and corporate profits, particularly with reference to the holding of the Tokyo Summer Olympics. The governmental negligence may have constituted a criminally indictable offence in causing the unnecessary deaths and injuries of many innocent civilians. Sociologists, criminologists, and socio-legal scholars refer to such governmental negligence as "white collar crime (WCC)," or more specifically "state crime" or "government crime," when committed by individuals or groups placed in entrusted positions of government or corporate hierarchies. Such systemic negligence may also constitute a violation of international law, specifically under the provision of "crimes against humanity (CAH)." The following section critically examines Japan's responses to COVID-19 in relation to possible violation of domestic and international law.

## III. White Collar Crime (WCC) and Crimes against Humanity (CAH) Claims against the Abe Administration

Considerable research and analysis have shown that the delayed pandemic response by the Japanese government resulted in serious public health crises in Japan. Despite the fact that airborne transmission of the novel coronavirus and its morbidity were already known, the Abe government tried to downplay the crisis in a bid to save the Tokyo Olympics (Kingston, 2020). The Abe administration delayed its declaration of a national emergency until the spread of coronavirus became undeniable in multiple metropolitan regions and beyond. Furthermore, the Japanese policy of limiting coronavirus testing kept citizens and policymakers from access to vital information, thus seriously hampering timely responses to the outbreak. As the number of infections and deaths surged, the Japanese government began to take action, but it was tragically belated. It became clear that the Japanese people had paid a huge price due to governmental negligence and failure to properly handle the outbreak from the outset, which offered stark contrast to the effective responses in some neighbouring Asian countries.

Because of the lack of government-sponsored, centralized coordination of strategies, testing, contact tracing, and subsequent proper coordination protocols, many prefectures and municipalities were forced to act on their own. Osaka Governor Hirofumi Yoshimura, for example, pleaded for the public to donate plastic raincoats because medical workers in Osaka had to wear trash bags as personal protective equipment (PPE) in treating COVID-19 patients (Lies, 2020). Local Hanshin Tigers Professional Baseball team and players donated 4,500 ponchos, much more than what the Abe government was able to provide (Japan Times, 2020a). At the Tokyo International Airport at Narita, many travellers were forced to stay in cardboard boxes and sleep on cardboard beds until they received test results (Chase-Lubitz, 2020). Many Japanese grew critical of Abe's crisis management, as reflected in a Mainichi Shimbun (i.e., Mainichi newspaper) poll indicating that 70% felt that Abe had waited too long to declare an emergency, thus losing precious time in managing and containing the outbreak (Iwashima, 2020). Another poll by Asahi Shimbun indicated that 77% believed that Abe should have declared a national emergency much sooner (Asahi Shimbun, 2020b).

The policy to pursue "herd immunity" in dealing with the pandemic had been one of the main topics of debate within the Japanese government as well as in the US, the UK, Sweden, and other countries. Then-President Trump had spoken positively about herd immunity (Aschwanden, 2020), and Dominic Cummings, chief of staff and principal strategist for UK Prime Minister Boris Johnson, stated at the end of February that the government's strategy was to pursue "herd immunity, protect the economy and if that means that some pensioners die, too bad" (Shaw, 2020). After Johnson was admitted to the ICU for his own coronavirus infection, the UK government immediately eliminated the policy (Burdeau, 2020). Sweden had chosen not to impose a hard national lockdown, opting in March 2020 to pursue "herd immunity" as the governmental policy, hoping that 40% of people would get the disease and develop protective antibodies (Reinberg, 2020). After the infections and death tolls began to far exceed those of its Scandinavian neighbours, the Swedish prime minister was forced to admit that "Sweden's [herd immunity] strategy was a preventable disaster. And I suspect it won't take generations for Swedes to demand a reckoning over what went so horrifically wrong" (Ghitis, 2020). Japan had also contemplated the possibility of pursuing a herd immunity strategy, as advocated by prominent scholars, including professor Tai Takahashi, who had previously served as a vice president for the Japanese Cabinet's Headquarters for Japan's Economic Revitalization (*Nihon Keizai Saisei Honbu*). Takahashi asserted that many Japanese people had already been exposed to the coronavirus and that herd immunity should be easily established in Japan, if indeed it had not already been so (Ohashi, 2020). After his announcement, however, the number of infections and COVID-19 deaths significantly spiked, illuminating the fact that the herd immunity he predicted had yet to be achieved in Japan (The Mainichi, 2020). The Japanese government refused to take a definite stand on the issue of herd immunity. Prime Minister Abe, responding to the accusations of delayed progress in fighting against COVID-19 in early April, said that achieving herd immunity was not the primary objective of his administration, though it could remain as a secondary or tertiary viable option (Newsweek, 2020). Without effective vaccination available, the imposition of herd immunity as a governmental policy could be seen as an unethical human experiment. If it were implemented in the US, for instance, some experts predicted the death of millions of Americans, extensive and needless civilian fatalities (Aschwanden, 2020). Some critical legal scholars pointed out that the policy of herd immunity, if broadly implemented by nation-states, could be punishable as a "crime against humanity" or even as genocide under international law (Lipton, 2000).

Meanwhile, the Abe administration had placed governmental priority on economic protection of its financial stakes and investments in the Olympics, rather than on the public health and general welfare of the national population. Critical scholars have equated such governmental negligence with "state crime" or "governmental crime," one of the white-collar crimes (WCC) committed by high-ranking state officials. The term "white-collar crime" was coined by American Sociological Association (ASA) President Edwin Sutherland in his 1939 address to the ASA annual conference. Sutherland challenged his colleagues to attend to this neglected inquiry in the area of criminology and criminal justice, suggesting that both sociologists and criminologists had been preoccupied with their examination of street crimes, i.e., criminal actions that take place in public places, and should also look into crimes committed by people in highly privileged positions in society. Analysis of WCCs has since led to the categorization of four distinct taxonomies: (1) occupational crime (i.e., embezzlement and theft committed by privileged people in large organizations); (2) corporate crime (e.g., selling harmful products, such as tobacco, the consumption of which has led to the deaths of nearly half a million people annually in the US);

(3) government (or state) crime (i.e., government actions that inflict human suffering, such as the Tuskegee medical experiment conducted by US Human Health Services in Alabama for four decades); and (4) state-corporate crime (e.g., extraordinary rendition, global assassination campaigns, PRISM projects, and other indiscriminate surveillance programs conducted through joint collaboration between corporate-business and state sectors). The Abe government's preoccupation with the Tokyo Olympic Games, their delayed response, lack of testing, PPE shortages, lax trace-quarantine procedures, and ineffective voluntary lockdowns, among many other factors, may have contributed to the significant public health crises, infections, and COVID-19 deaths in Japan. The lockdown measures, for instance, lacked any legal authority and were largely dependent on citizens' voluntary self-restraint, for the Japanese government failed to offer clear economic incentives to encourage and enforce robust public adherence (Shimizu et al., 2020).

As to the culpability of governmental and corporate elites and adjudication of state crimes, however, it is highly unlikely that Japanese prosecutors or courts would be willing to investigate and prosecute such allegations. Analysis of WCCs in Japan has revealed that Japanese prosecutors and courts have rarely prosecuted high-ranking state and corporate elites for their actions or decisions that resulted in deaths, injuries, and the suffering of many crime victims. Japan's judiciary has been particularly reluctant to adjudicate crimes committed by groups of ruling state and corporate elites. For example, Japanese prosecutors and judges refused to condemn powerful politicians and corporate oligarchs for the deaths, injuries, and forced uprooting of more than a million people because of the explosion of nuclear reactors at the Fukushima Daiichi Nuclear Power Plant in March 2011 (Johnson et al., 2020). An indictment was issued to three top CEOs of the nuclear plant operator, the Tokyo Electric Power Plant (TEPCO), Japan's largest and most powerful private corporation, for the deaths and forced evacuation of more than one million Fukushima nuclear victims, but the Japanese professional judges issued a not guilty verdict, despite strong public demands for the CEOs' prosecution and condemnation (Johnson et al., 2020). Another avenue for the rightful prosecution of Japan's WCC perpetrators would be at the International Criminal Court (ICC), which is responsible for prosecution of crimes under international law, namely "crimes against humanity."

*The Allegation of Crimes against Humanity in Japan, Brazil, and the US*

A government's "purposeful" negligence in dealing with a public health crisis may be an indictable offence under international law. For example, in July 2020, Brazil's trade unions asked the International Criminal Court (ICC) to charge President Jair Bolsonaro with crimes against humanity over his mismanagement of the COVID-19 pandemic (Prashad, 2020a). Specifically, the Brazilian Union Network (*UNISaude*), representing more than one million workers in 18 Brazilian states, the Federal District, and all other regions of the country, filed the case with the ICC, charging the Bolsonaro administration with being criminally negligent in its management of the COVID-19 pandemic, thereby risking the lives of healthcare professionals and peoples of the greater Brazilian society, including Afro-Brazilians, Indigenous populations, and other marginalized regional groups (Prashad, 2020a). Brazil's lawsuit has relied on the 1998 Rome Statute, which established four international criminal offences for prosecution, including (1) genocide, (2) crimes against humanity, (3) war crimes, and (4) the crime of aggression. The plaintiffs demanded that the ICC's decision should require the Brazilian government to refrain from acting without proper measures and should lead to governmental implementation of necessary steps to help reduce risks to healthcare professionals, Indigenous peoples, and the general populations (UNI-Global Union, 2020).

In North America, similar ICC adjudication was called for in relation to the Trump administration. Many human rights organizations, legal specialists, and academicians have called the ICC to adjudicate, under the provision of crimes against humanity, the Trump administration's "purposeful" negligence in the prevention of the spread of COVID-19 that led to the deaths of hundreds of thousands of Americans (Canter, 2020). The Trump administration had declared the national emergency on 13 March 2020, but Trump has denied personal responsibility for managing the pandemic, including his failure to provide sufficient coronavirus testing and effective protocol, stating instead that it was the fault of previous administrations because "we were given a set of circumstances and we were given rules, regulations and specifications from a different time" (Liptak, 2020). Journalist Bob Woodward's recorded interview with President Trump revealed that Trump had in fact been well informed as to the extremely highly contagion and potent lethality of COVID-19 as far back as January 2020. Just as the Abe administration's concern appeared to have been with the Tokyo Olympics rather than public health and safety, Trump may have feared that accurate information on the epidemic could have adversely affected the stock market, and he thus purposefully downplayed the severity of the situation, while declining to develop national strategies, such as using the Defense Production Act to produce a large quantity of medical masks, ventilators, and other medical equipment to save lives. Trump clearly failed to initiate proper negotiation with private and public industries to produce and secure PPE and life-saving machines, to implement a national program to encourage masks in public and mandate physical distancing to minimize spread of the virus, to secure availability of testing equipment and procedural protocols, and to provide available anti-viral drugs and life-saving medical devices to 50 states and healthcare institutions (Canter, 2020). Instead, the Trump administration discussed the use of herd immunity as part of a national strategy. His senior Health and Human Services (HHS) advisor Paul Alexander, for example, suggested that herd immunity could be achieved by letting young and middle-aged Americans become infected with COVID-19 in order to allow the spread of antibodies (Rummler, 2020), specifically arguing that "Infants, kids, teens, young people, young adults, middle aged with no conditions etc. have zero to little risk … so we use them to develop … we want them infected" (Diamond, 2020).

The Trump administration had already been sued for the allegation of crimes against humanity for its family separation immigration policy, which had been part of a "zero tolerance" approach intended to deter "illegal" immigration to the US. The family separation program led to over 2,000 immigrant children being separated from their parents (Dickerson, 2020). Ravina Shamdasani, a spokesperson for the UN High Commissioner for Human Rights, condemned this policy, declaring that it "amounts to arbitrary and unlawful interference in family life, and is a serious violation of the rights of the child … The use of immigration detention and family separation as a deterrent runs counter to human rights standards and principles" (Shamdasani, 2018). Former Nuremberg prosecutor Benjamin Ferencz also argued that Trump could be prosecuted for a crime against humanity for the family separation program. Ferencz, the only surviving Nuremberg prosecutor, was one of those prosecuting high-ranking German officers who were later executed for their crimes as a result of the Nuremberg Trials of the late 1940s (Sampathkumar, 2018).

Whether Abe's delayed responses to the pandemic and the public health crisis in Japan constitute crimes against humanity will require further investigation and elaboration. When the allegation of criminal negligence related to his administration's policies was implied, Abe abruptly resigned the prime minister position in August 2020 in the middle of the COVID-19 pandemic, citing his poor health, and ending his tenure as Japan's longest-serving prime minister in history. This was the second time that Abe had resigned from the position; in 2007, he also cited ill health after many months of political scandals in his cabinet and the electoral loss of his Liberal Democratic Party (LDP) (Onishi, 2007). Critical interrogations of alleged crimes against humanity would include the analysis of his cabinet's internal discussions, debates on balancing the economic merits of the Olympics against the costs of emerging public health crises, decision-making processes and procedures, and the extent of governmental evaluations and estimations of the costs and consequences of delayed governmental responses, including possible deaths and infections of Japanese civilians. The interrogation of the Abe administration's criminal culpability is particularly important in Japanese society today, for it has been predicted that

Japan and other Asian countries may face more frequent instances of deadly zoonotic virus infections and pandemics in the coming years as well as decades because of the threat of Anthropogenic catastrophes. It is thus vitally important to ensure that the government is held responsible for decisions and actions in times of national emergency.

## IV. Conclusions: Future Pandemics in the Anthropocene

This paper has examined the Japanese government's response to the COVID-19 pandemic and resulting public health crises. The Abe administration's delayed response to the pandemic has been highly criticized along many lines, including its contemplation of "herd immunity" and its downplaying of the COVID-19 outbreak for several months in its bid to save the 2020 Tokyo Summer Olympics. In the same period, other Asian neighbours were dealing more successfully with the outbreak, with their successes attributed to science-based approaches, an early installation of centralized preventive measures, effective social measures such as rent freezes and lease suspensions, and mandated mask use in public places and spaces. Also examined here are the successes resulting from the Cuba-China medical collaboration, as well as the international humanitarian contributions of Cuba, which has been sending doctors and healthcare workers not only to regions in Asia but also to Africa, Latin America, the Caribbean, Europe, and beyond since 1960. The final section has examined the possible allegation of "state crimes" (or government crimes) and "crimes against humanity," as committed by the Abe administration in relation to its failure to protect the public throughout the COVID-19 pandemic in Japan.

Many researchers have predicted future waves of other zoonotic virus pandemics, as the "natural barriers" once existing between human society and non-human habitats have been destroyed, making trans-species transmission of viruses more frequent and serious (Fukurai, 2003, Fukurai, Gabriel, & Liang, 2023). In order to combat the next round of pandemics, experts in Taiwan and New Zealand have suggested (1) strengthening a dedicated national public health agency to manage pandemic prevention and control; (2) formulating a generic pandemic plan; (3) investing in government resources and preventive infrastructure; (4) reviewing pandemic management and training programs; (5) developing systems for evaluating pandemic responses; and finally (6) establishing cultural acceptability for pandemic response measures, including mask use in public places and digital technology to aid contact tracing, among others (Summers et al., 2020). If and when the next pandemic hits, the Japanese government must listen carefully to expert advice, and deal far more effectively with potential outbreaks in Japan, its surrounding regions, and beyond.

REFERENCES

Abiven, K. (2020, April 2). In Cuba, medical students provide door-to-door virus care. *Jakarta Post*. https://www.thejakartapost.com/news/2020/04/02/in-cuba-medical-students-provide-door-to-door-virus-care.html.

*Asahi Shimbun*. (2020a, April 7). Abe declares state of emergency for seven prefectures. https://www.asahi.com/ajw/articles/13279041.

*Asahi Shimbun*. (2020b, April 21). Survey: 57% say Abe not showing leadership in COVID-19 crisis. https://www.asahi.com/ajw/articles/13315219.

Aschwanden, C. (2020, October 21). The false promise of herd immunity for COVID-19. *Nature*. https://www.nature.com/articles/d41586-020-02948-4.

Beusekom, M.V. (2020, April 9). Data do not back cloth masks to limit COVID-19, experts say. *CIDRAP*. https://www.cidrap.umn.edu/news-perspective/2020/04/data-do-not-back-cloth-masks-limit-covid-19-experts-say.

Brown, A. (2020, April 28). Uncertainties about Japan's COVID-19 data. *Safecast*. https://safecast.org/2020/04/uncertainties-about-japans-covid-19-data/.

Burdeau, C. (2020, March 17). Boris Johnson's talk of "herd immunity" raises alarms. *Courthouse News Service*. https://www.courthousenews.com/boris-johnsons-talk-of-herd-immunity-raises-alarms/.

Canter, L.S. (2020, November 14). Crime against humanity? *Times Union*. https://www.timesunion.com/opinion/article/Crime-against-humanity-15728167.php.

CBS News. (2013, September 7). Tokyo wins bid to host 2020 Olympics. https://www.cbsnews.com/news/tokyo-wins-bid-to-host-2020-olympics/.

Chang, M.H., Wei, W.T., & Ko, L. (2020, April 17). COVIDVIRUS/How an online post forewarned Taiwan about COVID-19. Central News Agency. https://focustaiwan.tw/society/202004170016.

Chaple, E.B., & Mercer, M.A. (2017). The Cuban response to the Ebola epidemic in West Africa: Lessons in solidarity. *International Journal of Health Services*, 47(1), 134–49. https://doi.org/10.1177/0020731416681892.

Chase-Lubitz, J. (2020, April 15). Cardboard boxes replace hotel rooms at Narita as Japan struggles with returnees. *Japan Times*. https://www.japantimes.co.jp/news/2020/04/15/national/cardboard-beds-narita-airport/.

Cubanobel.org. (2020, December 10). Nobel peace prize for Cuban doctors. https://www.cubanobel.org/.

CubaSi. (2020, October 25). Cuban doctors get award for the 60 years of Cuba-China relations. https://cubasi.cu/en/news/cuban-doctors-get-award-60-years-cuba-china-relations.

Denyer, S. (2020, April 15). Japan sets aside $22 million to buff government's global image amid pandemic struggles. *Washington Post*. https://www.washingtonpost.com/world/asia_pacific/japan-coronavirus-image-abe/2020/04/15/73bf1dee-7f00-11ea-84c2-0792d8591911_story.html.

Diamond, D. (2020, December 16). "We want them infected": Trump appointee demanded "herd immunity" strategy, emails reveal. *Politico*. https://www.politico.com/news/2020/12/16/trump-appointee-demanded-herd-immunity-strategy-446408.

Dickerson, C. (2020, January 21). Parents of 545 children separated at the border cannot be found. *New York Times*. https://www.nytimes.com/2020/10/21/us/migrant-children-separated.html.

Dooley, B., Motoko, R., & Inoue, M. (2020, February 29). In graying Japan, many are vulnerable but few are being tested. *New York Times*. https://www.nytimes.com/2020/02/29/world/asia/japan-elderly-coronavirus.html.

Excite News. (2020, May 3). 安倍首相はなぜ「五輪延期」踏み切れずコロナ対策が遅れたのか？ [Why Prime Minster Abe reluctantly admitted Olympics' postponement and delayed anti-corona strategies]. https://www.excite.co.jp/news/article/Cyzo_239700/?p=2.

Fawthrop, T. (2010. February 15). Cuba's aid ignored by the media? Aljazeera. https://www.aljazeera.com/news/2010/2/16/cubas-aid-ignored-by-the-media.

Fawthrop, T. (2020, April 24) Cuba's improbable medical prowess in Asia. *Diplomat*. https://thediplomat.com/2020/04/cubas-improbable-medical-prowess-in-asia/.

Faiola, A., & Brown, K. (2020, April 10). U.S. allies, encouraged by Washington, said goodbye to their Cuban doctors: As coronavirus surges, some are arguing for their return. *Washington Post*. https://www.washingtonpost.com/world/the_americas/coronavirus-cuba-doctors-trump-ecuador-brazil-bolivia/2020/04/10/d062c06e-79c4-11ea-a311-adb1344719a9_story.html.

Fukurai, H. (2022). The "Vaccine Genocide" of Indigenous nations and peoples, the intellectual property (IP) regime, "Vaccine Apartheid" and "Vaccine Untouchable." *Fourth World Journal, 21*(2), 53–67. https://papers.ssrn.com/sol3/papers.cfm?abstract_id=4361550.

Fukurai, H., Gabriel, R., & Liang, X.C. (2024, forthcoming). The COVID-19 crisis, herd immunity, and "Vaccine Apartheid" in the age of Anthropocene. *Asian Journal of Law and Society*.

Ghitis, F. (2020, December 17). Will there be a reckoning over Sweden's disastrous "herd immunity" strategy? *WPR*. https://www.worldpoliticsreview.com/articles/29296/in-sweden-herd-immunity-has-failed-tragically-will-there-be-a-reckoning.

Granma. (2020, February 3). Cuban doctors in China to reinforce medical command center established in our embassy. https://en.granma.cu/mundo/2020-02-03/cuban-doctors-in-china-to-reinforce-medical-command-center-established-in-our-embassy.

Green, M.H. (2014). Taking "pandemic" seriously: Making the Black Death global. *Medieval Globe, 1*(1–2), 27–61. https://doi.org/10.17302/TMG.1-1.3.

Hallingsworth, J. (2020, August 14). New Zealand imposes 12-day lockdown in its biggest city as it battles fresh outbreak. CNN. https://www.cnn.com/2020/08/14/asia/new-zealand-coronavirus-lockdown-intl-hnk/index.html.

Herszenhorn, D.M., Paun, C., & Deutsch, J. (2020, March 5). Europe fails to help Italy in coronavirus fight. *Politico.* https://www.politico.eu/article/eu-aims-better-control-coronavirus-responses/.

Human Rights Watch. (2020, May 26). Venezuela: Urgent aid needed to combat COVID-19. https://www.hrw.org/news/2020/05/26/venezuela-urgent-aid-needed-combat-covid-19.

Ingle, S. (2020, March 24). How Tokyo's 2020 hopes turned to ashes: But can rise again in 2021. *Guardian.* https://www.theguardian.com/sport/2020/mar/24/olympic-postponement-inevitable-but-finding-right-way-to-do-it-caused-delay.

Ivic, S. (2020). Vietnam's response to the COVID-19 outbreak. *Asian Bioethics Review, 12*(3), 341–7. https://doi.org/10.1007/s41649-020-00134-2. https://link.springer.com/article/10.1007/s41649-020-00134-2.

Iwashima, S. (2020, April 20). Majority in Japan don't approve Abe govt's coronavirus response: Mainichi Poll. *Mainichi.* https://mainichi.jp/english/articles/20200420/p2a/00m/0na/016000c.

*Japan Times.* (2020a, April 18). Tigers pitch in with 4,500 ponchos for health care workers in Osaka. https://www.japantimes.co.jp/sports/2020/04/18/baseball/japanese-baseball/tigers-pitch-4500-ponchos-healthcare-workers-osaka/.

*Japan Times.* (2020b, April 24). Two firms recall "Abenomasks" after complaints of stains, bugs and mold. https://www.japantimes.co.jp/news/2020/04/24/business/abenomask-recall/.

Johns Hopkins University. (2020). Coronavirus resource center. Retrieved 22 October 2020, from https://coronavirus.jhu.edu.

Johnson, D., Fukurai, H., & Hirayama, M. (2020). Reflections on the TEPCO trial: Prosecution and acquittal after Japan's nuclear meltdown. *Asia-Pacific Journal: Japan Focus, 18*, 2(1), 1–36.

Kingston, J. (2020, April 23). COVID-19 is a test for world leaders. So far Japan's Abe is failing: The pandemic has exposed feeble leadership on the part of Prime Minister Shinzo Abe. *Diplomat.* https://thediplomat.com/2020/04/covid-19-is-a-test-for-world-leaders-so-far-japans-abe-is-failing/.

Kyodo News. (2020a, February 27). PM Abe asks all schools in Japan to temporarily close over coronavirus. https://english.kyodonews.net/news/2020/02/c3c57bbce11d-breaking-news-govt-will-ask-all-schools-in-japan-to-shut-for-virus-fears-abe.html.

Kyodo News. (2020b, February 28). Hokkaido declares state of emergency over coronavirus. https://english.kyodonews.net/news/2020/02/69d1b85128b9-urgent-hokkaido-declares-state-of-emergency-over-coronavirus.html.

Kyodo News. (2020c, April 25). 2 Firms recall masks from PM Abe's distribution program. https://english.kyodonews.net/news/2020/04/084513e72dd4-update1-2-firms-to-recall-products-under-pm-abes-mask-handout-program.html.

Kyodo News. (2020d, May 25). Abe declares coronavirus emergency over in Japan. https://english.kyodonews.net/news/2020/05/a1f00cf165ae-japan-poised-to-end-state-of-emergency-over-coronavirus-crisis.html.

Kyodo News. (2020e, June 24). Report confirms record economic impact of rugby World Cup 2019. https://english.kyodonews.net/news/2020/06/6ea0af79362f-report-confirms-record-economic-impact-of-rugby-world-cup-2019.html?phrase=north%20korea&words=.

Kyodo News. (2020f, September 26). Japan FY 2021 budget requests likely to top 100 trillion yen for 7th year. https://english.kyodonews.net/news/2020/09/6f03e7c3f735-japan-fy-2021-budget-requests-likely-to-top-100-tril-yen-for-7th-yr.html.

Lies, E. (2020, April 15). Lacking protective gear, Japan's Osaka pleads for plastic raincoats. *Reuter*. https://www.reuters.com/article/us-health-coronavirus-japan-raincoats-idINKCN21X0RO.

Liptak, K. (2020, March 13). Trump declared national emergency – and denies responsibility for coronavirus testing failures. *CNN Politics*. https://www.cnn.com/2020/03/13/politics/donald-trump-emergency/index.html.

Lipton, R.J. (2000). *The Nazi doctors: Medical killing and the psychology of genocide*. Basic Books.

Macnaughtan, H. (2020, June 23). Japan, the Olympics and the COVID-19 pandemic. *East Asia Forum*. https://www.eastasiaforum.org/2020/06/23/japan-the-olympics-and-the-covid-19-pandemic/.

*The Mainichi*. (2020, December 17). Japan COVID-19 deaths increase dramatically amid rising infections among elderly. https://mainichi.jp/english/articles/20201217/p2a/00m/0na/014000c.

Mark, M. (2014, October 12). Cuba leads fight against Ebola in Africa as West frets about border security. *Guardian*. https://www.theguardian.com/world/2014/oct/12/cuba-leads-fights-against-ebola-africa.

*News on Japan*. (2020, June 12). Distribution of Y46.6 Billon "Abenomasks" likely to wrap up on Monday. https://newsonjapan.com/html/newsdesk/article/127457.php.

*Newsweek*. (2020, April 3). 安倍首相、感染防止策は「集団免疫」獲得を直接の目的とせず [Prime Minister Abe: The acquisition of "herd immunity" was not the primary objective]. https://www.newsweekjapan.jp/stories/world/2020/04/post-92982.php.

NHK. (2020, November 21). Go Toトラベル・イート運営見直しを表明・首相 [Prime minister announced operational reviews of "Go to Eat" campaign]. https://www.nhk.or.jp/politics/articles/statement/48773.html.

O'Connor, D. (2020, January 21). Japan PM Shinzo Abe presses forward with casino resorts in National Diet address. *Casino.Org*. https://www.casino.org/news/japan-pm-shinzo-abe-presses-forward-with-casino-resorts/.

Ohashi, Akiko. (2020, November 12). 新型コロナ「7段階モデル」で今冬の流行を予測 [Novel strain of coronavirus: "7 step models" predicting the pandemic in winter]. *Toyo Keizai*. https://toyokeizai.net/articles/-/386189?page=3.

*On Cuba News*. (2020a, February 4). Cuban doctors against coronavirus in China. https://oncubanews.com/en/cuba/cuban-doctors-against-coronavirus-in-china/.

*On Cuba News*. (2020b, August 18). Cuba sends a new team to Venezuela to fight COVID-19. https://oncubanews.com/en/cuba/cuba-sends-a-new-team-to-venezuela-to-fight-covid-19/.

Onishi, N. (2007, September 12). Prime minister of Japan to step down. *New York Times*. https://www.nytimes.com/2007/09/12/world/asia/12cnd-japan.html.

*Our World in Data*. (2020a). Coronavirus (COVID-19) testing. Retrieved 22 October 2020, from https://ourworldindata.org/coronavirus-testing#note-2.

*Our World in Data*. (2020b, June 30). Emerging COVID-19 success story: Vietnam's commitment to containment. Retrieved 22 October 2020, from https://ourworldindata.org/covid-exemplar-vietnam.

Padilla Torres, M.A., Saez, R.G., Soler, A.R., Bermudez, O.H., Bargaza, M., Chavez, G.S., & Sotero, L.M. (2020) *Beyond borders: International map of Cuban medical cooperation*. MEMO Publishers.

Park, J.M. & Takenaka, K. (2020, March 23). Governor of Japan's capital urges cooperation to avoid city lockdown. *Reuters*. https://www.reuters.com/article/health-coronavirus-tokyo-idINKBN21A19W.

Perez, Giovanni. (2020, May 25). Stay Japan: The official traveler. *A Japanese Travel Magazine*. https://issuu.com/stayjapanmedia/docs/2020_autumn_magazine_.

Phillips, T. & Giuffrida, A. (2020, May 6). "Doctor diplomacy": Cuba seeks to make its mark in Europe amid COVID-19 crisis. *Guardian*. https://www.theguardian.com/world/2020/may/06/doctor-diplomacy-cuba-seeks-to-make-its-mark-in-europe-amid-covid-19-crisis.

Phillips, D. & Augustin, E. (2018, November 23). Thousands of Cuban doctors leave Brazil after Bolsonaro's win. *Guardian*. https://www.theguardian.com/global-development/2018/nov/23/brazil-fears-it-cant-fill-abrupt-vacancies-after-cuban-doctors-withdraw.

Prashad, V. (2020a, August 11). Why a growing force in Brazil is charging that President Jair Bolsonaro has committed crimes against humanity. *Peoples Dispatch*. https://peoplesdispatch.org/2020/08/11/why-a-growing-force-in-brazil-is-charging-that-president-jair-bolsonaro-has-committed-crimes-against-humanity/.

Prashad, V. (2020b, August 25). Why Cuban doctors deserve the Nobel Peace Prize. *MR Online*. https://mronline.org/2020/08/25/why-cuban-doctors-deserve-the-nobel-peace-prize/.

Radio Havana Cuba. (2015, June 2). Cuban doctors attend to over 4,600 Nepal earthquake victims. https://www.radiohc.cu/en/noticias/nacionales/58038-cuban-doctors-treat-over-4600-nepal-earthquake-victims.

Reinberg, S. (2020, August 13). Sweden's COVID policy didn't create herd immunity. *WebMD*. https://www.webmd.com/lung/news/20200813/swedens-no-lockdown-policy-didnt-achieve-herd-immunity.

Rummler, O. (2020, December 16). Emails show former Trump health appointee advocated herd immunity strategy. *AXIOS*. https://www.axios.com/trump-herd-immunity-coronavirus-emails-ed33f289-e178-40a1-9bec-27c6d55c2f2e.html.

Sampathkumar, M. (2018, October 16). Last surviving prosecutor at Nuremberg Trials says Trump's family separation policy is "crime against humanity." *Independent*. https://www.independent.co.uk/news/world/americas/trump-border-crisis-nazis-nuremberg-trial-ben-ferencz-family-separation-migrants-un-a8485606.html.

Schumaker, E. (2014, December 2). We'll remember Cuba's Ebola compassion long after Castro's death. *Huffington Post*. https://www.huffpost.com/entry/cuba-medical-personnel-ebola-castro_n_58405570e4b0c68e047f5a87.

Shamdasani, R. (2018, June 5). Press briefing note on Egypt, United States, and Ethiopia. UN Human Rights Office of the High Commissioner. https://www.ohchr.org/en/NewsEvents/Pages/DisplayNews.aspx?NewsID=23174&LangID=E.

Shaw, M. (2020, March 23). "Herd immunity and let the old people die" – Boris Johnson's callous policy and the idea of genocide. *Discover Society*. https://archive.discoversociety.org/2020/03/23/herd-immunity-and-let-the-old-people-die-boris-johnsons-callous-policy-and-the-idea-of-genocide/.

Shimizu, K., Sakamoto, H., Mossalos, E., & Abel-Smith, B. (2020, August 18). Resurgence of COVID-19 in Japan. *BMJ*. https://www.bmj.com/content/370/bmj.m3221/rapid-responses.

Singh, S. (2020, January 29). Coronavirus airline & flight suspensions: Live updates. *Simple Flying*. https://simpleflying.com/worldwide-airlines-china-flights-suspension/.

Statista. 2020. Coronavirus (COVID-19) Deaths worldwide per one million population as of 3 March 2022, by country. Retrieved 3 March 2020, from https://www.statista.com/statistics/1104709/coronavirus-deaths-worldwide-per-million-inhabitants/.

Steen, E. (2020, October 5). Japan could reopen to tourists from spring 2021. *Timeout*. https://www.timeout.com/tokyo/news/japan-could-reopen-to-tourists-from-spring-2021-100520.

Sturmer, J. & Asada, Y. (2020, March 2). Coronavirus could see the Tokyo Olympics cancelled. Is Japan's handling of the outbreak to blame. ABC News. https://www.abc.net.au/news/2020-03-03/coronavirus-will-the-diamond-princess-sink-japans-olympics-dream/12015202.

Summers, J., Cheng, H.Y., Lin, H.H., Barnard, L.T., Kvalsvig, A., Wilson, N., & Baker, M.G. (2020). Potential lessons from the Taiwan and New Zealand health responses to the COVID-19 pandemic. *Lancet Regional Health: Western Pacific*, 4 (2020-100044), 1–6. https://doi.org/10.1016/j.lanwpc.2020.100044. https://www.thelancet.com/action/showPdf?pii=S2666-6065%2820%2930044-4.

Takahashi, R. (2020a, December 8). Support wanes for Go To Travel campaign as coronavirus continues to spread. *Japan Times*. https://www.japantimes.co.jp/news/2020/12/08/national/coronavirus-go-to-travel-study/.

Takahashi, R. (2020b, December 14). Japan to suspend Go To Travel Program nationwide from Dec. 28 to Jan. 11. Reuters. https://www.reuters.com/article/health-coronavirus-japan-travel/japan-to-suspend-go-to-subsidised-travel-programme-dec-28-to-jan-11-media-idUST9N2IA014.

Takenaka, K., & Slodkowski, A. (2020, April 19). As coronavirus infections mount, Japan at last expands testing. *Japan Today*. https://japantoday.com/category/national/as-coronavirus-infections-mount-japan-at-last-expands-testing.

TeleSUR. (2020a, February 2). Chinese doctors are using Cuban antivirals against coronavirus. https://www.telesurenglish.net/news/cuban-antiviral-used-against-coronavirus-in-china-20200206-0005.html.

TeleSUR. (2020b, December 22). Venezuela receives Chinese medical supplies to fight COVID-19. https://www.telesurenglish.net/news/Venezuela-Receives-Chinese-Medical-Supplies-To-Fight-COVID-19-20201222-0010.html.

Tiwari, V. (2020, March 31). COVID-19: Chinese medical team arrives in Venezuela to help battle pandemic. *Republic World*. https://www.republicworld.com/world-news/rest-of-the-world-news/covid-19-team-of-chinese-disease-control-doctors-arrive-in-venezuela.html.

UNI-Global Union. (2020, July 27). International Criminal Court case claims Brazilian government's COVID-19 response is a crime against humanity. https://uniglobalunion.org/news/international-criminal-court-case-claims-brazilian-governments-covid-19-response-is-a-crime-against-humanity/.

Wang, C.J., Ng, C.Y. & Brook, R.H. (2020, March 3). Response to COVID-19 in Taiwan: Big data analysis, new technology, and proactive testing. *JAMA*, *323*(14), 1341–2. https://doi.org/10.1001/jama.2020.3151.

World Bank. 2019. GDP per capita (Current US$). Retrieved 22 October 2020, from https://data.worldbank.org/indicator/NY.GDP.PCAP.CD.

Wyss, J. (2019, November 15). Bolivia's new government expels Cuban officials, recalls its diplomatic staff from Venezuela. *Miami Herald*. https://www.miamiherald.com/news/nation-world/world/americas/article237405369.html.

Young, J. (2020, March 24). Tokyo 2020 Olympics are postponed amid coronavirus pandemic. CNBC. https://www.cnbc.com/2020/03/24/japan-pm-abe-says-he-agreed-with-iocs-bach-on-the-idea-of-delaying-the-tokyo-olympics-for-one-year.html.

# South and Southeast Asia

# 3 Witnessing amidst Distancing: Structural Vulnerabilities and the Researcher's Gaze in Pandemic Times in Relation to Migrant Workers of India and Singapore

BY AMRITORUPA SEN AND JUNBIN TAN

*Two old and visually impaired men, about 60 years of age, walked from Delhi, India's national capital, towards their village in Uttar Pradesh (India). Both men had begged for a living on Delhi's streets for the past 40 years. People passing by them on those dust-filled streets would drop spare change into their open palms or provide them with small amounts of food. Now, during the COVID-19 pandemic, as people remained in their homes or hurriedly returned to their home cities, towns, and villages when lockdown was declared, only dust remained on Delhi's streets. Without passersby to feed them, the men found it impossible to live even as beggars and were forced to return home. The problem, and a serious one, is that home is 800 kilometres away. These men, with the few belongings they carry in a cloth knotted into a bag, and without monetary savings, could only travel home on foot. There was little option – either embark on an arduous journey home with blistered feet or wait for a slow death by starvation. COVID-19, as much as it managed to plague Indian cities with quarantining and social distancing protocols, was thus a less immediate concern for those with no means of livelihood and no shelter.*

 *From the computer screen in the safety of my home in a middle-class neighbourhood in West Bengal that doubled as my mother's teaching space – she taught mathematics to school-aged children who had to stay home during the pandemic – I watched as these men took cautious steps down the Delhi–Uttar Pradesh highway. By no means were they alone. Hundreds of thousands of others, albeit of slightly better economic status – odd-job workers, builders, factory workers, cleaners, and so forth – but who lost their job amid the lockdown, throttled their way from the city back to their village homes. Nonetheless, they were forced to march on, as the government declared that all public transportation had to be cancelled. The men and women continued in the heat and the rain. Their journeys took weeks, with only one another's voices for a sense of security. In the wake of their journeys, hundreds of media reports circulated: "Faceless and dispossessed: India's circular migrants in times of COVID" (Deshingkar, 2020), "The working person's right to life" (Chaudhuri, 2020), and others that revealed the distress of the migrant underclass in India. I looked on* these people as they marched and wondered how people were asymmetrically affected by COVID-19.

Amritorupa Sen

While some found themselves locked out of Delhi, others in Singapore were locked in. Bunk beds lined the walls of a tiny room. Towels hung from bed railings, creating a semblance of privacy. Men shifted left and right in their beds as they fiddled on their mobile phones, many shirtless as they tried to keep cool in Singapore's heat and humidity. As many as a dozen foreign workers – men from "developing" Asian countries, employed here in low-wage, labour-intensive jobs – shared one dormitory room. Men who lived in rooms on the same floor shared a dirty, cockroach-infested toilet. The balcony outside these rooms, where they used to gather to pass time, and from where they could see other unadorned buildings in this fenced-off compound, registered silence. These dorms were cordoned off after COVID-19 was found to be spreading among workers. Police stood guard and the men were told to stay in their rooms, as they awaited COVID-19 testing. Uncertainties about who was infected and the fear of whether one's pay and employment status would be affected loomed large, as they endured the unbearable heat and stuffiness of being confined.

In the city-state of Singapore, ordinary residents, save for some essential workers, were asked to stay home. In the safety of my apartment, where actual distancing is possible, I tuned to local news as reporters covered the surge in dorm infections, and state officials came forward to convince the public that the situation was under control. Yet, photos and videos of workers' dormitories circulated on social media and the Internet, which traced the problem to enduring neglect. Migrant workers had fallen through the cracks of state surveillance that was intent on curbing the spread of COVID-19, and whose success was internationally acclaimed. Cramped housing, rooted in decades of neglect of workers' well-being, also meant that physical distancing was unfeasible. While those who tested positive were moved to other facilities, others were locked in with their possibly infected roommates as they awaited testing. They were kept inside to protect us who were outside. Deeply entrenched inequalities came to light, which, for months, confronted our moral sensibilities, albeit with varied public responses – from remorse and guilt to the denial of one's complicity. How does one write, as a member of a society complicit in the migrant workers' plight, and whose privilege to remain in an enclosed safety is predicated on their segregation and being made vulnerable?

Junbin Tan

How can we conduct research, especially research that seeks to be politically engaged, when we are physically and socially distanced from our interlocutors? How do the relationships between us and our interlocutors take shape within broader structural inequalities, along the lines of social class, the urban-rural divide, and citizenship? Amidst the COVID-19 pandemic, we – your co-authors

Amritorupa Sen in West Bengal, India, and Junbin Tan in Singapore – in our homes and on our televisions and computers, watched as migrant workers all over India left their city workplaces for their rural village homes, and foreign migrant workers in Singapore were quarantined in the overcrowded, unsanitary quarters in which they live. Research has shown that the migrant underclass are disproportionately affected in past and present pandemics, especially among the underprivileged with fewer resources to protect themselves (Ciotti, 2020; Ghosh, 2020; see also Quesada, Hart, & Bourgois, 2011), among those who have become targets of xenophobia (Chuvileva et al., 2020; Dewanto, 2020; Pellegrino, 2020), and among those whose marginal status was exacerbated through policing, surveillance, and medical intervention (Castañeda & Lopez, 2020; Mason, 2012; Team & Manderson, 2020). Sangaramoorthy (2020) argues, "for many, COVID-19, like other conditions, is part of the broader context of inequity and suffering – an enduring crisis rather than a break from 'normal.'" COVID-19, as Team and Manderson (2020, p. 671), building on Quesada et al. (2011), suggest, "thrived on the structural vulnerabilities that worldwide shape access to quality and security of housing." We echo these scholars' views and further consider what it means for us to conduct research as individuals who belong to these social structures and are complicit – not only in this instance, but since time immemorial – in exacerbating the structural vulnerability of migrant workers.

Beginning in March 2020, when the pandemic wreaked havoc in the countries in which we live, we kept each other in conversation as we tried to make sense of the perils of the migrant workers in both contexts. We also lamented our inability and unwillingness, even if we had the option of being mobile, to leave our homes for research and other purposes. While at home, we only had access to the world beyond our four walls through mass and social media (and the occasional venturing out to obtain groceries and living essentials). We remained attuned to the news and to social media, which showed the fragility of lives, health, jobs, and economies outside. Research for this chapter combines our observations of information from these media with our experiences and reflections on engaging with them. That said, the media content framing the inequalities and vulnerabilities that migrant workers in Singapore and India face, we argue, is reflected uncannily in our research method – the lack of interaction with migrant workers in pandemic times. However, we also watched and listened as researchers sought to navigate the moving tensions between "noticing" (Tsing, 2015) and political "witnessing" (Thomas, 2019), between thinking about migrant workers through the media and thinking about how our writings might matter for the quotidian politics that limit their life chances. If we take the researcher's gaze as a form of relationality between the researcher and the subject, it will be clear, as we will see, that this relationality does not develop outside the social structures and discourses that subtend our relations –

as regular citizens – with the disenfranchised migrant workers whom we learn and write about. As researchers writing about structural "distance," we found ourselves inhabiting the privileged end of this very "distance" that we write about and critique.

On our television and computer screens, we are assaulted by moving images and mediated voices of the individuals and communities who remain exposed: left on the streets or cordoned off in workers' dormitories where the virus is known to circulate. To be clear, this chapter is not about the migrant workers *per se*, but about our relationship – as middle-class members of our respective societies, which you, our reader, may partially share – with the migrant underclass. As we struggled to protect ourselves – our health, our polities, and our economies – during the pandemic, the migrant workers faced the brunt of being exposed to infection. They were, in other words, "let die" (Foucault, 1976, p. 241). While the Foucauldian evocation is a useful heuristic, if "life" and its governance involve not only political apparatuses but also the public, then what do we make of our responsibility to individuals and communities whom we "let die" so that we can live? Attempts to look and listen from our enclosed safety cannot be neatly separated from our structural distance from the workers; their "letting die" is mired with our desire to make them live.

We approach this concern through analyses of the structural conditions that prevented our interactions with migrant workers, and our thoughts on living and doing research in the pandemic. We narrate our observations in separate sections – on Singapore through Junbin's gaze, and on India through mine. Each section's discussions will shed light on our structural relations with the migrant workers we write about in the absence of physical interaction as well as our sense of helplessness as we attempt to overcome this distance. While this project has our six years of friendship and intellectual conversations as its absolute beginning, it has a more concrete starting point in another project – a collection of short essays on COVID-19 in various Asian contexts – for a public forum that we undertook with our friends in universities across Asia.[1] These writing projects, our conversations with others whom we lived with during the pandemic who do not necessarily understand the migrant workers' plight, and other small deeds that we enacted represent our attempts to struggle against the systems of inequality that we gaze at, which we are complicit in, but which we hope to rectify.

**The Politics of the Gaze and of Writing**

We begin by contemplating Judith Butler's (2012) inquiries, such as "whether any of us have the capacity or inclination to respond ethically to suffering at a distance and what makes that ethical encounter possible?" and "what it means for our ethical obligations when we are up against another person or group," finding ourselves "invariably joined to those we never chose?" (2012, p. 134).

The concerns of distance, the impetus to respond, and the ambiguity of forms that political action can take resonate with our encounters with the images of migrant workers in pandemic times. Before we examine Butler's text, which we set in dialogue with the work of Jacques Lacan (1978), we turn our attention to two recent works whose discussions concern the politics of the gaze, and whose advice, reminders, and triggers for reflection are too important to ignore.

Anna Tsing, in *The Mushroom at the End of the World* (2015), advocates attentiveness to the "divergent, layered, and conjoined projects," caught in an "interplay of temporal rhythms and scales" (pp. 22–3), that human and nonhuman actors engage in as they live amid the ruins of modern capitalism. Noticing, in Tsing's terms, activates the senses. What Tsing notices, as she traces matsutake mushrooms' lingering smell across temporal and cultural registers, the durability of trader markets that flirt with but resist incorporation into capitalist systems, etc., are spaces of possibility that exist within regimes of domination. Inspired by the surprise that "alternative world-making projects" (Strathern, 1999) and "interplay across divergent projects" (Haraway, 2003) afford, she "traces out … interruptions of one kind of project by others" (Tsing, 2015, p. 293). To notice, in other words, is to "trace" acts and projects of resilience, and acts of refusal by our interlocutors that run up against hegemonic institutions (Simpson, 2014). Doing so not only recognizes our interlocutors' political agency, but also locates the definition of "political action" in their hands. Tsing remains silent on the prescription of action apart from noticing. Perhaps, for her, our role, in a limited but powerfully reflexive way, should first and foremost be to notice and to empower our readers to do the same; anything beyond this is presumptive.

We compare Tsing's argument with that of Deborah Thomas, who, in *Political Life in the Wake of the Plantation* (2019), accords greater significance to researchers' responsibilities.[2] To "bear witness," Thomas tells us, goes beyond the visuality of "eye-witnessing" and in doing so "destabilizes the ocular-centrism that had facilitated imperialism" and other hegemonic formations (Thomas, 2019, p. 2, citing Tait, 2011). Witnessing, for Thomas, is not limited to rights, status, and categories, but is instead "co-performative" and involves co-temporality, affect, and embodied practices (2019, p. 2). Beyond seeing, witnessing demands "response-ability" (2019, p. 2, citing Gordon, 2008) – our "attentive embodied care" for another individual and our responsibility to "respond to the psychic and sociopolitical dynamics in which we are complicit" (2019, p. 3). While "bearing witness" involves the attentive tracing (see also Tsing, 2015) of Tivoli Garden's residents' lives from the garrisons where they live to the Jamaican colonial plantations that continue to haunt contemporary politics, Thomas envisions her project as an explicitly political one. To that, we ask: what does it mean for us to be politically involved, and to claim responsibility especially in the lives of others? Thomas may deem these questions less relevant, since co-performative witnessing "destabilizes the boundaries between self and other, knowing and feeling, [and] complicity and accountability" (2019, p. 3), but we insist on recognizing a self-other dichotomy that exists prior to acts of witnessing, which we will have to confront when addressing witnessing as an ethical practice. Treading the slippery slope of what political involvement means, we also ask: what if circumstances do not permit direct, embodied involvement? Can one still "witness" in situations that call for physical distancing?

The pandemic forces us to confront these questions, of gazing, of complicity, of guilt, of responsibility, and of our (in)capacity to act while observing social distancing. "What does it mean to observe from a distance? What kinds of responses, and response-ability can/should one have, when gazing from afar?" Butler's questions resound alongside Thomas's and Tsing's respective viewpoints. However, for Tsing and Thomas, the motivations and the moods that underlie the gaze are not the focus of analysis. "Witnessing," in particular, when proposing that the witnessing of injustice must be accompanied by political action, takes both the gaze and its consequent action to be virtuous. Further to the opposition that we provide by comparing Tsing's work with Thomas', we offer another counterpoint between these works and Jacques Lacan's ideas, which brings to attention and calls into question the impulses that undergird one's inquisitive, but anxiety-inducing, and possibly voyeuristic gaze.

The younger Lacan's gaze was drawn to the tin can, or more precisely, to the blinding flashes of light that reflected as it floated on the sea:

> It's a true story. I was in my early twenties or thereabouts, and at that time, of course, being a young intellectual, I wanted desperately to get away, see something different, throw myself into something practical, in the country say, or at sea. One day, I was on a small boat, with a few people from a family of fishermen in a small port. At that time, Brittany was not industrialized as it is now. There were no trawlers. The fishermen went out in this frail craft at their own risk. It was this risk, this danger, that I loved to share. But it wasn't all danger and excitement – there were also fine days. One day, then, as we were waiting for the moment to pull in the nets, an individual known as Petit-Jean, that's what we called him – like his family, he died very young from tuberculosis, which at that time was a constant threat to the whole of that social class – this Petit-Jean pointed out to me something floating on the surface of the waves. It was a small can, a sardine can. It floated there in the sun, a witness to the canning industry, which we, in fact, were supposed to supply. It glittered in the sun. And Petit-Jean said to me – *You see that can? Do you see it? Well, it doesn't see you!* (Lacan, 1978, p. 95, italics in original)

Rather than being drawn *himself* to the object, as various analyses suggested (Copjec, 2015), it would be more accurate to say that Lacan's *gaze* was drawn to it – he was *compelled by*, as it was impossible not to notice, the can's glaring brightness. Krips (2010, p. 92) rightly noted, in our view, that Lacan's discomfort was "occasioned by a lurking political guilt at his own privileged position in relation to the working-class fishermen." Through displacement then, the tin can – a piece of waste that floated on the waves, that is, something unworthy – stood for and "brought to the surface … in the young Lacan a palpable and excessive anxiety, even shame, about who he is and what he was doing" (2010, pp. 92–3). Not only does one gaze at the object, the gaze also "'turns around', that is, reflexively turns back upon Lacan, at the same time as it switches from active to passive voice" (2010, p. 93), *but* – and this ought to be made clear – not as a self-initiated reflection but as one invoked by an encounter with an external. In the same way, we felt compelled to acknowledge the migrant workers' fates shown to us in the media. Guilt and shame, and the desire to react and reflect, followed from these mediated encounters. To reiterate our questions inspired by Butler (2012), *What does it mean, for us, to observe from a distance? What does it mean, for us, to feel compelled to respond in writing when gazing from afar?* What are the symptoms of our responses?

Butler (2012) addresses these questions dialogically and leaves us with more thoughts than definite answers. Her argument develops from an existentialist position that "everyone is precarious, and this follows from our social existence as bodily beings who depend upon one another," which she filters through a particularist argument that "our precarity is to a large extent dependent upon the organization of economic and social relationships, the presence or absence of sustaining infrastructures and social and political institutions" (2012, p. 148). Debunking the existentialist position, which remains an important pivot of her theorizing, she argues that "since [the existential claim] must be articulated in its specificity, it was never existential … precarity is indissociable from that dimension of politics that addresses the organization and protection of bodily needs" (2012, p. 148). As we encountered the images of migrant workers, it was as Butler had described:

> When any of us are affected by the sufferings of others, it is not only that we put ourselves in their place or that they usurp our own place; perhaps it is the moment in which a certain chiasmic link comes to the fore and I become somehow implicated in lives that are clearly not the same as my own … Some media representations of suffering at a distance compel us to give up our more narrow communitarian ties and to respond, sometimes in spite of ourselves, sometimes even against our will, to a perceived injustice … The kinds of ethical demands that emerge through the media in these times depend on this reversibility of proximity and distance … *Certain bonds are actually wrought through this very reversibility, however incomplete it is.* (Butler, 2012, p. 149, italics in original)

Butler's thoughts on political commitment and its conditions of possibility are particularly relevant for our work, as we think of the deep-rooted structural conditions of poverty and exploitation that undergird the lives of people we think, feel, and write about, but also – and Butler does not address this point as fully – how we are implicated in these unequal structural relations *prior to* the encounter. Despite the conditions that double as constraints to claims of empathy and understanding, we find inspiration and comfort in Butler's words that "certain bonds … however incomplete" can be wrought such that we can begin to participate politically in the lives of others (see also Butler, 2012, p. 149).

In the following, we peer beyond each layer as we narrate these stories, and the stories behind these stories, through partial readings of our encounters in India and Singapore. These stories *do not* tell all of what is going on "out there" in India and in Singapore – to claim to tell such stories is to be complicit in gazing – but are instead revelatory of our encounters in these places (Borneman & Hammoudi, 2009). We take inspiration from W.G. Sebald, who moves between personas and perspectives, and past and present, in *Austerlitz* (2001), as we move between the mediated encounters we narrate, our situatedness in the societies where we *and* the migrant workers live, and our thoughts on these encounters. Sebald's "periscopic" analytic and writing style (see also Silverblatt, 2007, p. 83), we propose, allows for us both to adopt, following ethical calls to our (in)action (see also Lacan, 1978), the positions and the politics of "noticing" (Tsing, 2015) and "witnessing" (Thomas, 2019) at different moments, when we are confronted with images that we never chose to see, of people with whom we are conjoined.

**Singapore: A Lockdown of Empathy?**

"Observing a different world" best described what I, Junbin Tan, felt of my position vis-à-vis the pandemic in April and May 2020, as I learned about migrant workers' predicament when news, images, and soundbites of their situations played continuously on television, on radio, and on my Facebook feed. We were all quarantined, but the conditions in which we lived, our risks of exposure to infection, and the anxieties we faced drastically differed. The image Whiteford (2020, p. 7) painted in "A Room with a View" that she "live[d] in a space in which the virus does not yet seem real," which gave her "a strange sense of both power and powerlessness, competing for [her] balance … [an] 'off balancing' that most engages and disturbs [her]," resonated with me. From my home near Little India, the silence of the streets on weekends – South Asian workers used to gather here on their day off – was a reminder of the pandemic. Just as Whiteford was struck by the contrast between the beauty of her garden view and the millions who had lost their jobs and were without medical insurance, I too was disturbed by the gap between my situation and migrant workers'. I was among the many Singaporeans who were safe at home, shielded from the few others in some neighbourhoods who might be infected. The act of gazing, from the safety of one's space, stands in contrast to what one gazes at, situated elsewhere. In crowded dormitories, since infection rates among workers were high, healthy workers in each room were as exposed to possibly infected others as they were safe from workers in other rooms. Beyond the distance that separates those spaces we inhabit, I sensed the distance – an effect of distancing – between our social worlds.

After months of rolling out health advisories, and weeks after imposing stay-at-home notices to citizens who returned from abroad, the Singapore government issued a nationwide stay-at-home order on 7 April 2020. The Circuit Breaker, as the government termed it, mandated physical distancing in the service of keeping households and individuals apart so that infection did not travel. We, ordinary residents, could no longer leave our homes except to buy food and essential items; wearing of surgical or government-issued masks outside was enforced; non-essential workers worked from home and some were asked to go on unpaid leave; schools conducted classes virtually; and gathering and dining spaces were closed. "Contact tracing" through mobile devices and scanning in and out at locations such as shopping malls, eateries, and hair salons was implemented. "Safety ambassadors" – many of them air crew or airport staff whose jobs were affected – were employed by the government and assigned to food centres and other areas to ensure that members of the public kept their distance (Lai, Tang, Kurup, et al., 2020).

In March and April, citizens abroad were flown back on government-chartered flights, placed on stay-home notice if they tested negative, and quarantined at designated sites if they were positive. I learned through social media that a friend who was pursuing graduate studies abroad returned, tested positive, and was quarantined at a resort. He lauded the government's effort in reacting quickly to the globalizing pandemic. Similar thoughts, views, and sentiments circulated in state-affiliated media channels, on social media, and in conversations. At the time, we were inundated with slogans like "Stay Strong, Stay Safe, Stay United" by the Ministry of Culture, Community and Youth (2020) as well as "together we keep Singapore strong" and "stay at home, do the right thing" by the Ministry of Health, all appearing on television commercials, radio channels, and posters placed in public spaces. Watching the pandemic develop globally, as infection and death figures skyrocketed, it was hard for me not to feel appreciative of state efforts, and comforted by and satisfied by the stability at home. I grew even more impressed when the government stepped up public health measures, distributed monetary reliefs and masks, and supported businesses affected by the lockdown.

In early April 2020, however, we – the government and members of the public – sobered up to case after case of COVID-19 outbreaks in migrant worker dormitories. Infection cases were traced to two dormitories, after which the government announced the quarantining of all 19,800 residents for 14 days, starting 5 April (Tang & Co, 2020). More dorms were declared isolation areas over the next two weeks, their residents placed on quarantine, and construction projects gradually placed on hold. On 20 April, the Ministry of Manpower announced that all construction work was to cease, and migrant workers were put on a 14-day stay-home notice until 4 May (Lai, 2020). Twenty-five dormitories were declared "isolation areas" and spaces were provided at vacant public housing, exhibition halls, military camps, and unused cruise ships to facilitate distancing. About 10,000 healthy workers in essential services – for example, maintenance and cleaning – "a skeleton staff to keep the country going," were housed separately (Y. Tan, 2020). Yet, at one point, over 1,400 confirmed cases were found in one day at the dormitories (Cai & Lai, 2020), and 19.4% of the residents at one 13,000-worker dormitory tested COVID-19 positive (Koh, 2020). The dormitories, as we described, were overcrowded and clearly unsuitable for distancing. By 6 May, exactly a month after the first dormitory lockdown, a total of 17,758 COVID-19 cases had been found among dormitory residents, constituting 88% of Singapore's 20,198 confirmed cases (Koh, 2020).

The local newspaper, the *Straits Times*, reported differences in outbreak between the workers in dorms and the rest of the population (Khalik, 2020), which the British Broadcasting Corporation (BBC) called "a 'pandemic of inequality' exposed" (Y. Tan, 2020). Migrant workers' poor housing conditions were most evident: while the Employment of Foreign Manpower Act states that employers must provide "acceptable housing," the relegation of these responsibilities to private, capitalist market forces meant that cost cutting usually resulted in less than adequate provision; workers were housed in commercially run, purpose-built dormitories (the subject of attention during COVID-19), temporary quarters on worksites, and, less frequently, in shop-houses and public flats (Ye, 2014). These overcrowded spaces, where as many as 60 people were housed in a single room (Hamid & Tutt, 2019, p. 529), provided the perfect conditions for the infection to spread. Commuting in packed vans to sites where they worked and working and resting alongside men from other dormitories hastened the spread (Y. Tan 2020). As news articles pointed out, and the state and the public had to admit, the seeds of the pandemic were sown many decades before the pandemic. To view this as *difference* in outbreak (Khalik, 2020), while already sobering, downplays that deeper *inequalities* exist, and persist, and go unresolved: that the workers, mostly from South Asia, work in dangerous or undesirable jobs in exchange for low wages (Abdullah, 2005; Ye, 2014), and deportation laws, enforced or used as a threat, "construct these workers as use-and-discard economic subjects" (Bal, 2015, p. 267; see also Hamid & Tutt, 2019). Videos of overcrowded dormitories and interviews with workers circulated,[3] confronting me and others with a parallel reality: we live alongside them as they build our apartments and maintain our environment, and we rely on and exploit their labour – which we see only in our peripheral vision. COVID-19

Migrant Workers of India and Singapore    95

Image 3.1. Migrant workers are pictured standing outside their dormitory rooms at Cochrane Lodge II, which was declared an isolation area because of the outbreak of COVID-19, in Singapore, 16 April 2020 (Reuters/Edgar Su).

brought the peripheral vision to the forefront, forcing us to address our neglect of this "oft-forgotten group," as Mokhtar (2020) termed it. Our neglect appears more apparent vis-à-vis the government's success in reducing infections in the general public, including Singaporeans like me returning from abroad.

Public attitudes often take shape in relation to news obtained through mainstream and popular media, and this relationship became more apparent during quarantine as we learned about the world only through the media. The disparities between migrant worker infection rates and the rest of the population, and the government's fast but hard-handed approach, were much talked about. Multiple discourses could be identified. The government admitted to their oversight in predicting the dormitory outbreaks, and Manpower Minister Josephine Teo promised that the government would deal with the housing problem that had long festered (Romero, 2020). The interventions made on 5 April and thereafter were also framed as "helping migrant workers"; the workers, Labour Chief Ng Chee Meng said, "helped build our homes, and contributed in many other areas. In time of crisis, it is only right that we too look after them" (Lai, 2020). Images of the government providing additional housing and quarantine areas – in exhibition halls (the EXPO), unused public housing, unbooked hotels, docked cruise ships, and so on – were broadcast on television and in the news. However, other perspectives also emerged, such as Ng's (2020) observation that "Singapore has treated the recent spike in COVID-19 cases among foreign workers as a separate outbreak," which echoes Mohan Dutta's argument that "the idea of reporting two different numbers [makes] the inequalities even more evident. One might even go so far as to say it's [an example of] 'othering'" (cited in Y. Tan, 2020). Y. Tan also narrated the story of Zakir Hossain Khokan, who stayed in a dormitory room with 11 others:

> "Day and night, we are just inside one room," [Khokan] says. "It's actually torturing our mind. It's like jail … Then we can't social distance because there's no space." Having already caught COVID-19, recovered, and gone back to work, [Khokan] thought his worst days were behind him. His dormitory was declared cleared of the virus in June. But last month a new cluster developed at the dorm, and like thousands of migrant workers, he was ordered back into quarantine. (Y. Tan, 2020)

Images of migrant workers remaining confined in dormitories and news of "a recent spate of suicides and attempted suicides" among them (Geddie & Aravindan, 2020; see also Romero, 2020) circulated alongside heart-warming "unauthorized" videos of workers, with surgical masks on, dancing alongside healthcare workers in protective suits at one care facility (Mohan, 2020). These multiple, sometimes contradictory, discourses were what I and other Singaporeans were exposed to and probably reflected on while staying home.

Our views, at least among my family and immediate social circle, were as varied and conflicted as the portrayals we saw. These ranged from empathizing with workers' risks of being exposed to COVID-19 to dismissing their suffering by saying that most of them were young, strong, and unlikely to fall seriously ill; from acknowledging structural inequality to a disavowal, pointing instead to how workers still receive salaries while under quarantine;[4] from seeing Singaporeans' – not only the government's – complicity in benefiting from the exploitation of cheap labour and condoning their poor living standards, which retired diplomat Tommy Koh criticized as Third World (Romero, 2020), to deflecting these questions that have become too apparent to be ignored. In our midst were some individuals, safely quarantined at home, who remained stricken with what I term the "lockdown of empathy," which preceded the COVID-19 pandemic. The structural distance that developed over years and decades, which separates migrant workers spatially and socially, discursively marked out in popular sentiment the distinction between us, the "make live," and them, the "let die" (Foucault, 1976, p. 241). Our emphasis on the government's and the public's attempts to get Singaporeans to return and to control the outbreak among "imported cases" was accompanied and contrasted with the overshadowing of a concern for outbreaks in workers' dormitories. I had found it difficult to reach out to migrant workers, and felt uncertain whether I should, since interacting with them might endanger my family; that these considerations were enmeshed with structural constraints attests to my belonging safely on one side of the divide. Apart from persuading others to identify with migrant workers' distress and speaking against vilifying them (and writing this chapter), I found that I could not do much – and I found "legitimation," in part, in physical distancing.

Our neglect, as Lacan would have it, is rooted in our Unconscious. Yet, our eventual confrontations with these outbreaks are also instances where occasions for reflection could develop, similar to that moment that Lacan described, when he was captured by the blinding flashes out at sea. I observed in conversations in April and May 2020 that some people more actively voiced their displeasure at how migrant workers were treated and began to acknowledge that they deserved better housing. Beyond "noticing" (Tsing, 2015), they could perhaps be seen as engaging in "witnessing 2.0" (Thomas, 2019) through their involvement in issues that they were confronted by, reflected on, and now cared about. In a pandemic, however, many found it difficult if not impossible to reach out to migrant workers, although many more vocalized the need to improve migrant workers' housing. It is apparent that the rampant COVID-19 outbreaks in workers' dormitories will be etched in our history. However, it remains to be seen how long our motivation to care for their rights to better housing and other living conditions will keep burning.[5] Or will we, after the pandemic, return to a state of lockdown of empathy? How adequately will the government's plans to improve the migrant workers' housing actualize (Romero, 2020), and are citizens and residents motivated to push for change? Are we willing to interact with and reach out to the workers after distancing measures are lifted, when we are no longer protected by distance?

## India: Re-visibilizing the Migrant Underclass

On 24 March 2020, Narendra Modi, India's prime minister, declared a 21-day nationwide lockdown suddenly and with urgency. Modi laid out the prescribed rules for prevention, foremost of which was "maintain social distance from others" and "stay at home." This, he stated firmly to all Indians, "is to save yourself, and your family" (CNN News 18, 2020). He repeated, with hands folded in earnest appeal, "stay wherever you are, do not go out." COVID-19 by then had started spreading. State governments along with hospital authorities were frantically preparing new medical set-ups for COVID-19 patients – more hospital wards and beds, more respirators and ventilators, and even medication that was thought to be used in preventing COVID-19. Fear and anxiety started to loom in the air. The nationwide lockdown was seen as an emergency step in the face of an imminent threat. However, as the state made its preparation, a large group of people – the migrant underclass – who are in fact Indian citizens, no less than other Indians, fell outside the orbit of its policy considerations.

When I, Amritorupa, awoke the following day, our middle-class neighbourhood was empty, shops were shut, and no humans could be seen in close vicinity. Usually on any given day in the morning, women from the nearby slums would be seen walking hurriedly along these roads, as they headed to work in several middle-class households, including my own, as domestic helpers. Domestic workers, who live in densely populated slum dwellings located amidst more developed, urbanized dwellings, had to interact with their employers and neighbours for their daily sustenance. The domestic workers' houses, usually one room, can barely accommodate their members. Workers share water and toilets with fellow slum dwellers; hence they are, particularly in government and healthcare workers' views, seen as dangerous "carriers" (Viswanath, 2020) of COVID-19, people whose movement had to be restricted. Most shopkeepers kept their shutters down, as the sale of non-essential goods ceased, including e-commerce home delivery services (Singh, 2020). Public transport did not operate, and public spaces such as malls and parks shut down operations (Hindu BusinessLine, 2020). Day by day, with increasing silence and decreasing noise, I realized what social distancing meant, and its emerging practical and social implications. The social and economic landscape around me was changing, and my only access to the world outside was through the media.

At a time when everyone was supposed to self-isolate, the most striking images on the news that caught the nation's attention were those of thousands of people, mostly men with their families, stranded in big cities, and waiting to go back home (*New Indian Express,* 2020). Helplessness showed on their faces, as much as confusion about the situation, which neither they nor we, the audience, would have fathomed before the start of the pandemic. There was a mass exodus of the migrant underclass – informal, daily-waged, poorly paid migrant workers whose workplaces in the cities had suddenly closed, and who found themselves without work. With no work and income, and with little savings, many of them embarked on their way back to their homes in other states. And with no provision of public transport, they started walking, walking hundreds of kilometres (Yashee, 2020). As these migrant workers walked, the rest of India witnessed what many called "the largest exodus since partition" on foot. Men, women, and young children were seen in the most distressing situation, walking immensely long distances along the highways, with little knowledge of how long it would take for them to reach home – a destination that was far off and as uncertain as going back to work (Venkatraman et al., 2020). The migrant underclass became visible to the rest of us, albeit in a different way, one in which looking at them was deeply intertwined with our own selves and identity. I could not distance myself from what I was seeing. Within the comforts of home, little did I think of those for whom home was not an easy reach. For the next three months, I found myself following their lives, through different media channels; reacting, questioning, and struggling with the discomfort of witnessing long-known forms of discrimination in a new light.

In India, the migrant underclass largely belongs in the informal sector. These migrant workers – unlike the neat frames that Junbin examined based on state categorization of the resident population along the lines of occupation, social class, and citizenship – are a diverse and dynamic group, often neither clearly segregated nor easily identifiable. According to the latest National Sample Survey Office (NSSO) data, liberalization skewed the economy even further, with the majority of jobs created in the last few decades emanating in the informal sector – largely unregulated and lacking social security or benefits (Salve & IndiaSpend.com, 2019). These workers are seen working in different sectors, from construction to factory work, domestic services, or the hotel industry. Encountering them in our daily lives is not unusual, since they live and work alongside us – more precisely, for us. I am reminded of Kasturi aunty, our domestic helper, who cooks and cleans in five households including ours. She came to West Bengal as a 10-year-old, from Orissa, a neighbouring state, and learned how to cook our local cuisine perfectly, since working for middle-class Bengali households was her most reliable source of livelihood. Krishna, the middle-aged man who delivers milk in our neighbourhood, took up the trade from his father. Together, father and son have been delivering milk for over 40 years and communicate with us fluently in Bangla, our local language. Like Kasturi aunty and Krishna, we meet several other migrant workers frequently and they are an intricate part of our daily routine. Amidst such daily interactions, what often goes unrealized, unregarded, and unacknowledged among privileged Indians is that the improvement in the quality of the urban life of the bourgeoisie occurred at the expense of working-class inhabitants, largely uneducated or lower-educated workers who migrated from villages to cities for survival (Breman, 2016).

Since lockdown, these migrant workers are missing from our sight. I could not get in touch with them, as their slums remain under police vigilance. Some vague news travels through our middle-class households, that many of them are preparing to go home. Meanwhile, in other states, such as from Maharashtra, Gujarat, for example, many workers who started walking back home were interviewed (Dutt, 2020). They feared the rise of the price of food during lockdown, that they would have no rent to pay, and that their employers had stopped paying them. Their vulnerability and helplessness confronted us from television screens as we sat watching, and it soon became apparent that their fears and anxieties were indeed warranted. A study conducted on 11,159 affected migrant workers during this period reported several facets of migrant distress. For instance, 78% of these workers had less than 300 rupees (approximately US$5) when they left their workplaces, and 89% of them had not received any payment from their employers. The study also highlighted that "there have been at least 195 documented lockdown-related deaths (compared to 331 COVID-19 related deaths), which include suicides, death during the journey that migrant workers made to their home states, and hunger, alcohol withdrawal and police brutality" (Adhikari et al., 2020, pp. 10–16). These migrant workers were reminded once again of who they are: exploited workers whose homes lie elsewhere and whose sustenance during the lockdown period depended on the "kindness" of the privileged classes. Despite being citizens of India in their host states, and in the low-paying jobs that they hold, they were considered second-class citizens before the pandemic *and* during the pandemic, given how the state, in trying to save the population and the economy, failed to give them the care and attention they deserved.

Journalists, who covered the issue extensively, walked and talked with the migrant workers on the highways and put up their videos on social media (Dutt, 2020). I saw some, as they walked long distances and crossed state borders, going without food, some without money, some walking barefoot, and many carrying infants. When they were asked what they were carrying in their bags, they mostly said biscuits and tools that they used for work. Some hoped that they might get some work on their way back. From a leading news forum, a journalist narrated, "It's not a morning walk. It's not a pilgrimage. The walk of migrant workers towards their home is a march against the stripping of their democratic rights" (Kumar, 2020; see also Venkatraman et al., 2020). Their appearance reveals their emotions shifting from confusion to fear, despair, anger, and a sense of surrender. COVID-19 threatens to affect every individual. But the ground reality reveals the politics of its spread. While some, like me, were shaken from work and ordinary routines, others, like the workers who were migrating to their rural home villages, were uprooted entirely from their workplaces and the places where they had perhaps lived for decades.

As we were settling down into the lockdown life and the terms "social distancing" and "social isolation" were becoming normalized, I followed the lives of the migrant workers more routinely in the months of April, May, and June 2020. News about the migrant workers affected our middle-class sentiments as we read, watched, and talked about it within the confines of our walls. Netizens actively shared posts, publicly revealing their emotions online. In response to a video showing a migrant worker's wife giving birth in the middle of the highway, a netizen commented, "I have been indifferent for so long. I am also to be blamed for what happened to them," while another wrote, "day by day we are ashamed of seeing the migrant workers [in such a situation]" (Team Mojo, 2020).

The pandemic made us aware of what we took for granted, our class position and the privileges that accompany it, and the material resources and social capital to which we have access. Moreover, it is because of the said social diacritica that I am able to contribute to the relevant sections of this chapter. In short, such lifeworld attributes make us aware of who we are and our position within the structural matrix. Concomitantly, they also make starkly visible who the migrant underclass is. In the face of a crisis when roads are emptied of people, walking on the empty roads itself becomes symbolic of their vulnerability and, hence, suffering. The different reactions to the migrant workers also make me cognizant of the distance between us – this distance which is, again, not only structural but also symbolic. The structural distance and the moment in time change the migrant underclass from an object of my sight to an object of its gaze. So, did the pandemic change how we see vulnerable groups like the migrant underclass? I have no answer. But the importance of the "gaze" lies in its ability to reflect on ourselves and our relations with others, that in the process of gazing at the other we become objects of our own sight (Lacan, 1978).

## Structural Inequalities and Vulnerabilities, or Reflections in and through the Gaze

Our portraits of India and Singapore, reflected through our gazes on what went on at these places during the height of the pandemic, converge and part ways at many points. In India as well as Singapore, it was apparent that the migrant underclass had fallen through the cracks in both governments' regulation of COVID-19. In Singapore, initial concerns were about returning Singaporeans, to the neglect of migrant workers, while in India, the lockdown was focused on protecting the health and well-being of people who could afford to be quarantined, and those whose livelihoods were already precarious were "let die" (Foucault, 1976). These effects of the pandemic were able to take root in the migrant communities because of existing inequalities, neglect, and vulnerabilities – a poor living environment for migrant workers in Singapore, and chronic poverty and unstable livelihood among India's rural-to-urban migrant underclass. In addition, the lockdown measures reflected a certain non-concern – of being "locked in" with possibly infected others or "locked out" of jobs, economies, and cities, and compelled to leave for home. While debates on whether lockdowns should be implemented loomed large (e.g., Ghosh, 2020; Grothe-Hammer & Roth, 2021; Monaghan, 2020), our two examples show that marginal groups remain disproportionately affected whether or not lockdown is implemented, since the conditions in which they live and how they are treated by the state and society at large are rooted in deeper, persistent inequalities that constrain them regardless.

Scambler (2020) argued that the pandemic is a "breaching experiment" that exposes a fractured society and the fragility of existing sociopolitical orders. While we largely agree with his observations, we argue that it is more accurate to describe the pandemic as a *natural* experiment that *reveals* or *exposes* the structural problems that he noticed. "Breaching" must be scrutinized: are we observing the breaching of norms and expectations, say, of an ordered society – which Scambler (2020), leaning on Garfinkel (1967), is concerned with – or the ruptures in livelihoods, living conditions, and lives of marginalized others in their biological, social, and political dimensions? The pandemic forced us to turn our gaze upon ourselves in the same gesture as we gaze upon the migrant workers who are locked in or locked out (Lacan, 1978), and as we become sensitized to the relations between us and socially distanced others. For the migrant workers, their lives are breached, but these breachings are rooted in social inequalities that are not unfamiliar – enduring "structural vulnerabilities," as Team and Manderson (2020) described; for us, as the inequalities hidden from our eyes through the compartmentalization of ordinary resident and migrant worker societies entered our vision; as we watched migrant workers trudge back into their villages in "the largest exodus since partition." Through this event, our sensibilities are "breached" (see also Scambler, 2020).

The core contention of this chapter, inspired by Lacan, is that the researcher's gaze is situated within, disciplined by, and proceeds from a privileged position due to broader structural relations between us and our subjects/interlocutors. We realized this when we found ourselves trapped in our homes and unable to do research – or unwilling to, even if we could, since this would mean exposing ourselves and our families to the dangers of infection. The roots of our non-interaction with the migrant underclass can be traced to pre-pandemic times. Migrant workers labour in the construction sector, maintenance services, domestic sector, and other low-wage, labour-intensive work both in India and Singapore, where their physical presence and the products of their labour are clearly visible, but these individuals are socially under-acknowledged. There is a collective disposition by others in our societies that relegates the migrant underclass as socially invisible, such that their poor housing conditions were a lesser concern in Singapore, which sees itself as a First World country (Romero, 2020), and their vulnerabilities were unrecognized by the Indian government that declared a lockdown with little consideration of its implications for migrant workers' lives (Deshingkar, 2020; Ghosh, 2020). It was under these conditions that we, observing from our privileged positions during the pandemic, found the precariousness of their lives and the suffering to which they were exposed startling and disorientating. In tracing our gazes beyond and before the pandemic, we become aware of the broader contexts that subtend the co-mingling of our lives and the migrant underclass we write about, and which we now begin to acknowledge.

Gazing from the enclosed safety of our homes, to quote Whiteford (2020, p. 7) again, is "of both power and powerlessness," of privilege as we found ourselves safely indoors, but also disturbing when we were unable to reach out to our intended interlocutors, the migrant workers, who were exposed to the pandemic. We identify with what anthropologists such as Blum (2020) discussed as the "privilege that enables (in-person, on-site) fieldwork" and the call to be "aware of the many ways anthropologists have always conducted research 'at a distance,'" but our concern exceeds purely methodological ones. While "fieldwork from afar" (Blum, 2020) is a legitimate concern, and an important one to say the least, we were also confronted with how we are still able to conduct research when socially or structurally "distanced" from intended interlocutors. That is, what does it mean to gaze at the migrant workers, especially in a mediated form that protects us from the dangers to which they are exposed? What does it mean when we are complicit in their vulnerabilities? How do we, and how should we, react upon realizing how we are implicating in the inequalities that impinge on their lives, when our gaze conjured a return gaze back on ourselves, as Lacan experienced in his encounter with the fishermen, which the floating tin can only metaphorically represents? The preceding sections address these questions on the entanglement of methodology with the research "object" for which it is employed, and of research with the broader political situation in which it is located. Just like Lacan, who could not help but gaze at the floating can, we were confronted by a prolonged sense of anxiety and helplessness. We envision that this nagging discomfort, which you might also experience, will nudge us to examine how we, as individuals complicit in existing inequalities, could be pushed to reflect on how we can participate in changing these systems.

However, what might political participation look like? Revisiting the debate that we staged between "noticing" and "witnessing," while we valued observing and "tracing" how individuals navigate, reproduce, but also struggle against inequalities (Tsing, 2015), we were also compelled and inspired to take up "response-ability" for the lives of migrant workers (Thomas, 2019, p. 2). While "attentive embodied care" and "co-performative," as Thomas had envisioned it, were impossible under pandemic conditions, we, your co-authors, responded by engaging in conversations on migrant worker situations in and across contexts in which we live. We reflected on what we saw. The conversations that we shared from April to July 2020 – when we started to write this chapter – have their roots in an earlier piece of writing for a public forum (Sen, 2020; J. Tan, 2020). Alongside other writers, we pooled our remunerations from the writing project, which Junbin helped to distribute as contributions to soup kitchens for migrant workers and slum dwellers in West Bengal, and to provide textbook subsidies to my students who were in need. We had also, when residing in our homes during lockdown and the Circuit Breaker, wrestled with unkind remarks and prejudices against migrant workers and called for recognition of our complicity. Our thoughts in this chapter represent an ongoing attempt to clarify our positions, not only as individuals but also as members of the communities in which we live, vis-à-vis the migrant workers. And we hope to inspire our readers to engage in this self-reflexive process of interrogating and thinking about similar issues that we do not see, even when they lie before us. In this sense, borrowing from Lacan, we respond to Butler's (2012, p. 134) question affirmatively, that we have an instinctual capacity – when startled and confronted by the discomforting glare of the glimmering can out at sea – "to respond ethically to suffering *at a distance*," even if we are concerned here with social instead of physical distance. This capacity to be discomforted and to respond by means of ethical reflection creates us as social individuals who "find ourselves invariably joined to those we never chose" (Butler, 2012, p. 134). The discomforting glare that unsettles our sensibilities, contained in the images of migrant workers' suffering that we saw during the pandemic, will situate us as "witnesses" (Thomas, 2019) insofar as we continue to reflect and consequently reorientate our interactions with migrant workers. And so, we locate a hope in public responses to these discomforting experiences and their potential in unsettling the inequalities and vulnerabilities that our gazes compel us to recognize.

*Acknowledgments*: The authors thank Princeton-Mellon Initiative in Architecture, Urbanism and Humanities for funding research for this project. Junbin Tan also thanks Tiffany C. Fryer and our students in the seminar Gender, Race, Empire at Princeton University in the year 2020, whose discussions on "bearing witness" sharpened the analysis in this chapter.

NOTES

1 See Sen (2020) and Tan (2020). The collection of essays was organized by Shu Hu and are written in or translated into Chinese.
2 Tan thanks Tiffany C. Fryer for inspiring the comparison between Tsing's (2015) "noticing" and Thomas's (2019) "witnessing." See also Fryer's upcoming monograph, tentatively titled *Things of War: Conflict and Heritage on Mexico's Maya Frontier*.
3 These include video footage that is officially disseminated on mainstream media, as well as "authorized" footage taken without official permission but which has circulated in social media and has sometimes been taken up by mainstream media, as in Channel News Asia (CNA) (Mohan 2020). The need for "authorization" attests to the regimentation in spaces where the migrant workers live, and the inability for outsiders to enter these spaces.
4 The Ministry of Manpower's (2020) "Advisory on Salary Payment to Foreign Workers Residing in Dormitories" mandates that "Employers must continue to pay foreign workers in dormitories their salaries during this period" and suggests that "As many foreign workers remit money home, employers are encouraged to pay salaries earlier than the due dates if their workers need to access physical remittance services provided at the dormitories."
5 Some individuals and communities initiated efforts to care for the migrant workers, such as food distribution by the COVID Migrant Support Coalition, an informal group consisting of volunteers; a Google Sheets that recorded the support and

services that migrant workers required (Sholihyn, 2020); and volunteers who tested workers quarantined in dormitories and hotels for COVID-19 (C. Tan, 2020). Yet, it appears that most of them belonged to communities who were already active in caring about migrant workers' welfare and interests.

REFERENCES

Abdullah, N. (2005). Foreign bodies at work: Good, docile and other-ed. *Asian Journal of Social Science*, *33*(2), 223–45. https://doi.org/10.1163/1568531054930785.

Adhikari, A., Narayanan, R., Dhorajiwala, S., & Mundoli, S. (2020, April 15). 21 Days and counting: COVID-19 lockdown, migrant workers, and the inadequacy of welfare measures in India. Stranded Workers Action Network. https://azimpremjiuniversity.edu.in/publications/2020/report/21-days-and-counting-covid-19-lockdown-migrant-workers-and-the-inadequacy-of-welfare-measures-in-india.

Bal, C. (2015). Dealing with deportability: Deportation laws and the political personhood of temporary migrant workers in Singapore. *Asian Journal of Law and Society*, *2*(2), 267–84. https://doi.org/10.1017/als.2015.17.

Blum, S.D. (2020, September 10). Fieldwork from afar: Pandemic insights. *Anthropology News*. https://www.anthropology-news.org/articles/fieldwork-from-afar/.

Borneman, J., & Hammoudi, A. (Eds.). (2009). *Being there: The fieldwork encounter and the making of truth*. University of California Press.

Breman, J. (2016). *On pauperism in present and past*. Oxford University Press.

Butler, J. 2012. Precarious life, vulnerability, and the ethics of cohabitation. *Journal of Speculative Philosophy*, *26*(2), 134–51. https://doi.org/10.5325/jspecphil.26.2.0134.

Cai, W., & Lai, K.K.R. (2020, April 28). Packed with migrant workers, dormitories fuel coronavirus in Singapore. *New York Times*. https://www.nytimes.com/interactive/2020/04/28/world/asia/coronavirus-singapore-migrants.html.

Castañeda, H., & Lopez, W.D. (2020, October 29). Immigrant communities in the COVID-19 pandemic: Old and new insights on mobility, bordering regimes, and social inequality. *Items: Insights from the Social Sciences*. https://items.ssrc.org/covid-19-and-the-social-sciences/disaster-studies/immigrant-communities-in-the-covid-19-pandemic-old-and-new-insights-on-mobility-bordering-regimes-and-social-inequality/.

Chaudhuri, S. (2020, May 12). The working person's right to life. *Economic Times*. https://economictimes.indiatimes.com/news/economy/policy/view-the-working-persons-right-to-life/articleshow/75683761.cms?from=mdr.

Chuvileva, Y.E., Rissing, A., & King, H.B. (2020). From wet markets to Wal-marts: Tracing alimentary xenophobia in the time of COVID-19. *Social Anthropology*, *28*(2), 241–3. https://doi.org/10.1111/1469-8676.12840.

Ciotti, M. (2020). Home-made biopolitics: India's migrant workers between bare life and political existence. *Social Anthropology*, *28*(2), 243–5. https://doi.org/10.1111/1469-8676.12820.

CNN News 18. (2020, March 24). Nationwide lockdown for next 21 days, declares PM Modi. https://www.youtube.com/watch?v=bzYjfXj24y8.

Copjec, J. (2015). *Read my desire: Lacan against the historicists*. Verso Books.

Deshingkar, P. (2020, June 16). Faceless and dispossessed: India's circular migrants in times of COVID-19. *Down to Earth*. https://www.downtoearth.org.in/blog/economy/faceless-and-dispossessed-india-s-circular-migrants-in-the-times-of-covid-19-71782.

Dewanto, P.A. (2020, August 20). Labouring situations and protection among foreign workers in Malaysia. *Heinrich Böll Stiftung, Southeast Asia*. https://th.boell.org/en/2020/08/20/labouring-situations-malaysia.

Dutt, B. (2020, April 22). A long walk: Migrant workers are still on the road amidst the 5th week of the lockdown. *Mojo Story*. https://www.youtube.com/watch?v=Ib74nWIi_ig&t=26s.

Foucault, M. (1976). *The history of human sexuality*. Éditions Gallimard.

Garfinkel H. (1967). *Studies in ethnomethodology*. Prentice-Hall.

Geddie, J., & Aravindan A. (2020, August 5). Spate of suicides among migrant workers in Singapore raises concern. *Jakarta Post*. https://www.thejakartapost.com/seasia/2020/08/06/spate-of-suicides-among-migrant-workers-in-singapore-raises-concern.html.

Ghosh, A. (2020, June 3). India's response to COVID-19 pandemic: A success story? *Medical Anthropology Quarterly*. https://www.academia.edu/43152040/Indias_Response_to_COVID_19_ Pandemic_A_Success_Story.

Gordon, A.F. (2008). *Ghostly matters: Haunting and the sociological imagination*. University of Minnesota Press.

Grothe-Hammer, M., & Roth, S. (2021). Dying is normal, dying with the coronavirus is not: A sociological analysis of the implicit norms behind the criticism of Swedish "exceptionalism." *European Societies*, *23*(sup1), S332–S347. https://doi.org/10.1080/14616696.2020.1826555.

Hamid, W., & Tutt, D. (2019). "Thrown away like a banana leaf": Precarity of labour and precarity of place for Tamil migrant construction workers in Singapore. *Construction Management and Economics*, *37*(9), 513–36. https://doi.org/10.1080/01446193.2019.1595075.

Haraway, D. (2003). Situated knowledges: The science question in feminism and the privilege of partial perspective. In Lincoln, Y.S., & Denzin, N.K. (Eds.), *Turning points in qualitative research: Tying knots in a handkerchief* (pp. 21–46). AltaMira Press.

Hindu BusinessLine. (2020, April 15). Lockdown to be lifted partially from April 20. https://www.thehindubusinessline.com/news/national/lockdown-to-be-lifted-partially-from-april-20/article31349594.ece.

Khalik, S. (2020, April 21). Singapore facing two separate outbreaks: In the community and in foreign worker dormitories. *Straits Times*. https://www.straitstimes.com/singapore/health/singapore-facing-two-separate-outbreaks-in-the-community-and-in-foreign-worker.

Koh, D. (2020). Migrant workers and COVID-19. *Occupational and Environmental Medicine*, *77*(9), 634–6. https://doi.org/10.1136/oemed-2020-106626.

Krips, H. (2010). The politics of the gaze: Foucault, Lacan and Žižek. *Culture Unbound*, *2*(1), 91–102. https://doi.org/10.3384/cu.2000.1525.102691.

Kumar, R. (2020, May 5). We, left behind in cities, are migrants, not them. NDTV. https://www.ndtv.com/blog/we-left-behind-in-cities-are-migrants-not-them-by-ravish-kumar-2223524.

Lacan, J. (1978). *The four fundamental concepts of psychoanalysis*. W.W. Norton and Company.

Lai, L. (2020, April 19). Coronavirus: Work permit and S Pass holders in construction sector will serve 14-day stay-home notice from Monday. *Straits Times*. https://www.straitstimes.com/singapore/work-permit-and-s-pass-holders-in-construction-sector-will-serve-14-day-stay-home-notice.

Lai, S.H.S., Tang, C.Q.Y., Kurup, A., & Thevendran, G. (2020). The experience of contact tracing in Singapore in the control of COVID-19: Highlighting the use of digital technology. *International Orthopaedics*, *45*(1), 65–9. https://doi.org/10.1007/s00264-020-04646-2.

Mason, K.A. (2012). Mobile migrants, mobile germs: Migration, contagion, and boundary-building in Shenzhen, China after SARS. *Medical Anthropology*, *31*(2), 113–31. https://doi.org/10.1080/01459740.2011.610845.

Ministry of Culture, Community and Youth. (2020, June 1). Stay safe, stay strong and stay united as circuit breaker measures ease. https://www.mccy.gov.sg/about-us/news-and-resources/press-statements/2020/jun/stay-safe-stay-strong-stay-united-as-circuit-breaker-measures-ease.

Ministry of Manpower. (2020, April 11). Advisory on salary payment to foreign workers residing in dormitories. https://www.mom.gov.sg/covid-19/advisory-on-salary-payment-to-foreign-workers.

Mohan, M. (2020, May 18). Woodlands health campus investigates unauthorised filming of COVID-19 patients in mass exercise at Singapore Expo. Channel News Asia (CNA). https://www.channelnewsasia.com/singapore/woodlands-health-campus-video-singapore-expo-covid-19-exercise-668906.

Mokhtar, F. (2020, April 21). How Singapore flipped from virus hero to cautionary tale. *Bloomberg*. https://www.bloomberg.com/news/articles/2020-04-21/how-singapore-flipped-from-virus-hero-to-cautionary-tale.

Monaghan, L.F. (2020). Coronavirus (COVID-19), pandemic psychology and the fractured society: A sociological case for critique, foresight, and action. *Sociology of Health & Illness*, *42*(8), 1982–95. https://doi.org/10.1111/1467-9566.13202.

*New Indian Express*. (2020, March 29). No work, no money: Thousands of stranded migrant workers walk back home as India under 21-day coronavirus lockdown. https://www.newindianexpress.com/galleries/nation/2020/mar/29/no-work-no-money-thousands-of-stranded-migrant-workers-walk-back-home-as-india-under-21-day-coro-102813.html.

Ng, E. (2020, May 12). Tale of 2 outbreaks: Singapore tackles a costly setback. AP News. https://apnews.com/article/understanding-the-outbreak-asia-pacific-singapore-malaysia-southeast-asia-fc9e282110b28dab1e1a7a23dd8a11df.

Pellegrino, M. (2020). COVID-19: The "invisible enemy" and contingent racism: Reflections of an Italian anthropologist conducting fieldwork in Greece. *Anthropology Today*, *36*(3), 19–21. https://doi.org/10.1111/1467-8322.12576.

Phua, R. (2020, April 10). NGOs launch initiatives to help migrant workers amid COVID-19 outbreak. Channel News Asia (CNA). https://www.channelnewsasia.com/singapore/covid19-migrant-foreign-workers-dormitory-food-coronavirus-763461 (see also https://twitter.com/channelnewsasia/status/1248462726596194304?lang=ga).

Quesada, J., Hart, L.K., & Bourgois, P. (2011). Structural vulnerability and health: Latino migrant laborers in the United States. *Medical Anthropology*, *30*(4), 339–62. https://doi.org/10.1080/01459740.2011.576725.

Romero, A.M. (2020, April 17). Josephine Teo explains 3-pronged approach to containing coronavirus outbreak in foreign workers' dormitories. *Independent*. https://theindependent.sg/josephine-teo-explains-3-pronged-approach-to-containing-coronavirus-outbreak-in-foreign-workers-dormitories/.

Salve, P., & IndiaSpend.com. (2019, May 10). Data check: 90% of jobs created in India after liberalisation were in the informal sector. *Scroll.in*. https://scroll.in/article/922863/data-check-90-of-jobs-created-in-india-after-liberalisation-were-in-the-informal-sector.

Sangaramoorthy, T. (2020, May 1). From HIV to COVID-19: Anthropology, urgency, and the politics of engagement. *Somatosphere*. http://somatosphere.net/2020/from-hiv-to-covid19-anthropology-urgency-and-the-politics-of-engagement.html/.

Scambler, G. (2020). COVID-19 as a "breaching experiment": Exposing the fractured society. *Health Sociology Review*, *29*(2), 140–8. https://doi.org/10.1080/14461242.2020.1784019.

Sebald, W.G. 2001. *Austerlitz*. Random House.

Sen, A. (2020, May 8). 保持社交距离之下的生活：来自印度加尔各答的观察 [Life under social distancing: Observations from Kolkata, India]. Mrs. Muses. http://zhishifenzi.com/depth/depth/8939.html.

SGUnited. (2020, May 26). https://www.sgunited.gov.sg/.

Sholihyn, I. (2020, April 6). Initiative launched on Google Sheets to collate support for migrant workers affected by COVID-19. *AsiaOne*. https://www.asiaone.com/digital/initiative-launched-google-sheets-collate-support-migrant-workers-affected-covid-19.

Silverblatt, M. (2007). A poem of an invisible subject. In Schwartz, L.S. (Ed.), *The emergence of memory: Conversations with W.G. Sebald* (pp. 77–86). Seven Stories Press.

Simpson, A. (2014). *Mohawk interruptus: Political life across the borders of settler states*. Duke University Press.

Singh, V. (2020, April 19). Coronavirus lockdown | E-commerce firms can't supply non-essential goods, says government. *The Hindu*. https://www.thehindu.com/news

/national/coronavirus-lockdown-e-commerce-firms-cant-supply-non-essential-goods-says-indian-government/article31380752.ece.

Strathern, M. (1999). *Property, substance, and effect: Anthropological essays on persons and things.* Athlone Press.

Tait, S. (2011). Bearing witness, journalism and moral responsibility. *Media, Culture & Society, 33*(8), 1220–35. https://doi.org/10.1177/0163443711422460.

Tan, C. (2020, July 2). Doctors, dentists, retirees and students among the 17,100 who have volunteered to help fight COVID-19. *Straits Times*. https://www.straitstimes.com/singapore/doctors-dentists-retirees-and-students-among-the-17100-who-have-volunteered-to-help-fight.

Tan, J. (2020, May 23). 学人说 |以"口罩"为鉴： 新冠疫情时期的自护与偏见 [Scholars say | Taking "masks" as a lesson: Self-protection and prejudice during the new corona epidemic]. Mrs. Muses. http://zhishifenzi.com/depth/humanity/9100.html.

Tan, Y. (2020, September 18). COVID-19 Singapore: A "pandemic of inequality" exposed. BBC. https://www.bbc.com/news/world-asia-54082861.

Tang, S.K., & Co, C., (2020, April 5). COVID-19: Nearly 20,000 foreign workers in quarantine in S11 Dormitory, Westlite Toh Guan. Channel News Asia (CNA). https://www.channelnewsasia.com/singapore/covid-19-nearly-20000-foreign-workers-quarantine-s11-dormitory-westlite-toh-guan-761976.

Team Mojo. (2020, May 14). Pregnant worker delivers baby by road, then walks 150 kms. "Can't someone put us on a bus?" *Mojo Story*. https://www.youtube.com/watch?v=9DIZ4d80z4I.

Team, V., & Manderson, L. (2020). How COVID-19 reveals structures of vulnerability. *Medical Anthropology, 39*(8), 671–4. https://doi.org/10.1080/01459740.2020.1830281.

Thomas, D.A. (2019). *Political life in the wake of the plantation: Sovereignty, witnessing, repair.* Duke University Press.

Tsing, A.L. 2015. *The mushroom at the end of the world: On the possibility of life in capitalist ruins.* Princeton University Press.

Whiteford, L.M. (2020). A room with a view: Observations from two pandemics. *Anthropology Now, 12*(1), 7–10. https://doi.org/10.1080/19428200.2020.1760630.

Venkatraman, T., Chauhan, S., Dey, S., & Mishra, R. (2020, May 16). In long walk back home, migrants battle hunger, scourge of COVID-19. *Hindustan Times*. https://www.hindustantimes.com/india-news/in-long-walk-back-home-migrants-battle-hunger-scourge-of-disease/story-TizRfUz69osJQ0Uqmm6jZN.html.

Viswanath, K. (2020, May 27). It is time to stop seeing domestic workers as COVID-19 "Carriers." *The Wire*. https://thewire.in/labour/covid-19-lockdown-domestic-workers.

Yashee. (2020, May 20). Why migrants are walking home: "We know of government schemes, they won't help." *Indian Express*. https://indianexpress.com/article/cities/delhi/why-migrants-are-walking-home-we-know-of-govt-schemes-they-wont-help-6418627/.

Ye, J. (2014). Migrant Masculinities: Bangladeshi men in Singapore's labour force. *Gender, Place & Culture, 21*(8), 1012–28. https://doi.org/10.1080/0966369X.2013.817966.

# 4 Social Distancing? "No Problem!": Explaining Thailand's Successful Containment of COVID-19

BY PIYA PANGSAPA

On 13 January 2020, Thailand became the first country outside of China to report a coronavirus infection (in a 61-year-old female tourist from Wuhan city). By mid-March, there were over 400 confirmed cases in the country, prompting the prime minister to declare a state of emergency on 26 March . A nationwide curfew was imposed on 3 April, followed by a nationwide ban on public gatherings, a nationwide ban on the sale of alcohol, a ban on all inbound international flights, and the implementation of lockdown measures across the country. The number of confirmed cases in Thailand as of this writing in late 2020 stands at 3,348, of which 3,150 have recovered, while the death toll of 58 remains unchanged since 2 June 2020. By the end of March, the government established the Centre for COVID-19 Situation Administration (CCSA), and a medical doctor (rather than a politician) was appointed as its spokesperson; meanwhile, over a million health volunteers were dispatched across the country to trace, identify, and isolate those thought to be at risk of contracting and/or spreading the virus. The health volunteers visited 11.8 million households in less than two months and identified over half a million at-risk people, who were advised to stay home and self-quarantine (Kasipat, 2020).[1] On 28 July 2020, Thailand was ranked first out of 184 countries in a global survey for the country's effective handling of COVID-19 and recovery from the crisis, its index score of 82.06 placing it on top of the global chart as an example of best practices in tackling the virus (Global COVID-19 Index (GCI), 2020).[2] This chapter will attempt to explain Thailand's successful containment of COVID-19, which may be attributed to a combination of factors, but its ability to control the spread of the virus would not have been possible without full public cooperation.

My chapter draws on Gelfand's tightness-looseness (TL) theory, which posits that cultures with stricter social norms are less tolerant towards individuals who deviate from those norms (Gelfand, Nishii, & Raver 2006; Jackson, 2020), and on Gelfand, Jackson, Pan, et al.'s (2020) study on the global pandemic, which points to the importance of cultural tightness and government efficiency in understanding why "some nations were more effective than others at limiting the spread of the virus" (Gelfand, Jackson, Pan, et al., 2020). In trying to explain why Thailand has been more effective at containing the virus than other countries, this chapter will examine the Thai government's response to the COVID-19 pandemic as well as the Thai public's response to the state's strong campaign of pandemic mitigation along with the pandemic itself. In the case of Thailand, we have a country that was shut down effectively through its "draconian measures" (Thitinan, 2020), difficult to do in most democracies, and with citizens who were willing, ready, and able to comply with restrictions on social activities, the closure of shops and services, including shopping malls, department stores, restaurants, bars, street food and beverage stalls, and all schools and universities. The country's immediate and sustained overreaction to the pandemic, reinforced by the media daily, instilled and consequently ingrained fear into the public psyche, so much so that public reaction to just *one* new case of infection was alarm and outrage.

On 23 July 2020, the World Health Organization (WHO) chose Thailand and New Zealand as model countries that had been successful in handling and curbing COVID-19 for a documentary film (Nation, 2020d; see also Nation, 2020b; Apinya, 2020h). In November 2020, Thailand Prime Minister Prayut Chan-o-cha shared lessons learned with the international community at the Third Paris Peace Forum, "Reflection on the principles and the priorities of the post-COVID world" (Nation, 2020c). In his online address via video conferencing, the prime minister attributed Thailand's success in containing the coronavirus to two main factors: "people centrality" and a strong public health system characterized by an effective medical system, competent healthcare professionals, and the support of over one million health volunteers stationed in villages throughout the country (*Bangkok Post*, 2020m; Nation, 2020b). Preliminary desk-based research on Thailand and COVID-19 as well as interviews conducted with several individuals also point to a cooperative public and a strong healthcare sector, which in 2019 ranked sixth in the world (*Bangkok Post*, 2019), and while Thailand's impressive containment of the coronavirus may be commendable, its success has come at a great cost.

**Government Response**

As already mentioned, the first confirmed case of COVID-19 outside of mainland China was reported in Thailand on 13 January 2020, in a female tourist from Wuhan, China, but only 19 confirmed cases of the virus were reported by the end of January (which included the first human-to-human virus transmission in Thailand). A total of only 40 cases were confirmed by the end of February (Apinya, 2020a; *Bangkok Post*, 2020c). The low numbers of infections may be explained by a sequence of actions taken by the Thai government immediately following the WHO statement of 13 January 2020, confirming the novel coronavirus in Thailand. These actions included the early screening and quarantining of passengers arriving from Wuhan at all major airports throughout the country, which subsequently raised public awareness about the virus, allowing hospitals and public health workers to prepare for the threat of the outbreak (Achadthaya, 2020; *Bangkok Post*, 2020f; World Health Organization, 2020). Thailand was also the first country in ASEAN to adopt Universal Precautions in preventing virus transmissions (Kavi, 2020b), which included the standard practice of hand hygiene, the use of personal protective equipment (PPE) such as face masks and face shields, and respiratory hygiene/cough etiquette. COVID-19 was soon designated a dangerous contagious disease, on 1 March 2020, by the Ministry of Public Health (MoPH) under the 2015 Communicable Diseases Act, a designation that would "facilitate officers to work more effectively" so as to prevent the situation from escalating to Phase 3 (Apinya, 2020d; *Bangkok Post*, 2020e; Nation, 2020a). Essentially, the declaration of COVID-19 as a dangerous contagious disease would give local officials the power to exercise discretionary authority in their respective jurisdictions under the purview of public health and safety.

Under the various provisions of the act, local disease control officers had the power to bring in confirmed or suspected COVID-19 cases for testing, treatment, observation, isolation, or quarantine; the act also allowed officers to do the same for travellers arriving from countries in Disease Infected Zones. Initially, foreign travellers arriving from 15 countries were required to present COVID-19-free health certificates as well as COVID-19 health insurance prior to boarding their flights to Thailand (*Bangkok Post*, 2020h). People were asked to refrain from unnecessary travel, and all travellers returning from high-risk areas were required to pass screening procedures at airports and self-monitor for symptoms. Anyone exhibiting a body temperature of 37.5 degrees Celsius (99.5 degrees Fahrenheit) or higher, accompanied by at least one symptom, was taken to a medical facility and tested, treated, diagnosed, isolated, and/or quarantined. Travellers not suspected of having COVID-19 were required to self-quarantine at their accommodation for a period of 14 days, during which time they were to record personal symptoms while being monitored by a surveillance officer. Travellers coming from areas with ongoing local transmission had to notify a disease control officer about the places they planned to visit so that their movements and contacts could be tracked. Thailand's National Communicable Disease Committee (NCDC) also had the authority to close down locations, order people they suspected to temporarily stop working, and/or deny them access to certain areas. Violators could face up to two years' imprisonment, a fine of up to 500,000 THB, or both. The strict enforcement of the stringent measures may explain why an outbreak was prevented early, even though some critics found government action to be "unnecessarily harsh" (*Bangkok Post*, 2020d; Khemthong & Rawin, 2020).

While there were only 40 confirmed cases of COVID-19 by the end of February 2020, the number of cases would escalate in the weeks to follow. On 6 March, a boxing match held at the Lumpini Boxing Stadium led to a spike in cases throughout the month, including the largest single-day rise in cases since January and three deaths (Apinya, 2020f). On 12 March, the government set up the CCSA and appointed a physician as its spokesperson. One day later, a mobile app called "SydeKick for ThaiFightCOVID" was introduced by the Digital Economy and Society (DES) Ministry to track and monitor individuals who were required to self-quarantine at home for 14 days. The app allowed public health officials and staff to monitor daily behaviour of quarantined persons and to ensure they remained at home. On 19 March, the Thai army began a nightly process of disinfecting roads in Bangkok, using trucks spraying disinfectant-water solution on streets "around trading zones, transport stations, ports, education facilities, outdoor activity areas and public gathering points" (Wassana, 2020). On 26 March, the prime minister declared a state of emergency, by which time there were over 900 confirmed cases in the country (1,651 cases by the end of March).

**The Good Doctor**

The appointment of Dr. Taweesilp Visanuyothin, rather than a politician, as the spokesperson for the CCSA is noteworthy. The physician played a key role in educating the public and communicating with the media about the pandemic. The appointment of this skilful, articulate, and charismatic doctor marked a "sea-change in government-media relations" (Kavi, 2020a), and, according to veteran Thai journalist Veera (2020), "the CCSA spokesman made it easier for the media and the public to understand this deadly disease in a way that his predecessors failed to do." Another veteran journalist, Kavi, spoke of the physician's extraordinary ability to respond intelligently to the consistent barrage of questions from journalists "with information backed by data," noting that this was a first in Thai media history (Kavi, 2020a). One respondent said that the doctor "gave a perception of trustworthiness and reassurance because he knew what he was talking about. He was a smooth and suave speaker, so people listened to him and followed his advice."[3] The doctor's daily appearance on every news station restored public confidence in the government so effectively that he was accused of being a government lackey, of projecting too positive an image for the government, and of making the Prayut government look too trustworthy (Kavi, 2020a; Mongkol, 2020c; Veera, 2020). Furthermore, the doctor's refusal to politicize the pandemic only enhanced his popularity: "He appears to have outshone the government's three spokeswomen … most if not all the cabinet ministers and, probably, even the prime minister" (Veera, 2020).

In addition to the daily briefings by Dr. Taweesilp, news about the novel coronavirus, information concerning preventive measures, and public service announcements motivating healthy behaviour dominated almost all broadcasting television stations and mobile video streaming services. One respondent pointed to the onslaught of "media and publicity as well as big brands and

KOLs (key opinion leaders) that helped spread awareness. There was just so much PR around coronavirus, posters, paid media, traditional, digital media, you name it." Another respondent felt that the country's overreaction to the virus was psychologically effective: "the media was getting the public to fear the virus more than most; hence our outrage at just one new case of infection, which doesn't happen anywhere else."[4] Several respondents[5] spoke positively about some of the government's measures, namely mandatory state quarantine and the close monitoring of those in self-quarantine:

- "There was strict imposition of mandatory quarantine that was initially paid for by the state (all arrivals were put up in luxury hotels and provided with three meals a day)."
- "There was strict monitoring of those in quarantine."
- "The nationwide curfew definitely helped curb the spread of COVID-19 and they tried to track people within the country."
- "Overall, I'm very impressed with the government and the people's handling of the pandemic."

In May, the CCSA introduced a mobile phone application, Thai Chana, as a way of mitigating the spread of COVID-19 by having people scan a QR code every time they checked in and out of convenience stores, supermarkets, shopping malls, restaurants, banks, office buildings, etc. In addition to all these measures, the government's early concern about a possible outbreak resulted in an over-calculation of the infection rate, a preemptive move that may have inadvertently helped to prevent the spread of the virus.

**Fear Mongering and the State of Emergency**

Even with the low numbers of confirmed cases by the end of February 2020, the Communicable Disease Division had estimated that Thailand could expect to see between 400,000 and 9.9 million infections by year's end. In March, the Disease Control Department announced that 33 million people (half the country's population) could be infected over the next two years in a worst-case scenario (*Bangkok Post*, 2020o; *Bangkok Post*, 2020b; Mongkol & Apinya, 2020b), and the government said it was preparing for a third-stage outbreak (Mongkol, 2020a). Meanwhile, thousands of people were fleeing the city to return to their home provinces (Mongkol & Apinya, 2020a). It is possible that such proclamations were intended to stress the severity of the situation and may well have caused fear and anxiety in people who would react by taking preventive measures seriously. In a poll conducted in March with people aged 15 and over, a majority of respondents (68.18%) said that they were concerned about the spread of the coronavirus (Apinya, 2020e; *Bangkok Post*, 2020g). At the same time (in February/March), over one million health volunteers were dispatched across the country to trace, identify, and isolate those thought to be at risk of contracting and/or spreading the virus (Apinya, 2020c; *Bangkok Post*, 2020a). The health volunteers had visited 11.8 million households and identified 660,000 at-risk people, who were advised to stay home and self-quarantine (Kasipat, 2020). Those identified included the nearly 60,000 returning from abroad and nearly half a million arriving in the provinces from Bangkok as well as those linked to the boxing match incident and other transmissions clusters such as bars and pubs.

On 21 March 2020, Bangkok authorities imposed a shutdown of various business establishments and shopping malls, and the state of emergency was declared on 24 March. On 3 April, a nationwide curfew was imposed, followed by a nationwide ban on public gatherings, a nationwide ban on the sale of alcohol, a ban on interprovincial travel, a ban on all inbound international flights, and the implementation of lockdown measures across the country. In September 2020, the government extended the state of emergency for the fifth time, with the emergency decree extended again in October and November 2020, at which time the government decided to extend the decree until mid-January 2021. And while Human Rights Watch condemned the Thai government for using the state of emergency as a pretext to "suppress fundamental freedoms," including criticism from the media and the general public (Human Rights Watch, 2020), the state's proactive approach in handling the coronavirus may explain the country's low rate of infections: there were only 3,448 confirmed cases of COVID-19 in September 2020, and a death toll of 58 that had remained unchanged since June. While government measures were quite thorough, extensive, and systematic, Thailand's ability to contain the spread of the virus would not have been possible without full cooperation from the public.

*Public Response*

In almost all the reports concerning Thailand's response to COVID-19, including those by the Ministry of Public Health and the WHO, there is no mention or acknowledgment of public cooperation as being an important factor in helping to contain the coronavirus. These reports only included recommendations and advice for the public on best practices for preventing infection and transmission. Only a report from the Ministry of Public Health indicated nationwide public cooperation in effective social measures as one of five key lessons learned from Thailand (Supakit, 2020). One of those measures that came with the shutdown (of almost all businesses as well as schools and universities) was *social distancing*, defined as the practice of staying home, physically away from others, to prevent the spread of COVID-19, and the encouragement of online communication (Maragakis, 2020). The general public readily complied with this measure, since staying at home was mandatory and a nationwide 10:00 p.m. to 4:00 a.m. night curfew was in effect. With the reopening of public places and the lifting of quarantine restrictions, however, *physical distancing*, the practice of staying 6 feet or 1.5 metres away from other persons in public, was difficult if not impossible to maintain, especially with the heavy congestion of public transport systems and the natural crowding of public places. The Department of Rail Transport therefore had to ease physical distancing measures on all train services, including the Bangkok Skytrain (BTS) and the Bangkok Subway (MRT), by allowing occupancy of all seats but requiring all passengers to wear face masks and to refrain from talking (Thai PBS World, 2020). Passengers also had to undergo a body temperature check before boarding, scan themselves in, and use hand sanitizers. Alternate seating arrangements were also implemented in public areas – restaurants put up plastic screen dividers for diners, adhesive tapes were used to mark an "X" between seats, physical distancing was enforced in waiting lines, and movie theatres isolated pairs of seats, to name but a few measures. All major department stores implemented maximum health and safety measures upon reopening (*Bangkok Post*, 2020l).

To gain perspective and insight on the public's response to the coronavirus and COVID-19 measures, a diverse cross section of people living and working in Bangkok were asked why *they* wore face masks, why *they* thought most people here did not mind wearing masks in public, and how might *they* explain Thailand's success in containing the coronavirus. Here are some of their responses:[6]

- The public doesn't trust in the government to control the virus, so we decided to protect ourselves from the virus.
- Thais were already wearing face masks to prevent the breathing in of air pollutants, the deadly PM2.5 dust particles, so asking us to put on face masks to prevent COVID infection was no big deal.
- Thai people showed unison and were committed to fight the virus. Schools and universities helped contain the virus too.
- Thais think of protecting themselves and others from the virus while Americans and Europeans think only people who are sick should wear face masks.
- Thais were cooperative and wore masks, used hand sanitizers, and large numbers of people stayed home and listened to the government.
- Thais culturally place high value on health/cleanliness and are repulsed by disease/filth. Being overly cautious is viewed as smart, not as "brainwashed." Negative social (not to mention health) repercussions outweigh benefits for risky behaviour.
- Thais have a culture of open-mindedness when it comes to looking at other countries for solutions. Instead of reacting with exceptionalism, "happens to someone else," Thais thought "Oh shit, this could be me!" and looked

to see what worked and what didn't work in countries where the virus had made further inroads.
- We saw an almost universal cooperation by the Thai people in wearing masks, socially distancing, etc. This indicates that, despite ongoing political disagreements, an overriding sense of togetherness (of being "one family") is still prevalent in Thai society.
- Thais are also more community-minded and recognize that the use of masks will cut down on the spread of COVID-19 to other people, whereas Western people think primarily of themselves and how they will overcome the virus when they contract it, but less about the virus's effects on the people around them.
- In explaining Thailand's success: Thais will come together and fully cooperate regardless of their background, as they understand the importance of being one as a society. I've noticed that whenever there was an announcement to do something, Thai people would react positively in the spirit of helping one another. When there weren't enough face masks, people would step in and manufacture them for free distribution or people would make their own masks from home and distribute – and this may be linked to the Thai Buddhist culture of giving and merit making. When someone doesn't have food to eat, people step in immediately to provide free meals, so we help each other. We don't feel like it's someone else's responsibility, the government's responsibility, but that it's *our* responsibility as a people as a society.
- I still find it hard to believe that there are only 58 deaths, but I will give credit when credit is due. There has been an overwhelming level of cooperation by Thais and non-Thais in general and that was probably key to our success. Everyone wore a mask voluntarily, there were hand sanitizers and temperature checkpoints everywhere. Even 7–11 staff members did their job (better than some police officers) and refused to let me inside when I'd forgotten to bring my mask. Overall, I'm very impressed with the government and the people's handling of the pandemic. It does indeed take a village. The heat may have also played a huge part in keeping the virus at bay.

Chris Parker, an American living in Thailand, featured a segment titled "Why Thailand Is Winning (and America Is Losing) the Fight against Covid 19" on his YouTube channel, and offers similar remarks with regard to mask wearing: "Thais and Asians in general have always done a really good job of understanding that we live in a society and certain decisions are made based on the betterment of that society as a whole." He observes of the Thai response to mask wearing:

> You will notice the topic of whether to wear a mask or not, it was never discussed here, wasn't ever discussed here; I don't think it was ever officially mandated by the government for people to wear masks, people just did it for the betterment of

society as a whole and now the issue has morphed into the top 10 things you need to do to wear your mask properly ... Everyone doesn't wear masks a hundred percent of the time, but they've been very, very reasonable about it, whenever they're in crowded places or in environments where they feel like it would help, most of them are wearing masks.

As for why he thinks Thailand is winning the battle: "In my opinion, it comes down to one thing: the will of the people. While Americans seem busy fighting for individual freedoms, the Thais are busy fighting as a society against a common enemy" (Parker, 2020). Indeed, within Thailand, Crispin (2020) noted that "resident Westerners were often the only mask-less outliers."

Casual conversations with people on the issue of mask wearing triggered similar reaction and commentary: expressions of incredulity, astonishment at the absurdity of people in Western societies who refuse to wear face masks in the midst of the pandemic. The collective eye roll was followed by an attempt to understand and explain cultural differences: East versus West mentality, different cultural values, differing attitudes – that is, the usual cultural stereotypes of Western individualism and self-centredness vis-à-vis the Eastern sense of community, unity, cooperation, and togetherness. Another commonly mentioned factor by respondents, but taken for granted, is the Thai way of greeting: a gesture involving no physical contact. To do the Thai *wai*, your palms are placed together at the chest, and you bow your head slightly until your fingertips touch your chin. The Thai *wai* was subsequently encouraged, recommended, and widely adopted in non-Asian countries (notably by world leaders) as a safe alternative in order to lower risk of transmission.

Interestingly, the general distrust in the government was frequently remarked on and is not unrelated to the fact that Thais were wearing face masks long before COVID-19. In 2019 Thailand was ranked the 28th most polluted country out of 98 countries in IQAir's World Air Quality Report (IQAir, 2019). The Thai government has consistently been criticized for lacking the political will to address the issue, whether in its failure to monitor polluting vehicles – the major source of smog and air pollution in the city – in its failure to require emission control standards, or in its failure to enforce pollution control regulations (*Bangkok Post*, 2020a; Supita, 2019). Poor air quality has meant that city residents have had to resort to wearing face masks outdoors. Subsequently, the wearing of face masks had become a regular part of daily life. An estimated 15 million people in Bangkok spent THB600 million (US$20 million) over a period of two months buying standard face masks in 2019 (Supita, 2019).

In March and April, 2020, the British market research and data analytics firm YouGov and Imperial College London conducted a study to look at COVID-19 preventive measures across six ASEAN nations – Singapore, Malaysia, Indonesia, Thailand, the Philippines, and Vietnam. Based on their online survey

**Thais most likely to wear facemasks in public and to use hand sanitiser**
Thinking about the last 7 days, how often have you taken the following measures to protect yourself or others from COVID-19? % that answered 'always' or 'frequently'

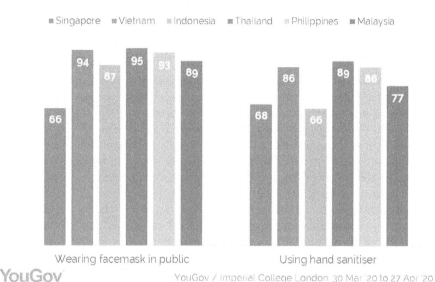

Figure 4.1. Mask wearing and hand sanitizer use as seen by YouGov and Imperial College, London study (Kim Ho, 19 May 2020).

of 12,999 adults aged 18 and over, the study found that Thais were most likely to use face masks in public (95%) and were also the most likely to use hand sanitizers, 89% (Ho, 2020). In fact, hygiene measures were taken seriously by the Thai public thanks to an "abundance of alcohol hand-sanitizing gels and hand-washing facilities" available everywhere (Gunawan et al., 2020). Only 66% of Singaporeans surveyed wore face masks, while Indonesians (also 66%) were the least likely to use hand sanitizers (see Figure 4.1). The controversial issue of wearing face masks is therefore the ideal conduit for discussing Gelfand's tightness-looseness (TL) theory in the Thai context.

### A Tight Society?

Gelfand's tightness-looseness (TL) theory posits that cultures with stricter social norms are less tolerant towards individuals who deviate from those norms (Gelfand, Nishii, & Raver, 2006; Jackson, 2020), and what we see in the Thai context is a culture of conformity and public sanction when it comes to face mask wearing. Two common remarks from respondents were: "You are the odd one out if you aren't wearing one" and "If I forget to put on my face mask someone will look right at me and gesture that I put it on." So, anyone out in public without their face mask on was subject to a glare, raised eyebrows, a gesture, or a polite plea. Since face masks must be worn before entering any type of local establishment and when using public transportation such as buses, vans, taxis, trains, and the metro system, this may explain the high percentage of mask wearing in Thailand as reported in the YouGov/UCL study. Thailand, the "Land of Smiles," is also known to be a non-confrontational society and the Thai people to be socially conforming – indeed, numerous articles, reviews, and travel guides have written about this facet of the Thai people and Thai culture.

In a BBC travel feature titled "Where People Don't Like to Say No," Laurel Tuohy writes, "As a collective culture, Thai people are taught to be more concerned with what's best for the group rather than what suits them personally" (Tuohy, 2017). Showing anger and extreme emotions in public is frowned upon in a country where people are expected to be gracious, pleasant, respectful towards one another, and accommodating (Tuohy, 2017; Williams, 2018). According to a communications professor at Kasetsart University, "Thais avoid confrontation because they live in a group-orientated culture. Showing emotion is considered immature or rude ... Thai society is highly conservative and traditional. It's a culture where showing gratification and emotion is controlled by strict social norms" (Tuohy, 2017). The respectful gesture of the Thai greeting in itself is an expression of a welcoming, friendly, and hospitable people, an outward reflection of a society that is "tolerant, understanding, civil, and peaceful" (Williams, 2018). These characteristics explain the subtlety and indirectness in verbal communication so as to avoid direct confrontation and any unpleasant situation. Thais do not like saying *no* because "social etiquette dictates that they say yes," according to Kasetsart University professor Rachawit (Tuohy, 2017).

In their work on the global impact of COVID-19, Gains (2020) along with Rustam and Griffin offer an interesting observation in regard to people's emotional responses to the lockdown, which were negative in Europe and other continents but not so in Asia, where people were "much more accepting of the situation" and where, in Asian cultures, it is more difficult to express negative feelings. This observation is a fitting depiction of Thai people in general, given that public sentiment towards the coronavirus and the lockdown was a matter-of-fact practical acceptance of reality. Schools and universities, supermarkets, and shopping malls closed, and local convenience stores provided hand sanitizers for customers, performed temperature checks, and enforced the wearing of face masks, which was mandatory on all public transit systems. Rather than interpreting such actions as a violation of individual rights, Thais saw these measures as common sense and necessary to protect themselves and others, allowing the state to effectively lock down the country. Social control and conformity in the Thai case were therefore salient in the collective response and behaviour of both Thais and most non-Thais as well as businesses and all other facets of society. According to Crispin (2020), such compliance is due in part to the government's "successful appeal to Thai nationalism … where stay-at-home and social distancing orders have been portrayed and obeyed as patriotic duty … seen in near universal adoption of mask-wearing," but not all Thais would agree with this characterization, since many wore masks simply to protect themselves. A survey of 26,000 respondents conducted by the Department of Mental Health in March found that 70% of Thais "understand and are engaging in the 'social distancing' campaign" (Apinya, 2020g).

A commonly used phrase in the Thai language is *mai phen rai*, which translates as "no worries," "it's all right," "don't mention it," or "not a problem," and it is this sentiment that is captured in this chapter's title. Thais for the most part complied with government-imposed social distancing with little opposition. "Harmonious relationships take precedence over being right or wrong, over personal agreement or dissent, even over professional progress," writes Tuohy (2017), and perhaps these social requisites for maintaining harmony explain the accommodating behaviour of citizens during the COVID-19 pandemic. In a non-confrontational society, it would not make sense to argue or get angry with someone who is asking you to put on your face mask. *Mai phen rai* is an expression associated with one's composure and maintaining self-control where in a difficult situation you are able to let go of the issue so as not to trouble the other person (Williams, 2018). *Mai phen rai* is related to the Thai attitude of *greng jai*, a uniquely Thai habit of rejecting help or rather a uniquely Thai concept of not wanting to inconvenience or trouble another person even if it poses an inconvenience to you (Nanticha, 2016). Note that there is no English equivalent translation for *greng jai*. The *greng jai* and *mai phen rai* attitudes tie into two frequently used Thai expressions: *sabai sabai* and *jai yen yen*. *Sabai sabai* means "relaxed" or "all is well," a phrase embodying a sense of calm, peace, and tranquillity, while *jai yen yen*, whose literal translation is "cool heart," is communicated to those feeling tense or agitated to remind them to "stay calm" or "remain cool." In the context of COVID-19, Thailand is a tight society where people keep checks on one another for the benefit of personal and public safety and where these everyday Thai expressions are used even more frequently.

Gelfand's notion of a tight society is one whose culture is "often correlated to its history of collective threat – such as natural disasters, war or infectious diseases" (Jackson, 2020), and Gelfand's tightness-looseness (TL) theory suggests that "strong norms and harsh punishments are effective for keeping groups organized and cohesive, especially in the face of ecological threats such as famine and warfare or growing social complexity" (Jackson, Gelfand, & Ember, 2020, pp. 2–4). In recent years, Thailand has had to deal with natural disasters such as the flooding crises in 2011 and 2017, the avian flu H5N1 outbreak in

2004, and "one of the world's largest ever botulism outbreaks" in the northern province of Nan in 2016 (Crispin, 2020). As the COVID-19 situation demonstrates, stringent measures and tough penalties imposed by the Thai government, coupled with public sanctions, have helped keep the pandemic at bay. For example, heavy penalties were imposed on those who hoarded face masks and other preventive products and on those who sold those items at inflated prices (Mongkol, 2020b). Violators of the nationwide curfew and bans also faced the maximum punishment by law; meanwhile the Thai police force had increased security patrols and screenings across the country to deter violations of containment measures (Apinya, 2020b; Mongkol, 2020b).

In their 2006 article, Gelfand, Nishii, and Raver introduced a multilevel theory of *cultural tightness-looseness*, defined as "the strength of social norms and the degree of sanctioning within societies" (p. 1225). The authors also specify two key components in their theory of societal *tightness-looseness*: "The strength of social norms, or how clear and pervasive norms are within societies, and the strength of sanctioning, or how much tolerance there is for deviance from norms within societies" (Gelfand, Nishii, & Raver, 2006, p. 1226). In the Thai COVID-19 context, what we see is public and state sanctioning upon those who violate "new normal" social norms and hence little deviance from those norms. Gelfand et al. (2020) posit those countries with traditionally looser cultures like Spain or Italy "will grow tighter as COVID-19 continues to develop and begin adopting the cooperative norms that we observe here in tight cultures with efficient governments" (p. 20). Gains's case study on the COVID-19 situation in Indonesia found that Indonesians demonstrated "a strong sense of belonging and family (which for many was strengthened by the situation)" and Indonesians also felt "the need for social conformity," which may explain the general acceptance of lockdown measures in Indonesia (Gains, 2020).

Moreover Jackson, Gelfand, and Ember (2020) point to variation in cultural tightness, that is, the strictness of cultural norms and harshness of punishments for norm violations, which in the Thai context would fall at the strong end of the scale. And while Thailand is not a homogeneous society in the same way that Japan is, there is a high degree of social homogeneity among Thais, despite the "great ethnic, religious, linguistic and cultural diversity among the people" (Pavin, 2010). If social heterogeneity is a major obstacle to cultural tightness (Jackson, Gelfand, & Ember, 2020), it can be argued then that social homogeneity is a key facilitator of cultural tightness, as evident in the Thai case and COVID-19. Gelfand et al.'s (2020) study on the global pandemic points to the importance not only of cultural tightness but also of government efficiency in understanding why certain countries were more effective in combatting COVID-19's infection rates. It would be interesting to see how the concept of cultural tightness and government efficiency played out in other Southeast Asian nations vis-à-vis their respective levels of effectiveness in containing COVID-19.

Image 4.1. Sign atop cabinet states "Cupboard of Happiness" (translation of *Too Pen Sook*); placard inside front latch asks visitors to be considerate and take only what is needed (all three photographs by Nichapa Thichakornsakul, 12 August 2020).

Image 4.2. Those in need visit the cupboard.

Image 4.3. Items inside this particular Cupboard of Happiness include dozens of eggs, ramen noodle packs, cups for noodle preparation, bottles of vegetable oil, and canned tomatoes.

Cultural tightness thus segues nicely into what COVID-19 in the Thai example brought forth in terms of a sense of community and a sense of togetherness and belonging. In a predominantly Buddhist country, showing kindness and generosity is often connected with gaining merit and more positive karma (Williams, 2018), and acts of kindness and generosity were evident. Beginning in May, community pantries or "Pantries of Sharing" or "Pantries of Happiness" (two Thai literal translations being "Cupboards of Happiness" or "Cupboards of Kindness") started appearing across the country, initiated by a group of good Samaritans intent on helping people who were in need during the pandemic (*Bangkok Post*, 2020i; *Bangkok Post*, 2020k; *Bangkok Post*, 2020j; Wasant, 2020). The cupboards are placed on roadsides in public areas and are stocked with necessities like rice, eggs, canned fish, instant noodles, energy drinks, fruit juice, and drinking water. Anyone could come up to the pantry and take what they needed, and the pantry could be restocked by anyone (see images 4.1, 4.2, and 4.3). A similar initiative started by a university student, called the "Goodwill Account," gave out coupons that could be taken to participating vendors in exchange for a meal and/or drink (Wasant, 2020). In addition to full public cooperation and a community support system such as that of the happiness pantries, the coordinated effort of health volunteers early on was critical in containing spread of the virus within the country.

**A Strong Public Health System**

In August 2020, all local districts were instructed to monitor the health of those who completed their 14-day quarantine to make sure that they were clear of infection (after two Thai women who completed their quarantines later tested positive). The outcome of such thorough measures taken by the state, including the monitoring and tracking of people and patients, was Thailand's ability "to flatten the COVID-19 curve with limited local transmission" (Greater Mekong Subregion, 2020). In September, there were only 3,448 confirmed cases of COVID-19 (of which 3,150 had recovered) and a death toll of 58 that had remained unchanged since June. The Greater Mekong Subregion (2020) report, "Thailand shares best practices and lessons learned," pointed to several innovative public health strategies that were implemented to control the disease, including social campaigns to "stay home, stop the virus, save the nation" as well as the provision of free treatment to all COVID-19 patients, including non-Thai citizens. Thailand's sub-national public health capacity effectively deployed the Rapid Response Teams and Village Health Volunteers, extensive networks tasked with identifying suspected cases, ensuring immediate isolation, treatment, quarantining, and track and tracing, to minimize spread of the virus.

Thailand's ability to control the spread of the coronavirus, however, has come at great cost, according to studies by the Asia Foundation (Parks et al.,

2020) and the World Bank (2020) that examined the socio-economic impact of COVID-19 on micro, small, and medium-sized businesses, the Thai workforce (which included vulnerable workers), and the formal and informal economy, as well as how these affected the overall livelihood of the Thai people. The impact of the pandemic was most severe on the Thai tourism industry, which in 2019 accounted for nearly 20% of the country's GDP and 21.4% of total employment. The country relies heavily on tourists from China, who make up a significant percentage (30%) of visitors to the Kingdom. Indeed, 11 million of the nearly 40 million tourists who visited Thailand in 2019 were from mainland China (Thaiwebsites.com, 2020; Statista, 2020). At the onset of the outbreak in January, Thailand had welcomed more than 3.5 million foreign tourists alone. Despite the devastating impact upon the tourism sector, an online opinion poll carried out in July that surveyed 1,459 people throughout the country found that most Thais (94.51%) would continue to ban foreigners from entering to prevent a second wave. Moreover, survey respondents also indicated that screening and quarantine processes should be made more stringent (*Bangkok Post*, 2020n).

## Explaining Thailand's Impressive Containment of COVID-19

Thailand's remarkable containment of COVID-19 during the pre-vaccination phase has been met with global praise owing to a series of actions that were taken, such as the country's point-of-entry screening and quarantining systems, an efficient and robust public healthcare sector, use of information and communications technology (ICT) in contact tracing, effective risk communication that involved the raising of awareness through social media on the dangers of COVID-19, and, most notably, social cooperation and participation in observing preventive measures against COVID-19 (Apinya, 2020b; Suvimon, 2020). Writing about the key secrets to Thailand's success, Kavi also cites the nation's health security system, a resilient Thai public, quality leadership, and fact-based communication, and says that Thailand "pursued the middle path which was neither too stringent nor too loose" regarding preventive measures (Kavi, 2020b). With regard to leadership, Crispin (2020) credits the prime minister, whose "strongman move" empowered "bureaucrats and medical professionals to command, lead and communicate the government's response to COVID-19." In his article "Thailand's Unsung COVID-19 Success Story," Crispin (2020) attributes Thailand's "so far mild COVID-19 experience … to a uniquely Thai mix of factors" – Thai culture, effective risk communication, emergency rule, centralized crisis management, the phasing in of hard lockdown measures, and the country's universal public healthcare system. Kavi (2020b) would call this uniquely Thai mix *Thai exceptionalism*, which he defines as "any force that enables the Thai society and its people to perform some extraordinary work."

Gelfand et al. (2020) argue that the interaction of cultural tightness and government efficiency is key in explaining differences across cultures. Additionally, Jackson notes how "countries that are loose and have inefficient governments struggled to flatten curves and death rates in the early stages of the pandemic" (Jackson, 2020), with the US being a good example. Thailand on the other hand has demonstrated a tightening of measures, seen as necessary in keeping the pandemic under control, while cultural factors, including social cohesion and social cooperation on the part of the Thai public, have also contributed to the country's effective handling of the virus. Thailand's response to COVID-19 further illuminates Gains's and Gelfand's analyses on why people in Asia were more compliant in lockdown measures and why some nations contained COVID-19 more effectively than others. Thus, in the context of Thailand's pandemic, Thai civil society appeared to be self-regulating, ensured social harmony, and promoted social cohesion independent of state sanction.

Thailand's success in this respect can be summarized into three main points succinctly captured by Parker (2020):

- *Early and widespread action by national and local governments* to combat COVID-19,
- *Early action by business literally overnight* (referring to action taken by shopping malls and local convenience stores in every town that implemented temperature checks and provided hand sanitizer upon entry to their premises), and
- *Masks and social distancing*, which according to Parker is the "single biggest reason why Thailand is winning this battle."

Almost all new COVID-19 cases reported in the country between September and October 2020 were returnees from the Middle East and foreign travellers arriving from Europe, North America, Africa and the Middle East, South Asia, and East Asia (three new COVID imports). On 7 September, Thailand's first locally transmitted case after over 100 days was detected in a male inmate (Thanthong-Knight et al., 2020), and 18 September marked the country's first new death since 2 June (a Thai returnee from Saudi Arabia), bringing the total number of deaths to 59 and the total number of cases to 3,583 by 3 October.

On 16 July, the *New York Times* published an article titled "No One Knows What Thailand Is Doing Right, but So Far, It's Working," questioning whether Thailand's low rate of coronavirus infections could be attributed to culture, genetics, face masks, or a combination of all three (Beech, 2020). Beech points out that Thais were quick to adopt wearing face masks and maintained social distancing measures early on, but she also raises a question concerning the immune systems of Thai people and of people in the countries of the Mekong River basin, namely, Myanmar, Cambodia, Laos, and Vietnam, given the low rates of infection in the region. (Vietnam had no deaths and had no case of community transmission in three months while Myanmar reported 336 cases, Cambodia 166 cases, and Laos only 19 cases at the time of her writing.) Further studies will be needed to determine whether there is a genetic explanation for resistance to COVID-19.

Gelfand et al.'s study (2020) concludes that "Nations with high levels of cultural tightness and government efficiency have slower growth and mortality rates compared with nations that have one of these factors or neither" (Gelfand et al., 2020, p. 18). While Thailand would fall into the former category, the authors do acknowledge that their analysis "only partly explains cross-cultural variation in COVID-19 case growth and mortality rates," and that "other factors such as political leaders," personal beliefs about the seriousness of COVID-19, the nature, extent, and timing of countermeasures such as quarantining, and the quality of governments' communications about the virus may also be associated with responses to COVID-19 (Gelfand et al., 2020). These "other factors" are indeed relevant to the Thai case and are addressed herein, and while the observation that tighter cultures are "more effective in the early stages of threats" is accurate, the authors posit that looser cultures may be "more effective at the later stages of the pandemic, when innovation is needed" (Gelfand et al., 2020, p. 20).

Perhaps there are certain aspects of or lessons learned from the Thai COVID-19 experience that can be considered as best practices for countries that need to flatten their COVID-19 curve. They would include a more proactive and effective government that can provide fact-based communication and that can effectively inform its citizens about risks and encourage cooperative behaviours without being too harsh or extreme (since "strong norms to fight COVID-19 do not imply that governments should become autocratic" [Gelfand et al., 2020]). While the kinds of measures taken by the Thai government may have been harsh and extreme, they were imperative. But irrespective of the government's actions, it appears that the country's successful containment of the novel coronavirus was in large part a consequence of a culturally tight society where strong social norms endured and indeed encouraged and engendered cooperative behaviour. Perhaps, the country's success in this respect can be attributed to a uniquely Thai mix of factors.

NOTES

1 It is conventional practice to cite Thai first names in parenthetical citations and to alphabetize by first names in the references.
2 The score comes from the GCI, which bases 70% of its calculation on big data and daily analysis from 184 countries and 30% on the Global Health Security Index, an assessment of global health security in 195 countries prepared by the Johns Hopkins Center for Health Security (Nation 2020b).

3  From email correspondence with respondent, 6 September 2020. Respondents to remain anonymous.
4  Selected excerpts from email correspondence with respondents, 6 September 2020.
5  Email correspondence, 6 September 2020. Respondents to remain anonymous.
6  Email correspondence, 6 September 2020. Respondents to remain anonymous.

REFERENCES

Achadthaya, C. (2020, January 14). Phuket airport screening Wuhan arrivals for coronavirus. *Bangkok Post*. https://www.bangkokpost.com/thailand/general/1835629/phuket-airport-screening-wuhan-arrivals-for-coronavirus.

Apinya, W. (2020a, January 22). First Thai infected with coronavirus. *Bangkok Post*. https://www.bangkokpost.com/thailand/general/1841199/first-thai-infected-with-coronavirus.

Apinya, W. (2020b, February 11). New clinics to screen virus. *Bangkok Post*. https://www.bangkokpost.com/thailand/general/1854774/new-clinics-to-screen-virus.

Apinya, W. (2020c, February 21). Thailand expands virus detection. *Bangkok Post*. https://www.bangkokpost.com/thailand/general/1862789/thailand-expands-virus-detection.

Apinya, W. (2020d, February 25). Virus gets "dangerous disease" tag. *Bangkok Post*. https://www.bangkokpost.com/thailand/general/1864809/virus-gets-dangerous-disease-tag.

Apinya, W. (2020e, March 18). At least 400,000 infections expected. *Bangkok Post*. https://www.bangkokpost.com/thailand/general/1880940/at-least-400-000-infections-expected.

Apinya, W. (2020f, March 21). 500 at boxing match risk. *Bangkok Post*. https://www.bangkokpost.com/thailand/general/1883170/500-at-boxing-match-risk?cx_placement=article#cxrecs_s.

Apinya, W. (2020g, April 2). 70% [of] Thais stick to social distancing. *Bangkok Post*. https://www.bangkokpost.com/thailand/general/1891290/70-thais-stick-to-social-distancing.

Apinya, W. (2020h, July 23). Thailand to be star player in COVID success story. *Bangkok Post*. https://www.bangkokpost.com/thailand/general/1955827/thailand-to-be-star-player-in-covid-success-story-doco.

*Bangkok Post*. (2019, September 9). Thailand's healthcare ranked sixth best in the world. https://www.bangkokpost.com/thailand/general/1746289/thailands-healthcare-ranked-sixth-best-in-the-world.

*Bangkok Post*. (2020a, January 21). Govt takes heavy flak for toxic smog response. https://www.bangkokpost.com/thailand/general/1840074/govt-takes-heavy-flak-for-toxic-smog-response.

*Bangkok Post*. (2020b, January 24). Thailand on alert to prevent spread of virus. https://www.bangkokpost.com/thailand/general/1842144/thailand-on-alert-to-prevent-spread-of-virus.

*Bangkok Post*. (2020c, January 31). Human transmission of coronavirus confirmed in Thailand. https://www.bangkokpost.com/thailand/general/1847884/human-transmission-of-coronavirus-confirmed-in-thailand.

*Bangkok Post*. (2020d, February 20). Tougher COVID-19 measures take effect Sunday. https://www.bangkokpost.com/thailand/general/1868439/tougher-covid-19-measures-take-effect-sunday.

*Bangkok Post*. (2020e, February 21). COVID-19 to get a "dangerous" label. https://www.bangkokpost.com/thailand/general/1862004/covid-19-to-get-dangerous-label.

*Bangkok Post*. (2020f, March 10). Govt ramps up COVID-19 measures. https://www.bangkokpost.com/thailand/general/1875139/govt-ramps-up-covid-19-measures.

*Bangkok Post*. (2020g, March 15). Majority concerned about COVID-19 poll. https://www.bangkokpost.com/thailand/general/1879255/majority-concerned-about-covid-19-poll.

*Bangkok Post*. (2020h, March 19). Travellers from 15 countries must have COVID-free certificates. https://www.bangkokpost.com/thailand/general/1882185/travellers-from-15-countries-must-have-covid-free-certificates.

*Bangkok Post*. (2020i, May 10). Community pantries help virus-hit needy. https://www.bangkokpost.com/thailand/general/1915680#cxrecs_s.

*Bangkok Post*. (2020j, May 12). "Pantries of sharing" fan out nationwide. https://www.bangkokpost.com/thailand/general/1916380/pantries-of-sharing-fan-out-nationwide.

*Bangkok Post*. (2020k, May 12). Pantries proliferate. https://www.bangkokpost.com/photo/1916256/pantries-proliferate#cxrecs_s.

*Bangkok Post*. (2020l, May 18). Siam Paragon, Siam Center, Siam Discovery and ICON Siam reopen with maximum health and safety measures and integrated innovations. https://www.bangkokpost.com/thailand/pr/1920076/siam-paragon-siam-center-siam-discovery-and-iconsiam-reopen-with-maximum-health-and-safety-measures-and-integrated-innovations.

*Bangkok Post*. (2020m, June 9). A million volunteers help curb coronavirus infections. https://www.bangkokpost.com/business/1931824/a-million-volunteers-help-curb-coronavirus-infections.

*Bangkok Post*. (2020n, July 19). 95% say "no foreigners" to prevent 2nd COVID wave: Poll. https://www.bangkokpost.com/thailand/general/1953916/95-say-no-foreigners-to-prevent-2nd-covid-wave-poll.

*Bangkok Post*. (2020o, September 16). COVID shock worse than expected. https://www.bangkokpost.com/business/1986139/covid-shock-worse-than-expected-adb-says.

Beech, H. (2020, July 16). No one knows what Thailand is doing right, but so far, it's working. *New York Times*. https://www.nytimes.com/2020/07/16/world/asia/coronavirus-thailand-photos.html.

Crispin, S.W. (2020, May 26). Thailand's unsung COVID-19 success story. *Asia Times*. https://asiatimes.com/2020/05/thailands-unsung-covid-19-success-story/.

Gains, N. (2020, August 2020). From confusion to frustration to anger: A pandemic case study from Indonesia (with Riri Rustam & Craig Griffin). *Tapestryworks*. https://tapestry.works/2020/08/20/from-confusion-to-frustration-to-anger-a-pandemic-case-study-from-indonesia-with-riri-rustam-craig-griffin/.

Gelfand, M., Jackson, J.C., Pan, X., Nau, D., Dagher, M., Van Lange, P., & Chiu, C.Y. (2020). Preprint. The importance of cultural tightness and government efficiency for understanding COVID-19 growth and death rates. *Psyarxiv*. https://psyarxiv.com/m7f8a/.

Gelfand, M., Nishii, L., & Raver, J. (2006). On the nature and importance of cultural tightness-looseness. *Journal of Applied Psychology*, 91(6), 1225–44. https://doi.org/10.1037/0021-9010.91.6.1225.

Global COVID-19 Index (GCI). (2020). The GCI dashboard. https://covid19.pemandu.org/.

Greater Mekong Subregion. (2020). Thailand shares best practices and lessons learned. https://greatermekong.org/thailand-shares-best-practices-and-lessons-learned-covid-19-pandemic.

Gunawan J., Yupin A., & Marzilli, C. (2020). "New normal" in COVID-19 era: A nursing perspective from Thailand. *JAMDA*, 21(10), 1514–15. https://doi.org/10.1016/j.jamda.2020.07.021.

Ho, Kim. (2020, May 19). Thais most likely to wear facemasks in ASEAN. YouGov. https://th.yougov.com/en-th/news/2020/05/19/thais-most-likely-wear-facemasks-asean/.

Human Rights Watch. (2020, October 15). Thailand: Emergency decree pretext for crackdown. https://www.hrw.org/news/2020/10/15/thailand-emergency-decree-pretext-crackdown.

IQAir. (2019). 2019 world air quality report. Goldach, Switzerland.

Jackson, J.C. (2020, July 1). Will COVID-19 lead to "tighter" societies? Research. umd.edu. https://bsos.umd.edu/featured-content/will-covid-19-lead-%E2%80%9Ctighter%E2%80%9D.

Jackson, J.C., Gelfand, M., & Ember, C.R. (2020, July 1). A global analysis of cultural tightness in non-industrial societies. *Proceedings of the Royal Society B*. https://doi.org/10.1098/rspb.2020.1036.

Kasipat, W. (2020, April 12). 660,000 COVID-19 "high-risk" people under watch of health volunteers. Thai PBS World. https://www.thaipbsworld.com/tag/11-8-million-households/.

Kavi, C. (2020a, April 28). Watershed Thai media moment? *Bangkok Post*. https://www.bangkokpost.com/opinion/opinion/1909000/watershed-thai-media-moment-.

Kavi, C. (2020b, June 23). 7 secrets of Thai COVID-19 success. *Bangkok Post*. https://www.bangkokpost.com/opinion/opinion/1939416/7-secrets-of-thai-covid-19-success.

Khemthong, T., & Rawin, L. (2020, May 8). Health before rights and liberties: Thailand's response to COVID-19. *Verfassungsblog*. https://verfassungsblog.de/health-before-rights-and-liberties-thailands-response-to-covid-19/.

Maragakis, L.L. (2020, July 15). Coronavirus, social and physical distancing and self-quarantine. *Johns Hopkins Medicine.* https://www.hopkinsmedicine.org/health/conditions-and-diseases/coronavirus/coronavirus-social-distancing-and-self-quarantine.

Mongkol, B. (2020a, March 18). PM: Third-stage COVID-19 would justify countrywide lockdown. *Bangkok Post.* https://www.bangkokpost.com/thailand/general/1881445/pm-third-stage-covid-19-would-justify-countrywide-lockdown.

Mongkol, B. (2020b, April 2). Cops to get tough with rule breakers. *Bangkok Post.* https://www.bangkokpost.com/thailand/general/1891230/cops-to-get-tough-with-rule-breakers.

Mongkol, B. (2020c, April 12). Taweesilp slammed as "govt lackey." *Bangkok Post.* https://www.bangkokpost.com/thailand/general/1897915/taweesilp-slammed-as-govt-lackey.

Mongkol, B., & Apinya, W. (2020a, February 20). Thailand "ready for worst" COVID-19 scenario. *Bangkok Post.* https://www.bangkokpost.com/thailand/general/1861929/thailand-ready-for-worst-covid-19-scenario.

Mongkol, B., & Apinya, W. (2020b, March 28). Mass exodus may cause COVID-19 numbers to soar. *Bangkok Post.* https://www.bangkokpost.com/thailand/general/1887995/mass-exodus-may-cause-covid-19-numbers-to-soar.

Nanticha, O. (2016, February 23). It's not okay: What Thais really mean by "Mai Pen Rai." *Prachatai.* https://prachatai.com/english/node/5877.

*Nation.* (2020a, February 24). COVID-19 declared a dangerous contagious disease. https://www.nationthailand.com/news/30382747.

*Nation.* (2020b, July 28). Thailand tops global ranking for handling of COVID-19 crisis. https://www.nationthailand.com/news/30392083?utm_source=category&utm_medium=internal_referral.

*Nation.* (2020c, November 6). PM to share lessons from Thailand's success in containing COVID-19 with Paris forum. https://www.nationthailand.com/news/30397514?utm_source=homepage&utm_medium=internal_referral.

*Nation.* (2020d, November 6). WHO picks Thailand among 4 countries as example of effective COVID-19 handling. https://www.nationthailand.com/news/30397523?utm_source=homepage&utm_medium=internal_referral.

Parker, C. (2020, July 24). *Why Thailand is winning (and America is losing) the fight against Covid 19. Retired Working for You.* Retrieved 25 December 2020, from https://www.youtube.com/watch?v=Cpl3ov9IYIc.

Parks, T., Chatsuwan, M., & Pillai, S. (2020, September 8). Enduring the pandemic: Surveys of the impact of COVID-19 on Thai small businesses. *Asia Foundation.* https://asiafoundation.org/publication/enduring-the-pandemic-surveys-of-the-impact-of-covid-19-on-thai-small-businesses/.

Pavin, C. (2010, July 1). The myth of Thai homogeneity. *Asia Sentinel.* https://www.asiasentinel.com/p/the-myth-of-thai-homogeneity.

Statista. (2020, January). Total number of tourist arrivals from China to Thailand from 2014 to 2019 (in millions). Retrieved 25 December 2020, from https://www.statista.com/statistics/1048386/thailand-tourist-arrivals-from-china/.

Supakit, S. (2020, August). *Thailand's experience in the COVID-19 response*. Ministry of Public Health. https://greatermekong.org/sites/default/files/Thailand%20s%20experience%20in%20the%20COVID-19%20response_0.pdf.

Supita, R. (2019, November 17). The pollution paralysis: Thailand's structural inability to clean up its air. *Bangkok Post*. https://www.bangkokpost.com/thailand/special-reports/1796019/the-pollution-paralysis-thailands-structural-inability-to-clean-up-its-air.

Suvimon, S. (2020, December 11). Point of entry screening and quarantine systems enabled Thailand to control COVID-19. World Health Organization (WHO). https://www.who.int/thailand/news/feature-stories/detail/point-of-entry-screening-and-quarantine-systems-enabled-thailand-to-control-covid-19.

Thai PBS World. (2020, September 1) No more enforced social distancing on Thailand's public rail networks. https://www.thaipbsworld.com/no-more-enforced-social-distancing-on-thailands-public-rail-networks/.

Thaiwebsites.com. (2020, December). Tourism statistics Thailand 2000–2020. Retrieved 25 December 2020, from https://www.thaiwebsites.com/tourism.asp.

Thanthong-Knight, R., Chuwiruch, P., & Yuvejwattana, S. (2020, September 3). Thailand's 100-day virus free run ends with new local case. *Bloomberg*. https://www.bloomberg.com/news/articles/2020-09-03/thailand-reports-first-local-case-after-100-day-virus-free-run.

Thitinan, P. (2020, April 17). Coronavirus blues and clues in Thailand. *Bangkok Post*. https://www.bangkokpost.com/opinion/opinion/1901360/coronavirus-blues-and-clues-in-thailand.

Tuohy, L. (2017, January 25). Where people don't like to say no. BBC. http://www.bbc.com/travel/story/20170123-where-people-dont-like-to-say-no.

Veera, P. (2020, April 13). Taweesilp's limelight fully earned. *Bangkok Post*. https://www.bangkokpost.com/opinion/opinion/1898470/taweesilps-limelight-fully-earned.

Wasant, T. (2020, May 16). Keep govt out of "Happiness Cupboards." *Bangkok Post*. https://www.bangkokpost.com/opinion/opinion/1918968/keep-govt-out-of-happiness-cupboards.

Wassana, N. (2020, March 19). Army disinfects Bangkok roads nightly. https://www.bangkokpost.com/thailand/general/1881470/army-disinfects-bangkok-roads-nightly.

Williams, S. (2018, April 6). 10 values to live by in Thailand. *Culture Trip*. https://theculturetrip.com/asia/thailand/articles/10-values-to-live-by-in-thailand/.

World Bank. (2020, June 30). Major impact from COVID-19 to Thailand's economy, vulnerable households, firms: Report. https://www.worldbank.org/en/news/press-release/2020/06/30/major-impact-from-covid-19-to-thailands-economy-vulnerable-households-firms-report#:~:text=While%20Thailand%20has%20been%20successful,tourist%20arrivals%20since%20March%202020.

World Health Organization (WHO). (2020, October 14). Thailand's review of the health system response to COVID-19. https://www.who.int/thailand/news/detail/14-10-2020-Thailand-IAR-COVID19.

# 5  Trust in Numbers? The Politics of Zero Deaths and Vietnam's Response to COVID-19

BY AMY DAO

My phone buzzed at 5:36 a.m. on 27 January 2020. I opened up Zalo – a Vietnam-based chat application. With about 100 million users worldwide in 2019, Zalo is used by nearly all Vietnamese. Along with Facebook, Zalo helped me stay in touch with family and the friends I made while carrying out fieldwork between 2012 and 2019. I did not expect that the "person" on the other side would be Vietnam's Ministry of Health. The message linked me to a page entitled "Measures to reduce the risk of contracting the novel coronavirus." Most notably, it was co-written by the Ministry of Health and the World Health Organization (WHO). What followed was a series of infographics and instructional videos detailing proper behaviours to curb transmission. The landing page ended with a list of recent Vietnamese news articles related to what was then only known as "nCov" or the novel coronavirus. News coverage about COVID-19 during the early months of 2020 from the United States and English-based media was sparse and felt largely confined to the Asia-Pacific region. Little did many know that the first person with COVID-19 symptoms had already presented at a US clinic on 19 January 2020 (Holshue et al., 2020) – only days before the text message arrived in my chat inbox.

Weeks passed as I continued to track reporting on SARS-CoV-2, the virus that causes COVID-19. South Korea, Italy, the United Kingdom, and the United States began experiencing increasing case numbers, hospitalizations, and deaths. Horrific stories filled my social media feeds: hospitals inundated with critically ill patients, healthcare workers making impossible decisions about who would live and die because of limited ventilators, people unable to be with loved ones at the end of their life. Although my friends and family in Vietnam were worried, cases spiked only as travellers returned to Vietnam. Hotspots and outbreaks were immediately addressed via lockdowns, contact tracing, testing, and quarantine. Vietnam did not record its first death from COVID-19 until 31 July 2020, nearly eight months after the coronavirus was reported from Wuhan, China in December 2020. The low case numbers and zero death count up until that point led many public health practitioners, researchers, and journalists to document and explain the country's success during the pre-vaccination phase.

Praise for Vietnam was not without controversy, however. As more news articles and NGO reports relayed the unfolding situation, the same questions appeared over and over in Vietnamese and Western English media: How could Vietnam – a low- to middle-income country with a population of 95 million, geographically sharing a 1,600-kilometre border with China – successfully manage its coronavirus situation? How could a country with a health system deemed as underdeveloped by international and national critics protect its population from the worst damages of the virus? With climbing cases, hospitalizations, and deaths in much of Europe and North America, reporting on Vietnam's numbers was often laced with doubt.

In this chapter, I analyse moments of incredulity across journalistic and social media reporting about Vietnam's COVID-19 numbers. I focus specifically on the ways that Vietnam's low COVID-19 case numbers and zero deaths were presented in the news. Following anthropologists and other social scientists studying quantification, I consider numbers as "globally circulating knowledge technologies" (Adams, 2016). Numbers transform social phenomena into information. Our trust in numerical representation is thus a hallmark of Western modernity (Porter, 1996). One only needs to look at how governments and organizations use numerical values to make decisions, compare, and rank to see that widespread recognition of statistics provides a common language of communication within and across institutions. It is easy to forget, however, that numbers do not speak for themselves; they require interpretation and explanation. The extent to which numbers are trusted or not is where politics is made visible. Thus, I look at the interpretation of COVID-19 numbers by English and Vietnamese media, and the circumstances in which they are seen as trustworthy or transparent.

Edward Said famously argued in *Orientalism* that the West comes to know itself and gain its power from its knowledge and construction of the East (Said, 1978). Representations of the East as exotic and foreign led to the production of Europe as civilized, and therefore entitled to a position of political and cultural authority. But what happens when the script is flipped by all accounts and indicators? Following Said's framework, I argue that news stories about Vietnam's zero death count reveal Western attempts to maintain global power over authority and legitimacy. Whereas metrics are purported to replace political debate with technical solutions and provide a level of transparency, the Vietnam case study shows that even in circumstances where numbers are reported, colonial politics continue to shape our interpretative understanding of numbers as pieces of information. Who is believed and how they are believed reflect ongoing relationships in which countries of the "Global North" continue to hold epistemic power over countries in the "Global South" in the field of global health. I highlight three ways in which the impact of the colonial project continues to surface via representations in reporting on COVID-19 numbers: biological racialization of Vietnamese and other Southeast Asian nations to explain the country's effective response to the pandemic, the centring of Western-affiliated expertise to verify and provide confidence in Vietnam's numbers, and the use of exoticizing and unverified explanations as reasons to distrust Vietnam's COVID-19 counts.

My arguments are drawn from online articles published in English- and Vietnamese-language news sources. Articles were collected between 1 January 2020, and 30 September 2020, as they surfaced on my Google Alerts and Twitter feed. When needed, I searched newspaper databases using key words such as "Vietnam," "COVID-19," and "zero deaths." As the pandemic advances forward even at the time of this writing in late fall 2020 during the pre-vaccination phase, this chapter is necessarily incomplete, but its conclusions point to longstanding insights from post-colonial approaches to power and cultural domination. I show how even in the context of successes of the Global South, the lasting impact of colonialism continues to uphold global hierarchies of knowledge in a subtle and insidious fashion.

To be clear, this chapter is not criticizing the importance of sharing knowledge about how a country successfully managed its coronavirus outbreak. I believe in the power of sharing lessons learned across all countries when facing a common problem (Dao & Nichter, 2016). It is the manner through which the information is conveyed, how it centres Western institutions and authority as a form of verification and validity and remakes divisions between East and West, which I wish to illuminate. In this way, I am inspired by recent calls by a new generation of public health scholars to decolonize global health (Abimbola & Pai, 2020; Büyüm et al., 2020; Glass, 2020; Green, 2019; Guinto, 2019; Pailey, 2020). The first step in "decolonizing" or dismantling oppressive systems of knowledge is to recognize the patterns that help maintain the global status quo. I aim to contribute to this literature by offering an analysis that asks how knowledge about Vietnam's COVID-19 numbers is constructed and what the political implications of the country's cultural representation during the pandemic are in relation to imperialism.

## Numerical Truth Telling and COVID-19

If our current era is characterized as an information society – wherein the availability of information is essential for good and just governance – then metrics, indicators, and a commitment to transparency are its political currency. "Data has been the only way to truly understand the scale and impact of COVID-19" is the opening argument of the Visual and Data Journalism Team of the *Financial Times* (FT Visual & Data Journalism Team, 2020). By data, they refer to quantified information on COVID-19 deaths, hospitalizations, cases, testing rates, and changes in a country's GDP. Through these numbers, the *Financial Times* team can relay stories about the hardest-hit countries, spikes in cases over time and space, how countries fared in comparison to one another, and the impact of responses on a country's economy. As quantified representations of suffering, these numbers, when tied to specific countries, prompt interpretation about the development of the coronavirus in national and international terms.

Not only do numbers facilitate qualitative judgments about what is happening in the pandemic, they also guide highly consequential decision-making processes about what to do during the pandemic. Flights and cruises were cancelled, universities and public schools closed for face-to-face classes, and events such as concerts and sports postponed. Each of these alterations in everyday life was the result of state-mandated restrictions on social and economic activities based on the ratio of coronavirus infections and hospitalizations among the population (often expressed as number of cases per 100,000 people in order to compare with other geographic areas). The early portion of the pandemic saw public circulation of a graph imploring the citizens to "flatten the curve," since the number of those sick enough to need hospitalization could either overwhelm the healthcare system or not depending on the time frame of COVID-19's dispersion. The objective was to change behaviour and slow the spread enough to keep hospitalizations below the number of available ventilators and hospital beds. As of the final draft of this chapter, people continue to check websites such as Worldometers, the Johns Hopkins Coronavirus Resource Center, Our World in Data, and local county dashboards to follow the live statistics of the pandemic.

Anthropologists and other social scientists studying quantification have long questioned the "truth value" of numbers and shown the powerful ways in which they are enlisted in governance (Adams, 2016; Davis, 2020; Merry, 2011; Merry, Davis, & Kingsbury, 2015). In studying the power of metrics, Adams argues that "Specific numbers can certainly move policy, confer political allegiance, guarantee funding, even bring about health. But they do so not simply by claiming truth about the empirical world. They do so because of the ways they are 'produced' and the ways they are circulated. These productivities and circulations are the stories that precede and exceed numerical forms of truth-telling" (Adams, 2016, p. 9). Much work has focused on what Adams refers to as the "production" of numbers from construct to count. In this chapter, I look at the way in which zero deaths was contextualized via media stories that circulated as a response to this numerical representation.

To these insights, I add that numbers, as representations and information, require interpretation by human spectators. They need an audience or third parties to ultimately decide if the numbers are in fact a reflection of realities on the ground. Implicit within the application of numbers for governance is an assumption that what is signified by numerical values matches the social phenomena it describes. While the aura of transparency that numbers possess is the product of a historical process (see also Porter, 1996), I argue that one of the lasting effects of colonialism is to steer interpretation and trust in numbers in particular directions that uphold the supremacy of high income countries (HICs).

To trust in numbers is to believe in the transparency of those numbers. With increasing globalization and communication between countries, transparency of information (here in the form of indicators or metrics) becomes a virtue in and of itself. Transparency has become a symbol of integrity (Ballestero, 2018), and thus has many benefits: some have argued that transparency helps solidify relationships and builds economic and political trust; others purport that it is the cornerstone of democracy and a prerequisite for a healthy economy (Hetherington, 2011, p. 6). For these reasons, being transparent could bring rewards in terms of fiscal or social gain. For example, economists have tried to make a causal link between economic and political transparency and an increase in foreign direct investments (Drabek & Payne, 2002; Zhao, Kim, & Du, 2003). Other international organizations, such as the World Economic Forum, argue that more transparency will help the global economy grow because approximately 15% of global GDP is lost to corrupt practices of procurement (Konanykhin, 2018). Similarly in global health, the ability to use numbers to show impact can provide access to funds and social capital (Adams, 2016; Nelson, 2010). Yet transparency is also a second-order observation often assessed by someone on the "outside." The case of Vietnam's COVID-19 numbers shows that transparency is not a given. For Vietnam and countries like Vietnam, to make a legitimate claim of transparency engenders a process that cannot be done with metrics alone. Countries in the Global South must rely on associations to verify their numbers and these associations are linked to global hierarchies of power and colonial histories. What practices are necessary for the production of transparency and belief in numbers? Who can claim numbers as "indelibly factual"? Who has the privilege of not having their numbers questioned?

**Background: A Brief Overview of Vietnam's Strategies to Contain COVID-19**

Vietnam used a combination of strategies for detecting, containing, and communicating the risks of the virus. By 15 January 2020, the government had assembled a National Steering Committee involving all the major ministries and several international organizations to plan for and run hypothetical scenarios in the event of increasing caseloads as well as to organize drill scenarios for how to handle as little as dozens to as many as thousands of cases. The government was aware of limitations in the country's health system. As Deputy

Prime Minister and the head of the National Steering Committee Vu Duc Dam stated, the leaders of Vietnam knew they could not embark on the same course of action as developed countries because of limitations in resources and infrastructure (Exemplars of Global Health, 2020). Thus, they had to ensure that they were prepared for the worst-case scenario. The government operated on the philosophy that "human lives and human health shall be most prioritized and no one shall be left behind" (Exemplars of Global Health, 2020).

Vietnam's first wave of coronavirus outbreaks occurred from the end of January to April 2020. Two days after discovering the first case of local transmission on 23 January 2020, the government halted all flights from Wuhan, China, the epicentre of the outbreak. During the middle of Lunar New Year holiday in mid-February, as people migrated back to their hometowns to spend the holiday with their families, the government suspended all school activities. This suspension would last until April 2020, when it was clear that there were no community transmissions.

Most cases were imported as Vietnamese nationals and non-nationals who had been travelling abroad entered the country. Despite the use of health declarations at the ports of entry, cases continued to rise. In response, the government suspended all flights to Vietnam beginning on 22 March 2020. A national period of quarantine was implemented from 1 April to 15 April, wherein restaurants, bars, and other services were closed to curb any community spread. Government officials stated that, without the same number of resources to test many citizens as South Korea or Japan, Vietnam had to take more economic measures for managing potential asymptomatic carriers, which included targeted testing and isolation. With testing, the government was quick to prevent imported cases from spreading. When positive cases were found, the state engaged in aggressive contact tracing. People who were in direct contact with someone with a positive coronavirus test were tested. Those who were in close contact with someone who had direct contact with a positive test case up to the third level were also tested once testing capacity increased. Vietnam performed more tests per confirmed case than any other country around May, even though the total number of tests was low (Pollack et al., 2020). In this way, testing was very strategic with limited resources. Vietnam was the fourth country to isolate the virus in the lab. Vietnamese scientists developed four types of tests soon afterward that met quality standards established by the WHO.

The country engaged in proactive containment. A national mask mandate was immediately implemented in February 2020, and the state put a moratorium on exports of masks. When clusters of positive cases were detected, neighbourhoods or buildings were gated off and locked down. For example, the commune of Son Loi in Vinh Phuc Province – about 27 miles from Hanoi – was put on lockdown when 11 coronavirus cases were found. The streets to Son Loi were gated off and sanitized, and necessary goods such as food were delivered by the military to the 10,000 residents of the commune as they underwent a 20-day period of isolation and restricted movement. As the government began to allow Vietnamese nationals, permanent residents, and other diplomats to re-enter the country, it prepared quarantine wards run by the military to house all who were required to undergo the 14-day isolation period before being able to reintegrate into the larger community.

Communication was also timely and transparent. The government stated that inclusion of the public was part of their strategy to keep cases from spreading. The Ministry of Communication utilized the media to disseminate information to ensure that people were informed about the pandemic, the risks, and mechanisms for infection. Anyone in the country could download two phone applications created by the government to receive information about positive cases nearby. New cases were also reported in the news, using case numbers and detailed descriptions of the person's whereabouts in the last few days so that anyone who suspected they might have been in contact could self-isolate. Cases were also displayed on the Ministry of Health's website, and special notices were given to each local ward where there were residents who tested positive. Their names were kept anonymous, but their city of residence was reported. Other forms of communication involved the use of visual and audio propaganda to teach people how to wash their hands, keep their social distance, and keep their respective areas sanitized. Most notably, a music video Ghen Cô Vy ("Jealous Coronavirus") was released on 23 March 2020, and went internationally viral as a song and dance that was copied on Tik Tok. The song served as a public service announcement to prevent viral transmissions and was created in partnership with the country's Ministry of Health. Laws were put in place that levied fines on people who spread false information about the virus on social media (Nguyen, 2020).

Vietnam's management of the coronavirus epidemic was lauded for being swift and proactive. The government was also praised for practising transparency in sharing its coronavirus numbers and holding regular meetings with international institutions such as WHO, the United States Center for Disease Control, and the World Bank. Deputy Prime Minister Dam was quoted as saying, "in their discourse, I usually tell them that I very much want to hear detailed, candid and critical opinions from them as from my own advisors, so that together we may arrive at the most appropriate way to fight the pandemic" (Exemplars of Global Health, 2020). Given its success, many think tanks and development organizations have held up the country as an example and viable model for other lower-resourced countries. Many attributed Vietnam's success to its historical experience of the SARS outbreak of 2003, which laid the groundwork for preparedness for the 2020 coronavirus pandemic. Indeed, Vietnam was the first in the Asia-Pacific region to be declared SARS free by WHO. At the time Vietnam was also praised for its handling of the virus.

## Erasing and Essentializing Vietnam's Success

"If you are a white middle-aged man, it is hard to re-program yourself. When this [pandemic] started we were still at the point of counting ventilators. Once COVID-19 took hold of Vietnam, I was sure this country was going to suffer. At least Britain was developed," writes Steve Jackson in a post on Medium published on 21 June 2020, and then reprinted in the Vietnamese online news site VN Express on 23 June (Jackson, 2020). Jackson is a British citizen based in Hanoi who writes for a local NGO. The essay describes his experience of the pandemic in Vietnam as he simultaneously watches how it unfolds in his home country of Britain. He begins by wondering if he should go back to Britain, but sentiments change as his thoughts are overturned when he is presented with evidence that Vietnam is responding to the crisis much more efficaciously than Europe. A repeated motif he uses at various points to reinforce Vietnam's success is the zero-death count: "Vietnam was still at zero deaths ... We were waiting for the explosion. The point where everything was beyond controlling. It never came ... Deaths were still at zero ... In Britain, the official COVID-19 death toll is 42,000. In reality, it's likely to be 50% higher. In Vietnam it is still zero ..." (Jackson, 2020).

Jackson praises the Vietnamese government, the military, and a culture of national solidarity as factors which pulled the country through the pandemic. But the sentiment he expresses at the beginning of the essay is emblematic of more widespread attitudes of those across Europe and North America. These are the cultural biases and ongoing colonial ideologies which lie underneath many of the headlines that hail Vietnam as a surprise success story. Jackson himself comments on the global media's widespread incredulity about Vietnam's numbers:

> Occasionally international media, think tanks, academics, etc. question those numbers. Vietnam is hiding something, right? Once they are set straight they adjust their questioning. "There must be at least one or two?" While tens of thousands die in their own countries there are people fixated on individual cases in Vietnam. (Jackson, 2020)

Jackson's article highlights an important explanation for the incredulity of English-language reporting: the discomfort of failure in the Global North is acknowledged by numeric comparison to Global South success.

Scholars in the post-colonial tradition have long argued that essentializing differences between the "East" and "West" was a strategy for European domination of the world. As discussed at the outset of this chapter, Said's articulation of Orientalism outlines how European superiority was based on the denigration of colonized populations not only through legal or economic means but also through cultural domination. These include representations and discourses that constructed the "Other" in a way that continued Western hegemony over knowledge. In this section, I show how Western epistemic hegemony continues in two ways: first, by underreporting Vietnam's success in the English-speaking media during the early months of the pandemic; and second, via the use of biological essentialism to explain Vietnam's low COVID-19 numbers and zero deaths. To make arguments based on a group's supposed shared biological features is to overlook the policies and labour carried out by Vietnamese leaders and the population in containing the spread of the coronavirus.

Reporting on Vietnam's success by several journalists points out the silence or erasure of Vietnam's efficacious pandemic response in Western-based media, in favour of covering other countries who have had comparable successes such as South Korea, Japan, Thailand, and New Zealand (Hirsch, 2020; Tatarski, 2020; see also Ward, 2020 for commentary on the overlooked success of African countries). For example, a *Washington Post* article published on 29 April 2020, about Vietnam's response points out this omission: "Vietnam has 'basically put the pandemic under control,' Prime Minister Nguyen Xuan Phuc said Tuesday. But this effective control has yet to be greeted with the global plaudits that many other nations have received, perhaps because Vietnam does not fit neatly with other success stories" (Taylor, 2020). This erasure was also expressed by Peter Piot of the London School of Hygiene and Tropical Medicine in a webinar by the Exemplars of Global Health titled "COVID-19 Exemplars: The Power of Positive Outliers, and What We Can Learn from Their Success." The webinar was held on 20 July 2020, and the first country featured on it was Vietnam. As Piot makes his introductory remarks, he says, "and then there's Vietnam, perhaps less well publicized in the West, certainly, but they have had no deaths, and looking at tests per cases, which is one of our indicators, Vietnam is a real outlier because, in fact, in May they stopped reporting because there were no cases. So that's the best situation to be in" (PAHO TV, 2020). Exemplars in Global Health is an initiative by Bill and Melinda Gates that looks at different health outcomes across countries of comparable income levels. The objective is to provide clear explanations for why a country is doing well and evidence for best practices for responding to common health system problems such as under-five mortality, healthcare worker shortages, and more recently, emergency epidemic preparedness and response. Although Vietnam has been recognized by the global health community, its effective response to COVID-19 has been largely ignored in English-language media.

Remnants of colonial attitudes are reflected in how Vietnam's success is explained through essentialist biological terms. Rather than attributing the low COVID-19 numbers to the implementation of strategies to curb the virus's spread, some news reporters featured viewpoints from sources who believed that people of Vietnamese or Southeast Asian descent had biological, and thus racialized, advantages against the coronavirus. None offered much further evidence beyond speculation. An article in the *Telegraph* titled "Vietnam miracle escape from COVID may be down to 'natural immunity'" was published on 1 August (Smith, 2020), well after the policies and practices of Vietnam had been documented by public and academic presses. About half of the experts interviewed in the article attempted to steer the explanation away from biology and towards the Vietnamese government's quick response to track, isolate, and contain the virus through communication strategies, mask wearing, restricting movement at the borders, and quarantine. But the other half of the report – and the main focus of the article – was to present the Vietnamese people as "not as immunologically 'naïve,'" and thus less medically affected by COVID-19 (Smith, 2020). Similarly, writers of a *New York Times* article investigating Thailand's successful management of coronavirus ask, "Is there a genetic component in which the immune systems of Thais and others in the Mekong River region are more resistant to the coronavirus? ... Thailand's low rate of infection appears to be shared by other countries in the Mekong River basin. Vietnam has not recorded a single death and has logged about three months without a case of community transmission. Myanmar has confirmed 336 cases of the virus, Cambodia 166 and Laos just 19" (Beech & Dean, 2020). By focusing on biological justifications, they credit the ability to contain the spread of the virus to genes or t-cells while discrediting the actual work and policies implemented to manage the pandemic. Biological justifications also absolve European and US accountability in their failure to contain the coronavirus, while reinforcing categories of racial difference. Racialization that masquerades as science is a feature of colonial thought.

**Where Are the Bodies? The Auditing of Vietnam's Death Count**

Our epistemic culture assumes visibility to be a conduit for knowledge (Strathern, 2000, p. 310). One response to the doubt raised by many who questioned Vietnam's zero death count was to find the bodies. In other words, to see mortality and morbidity in the flesh provides visible evidence that could prove or disprove Vietnam's claims. Although the epidemiological data about COVID-19 patients, hospitalizations, and deaths were available and updated daily on the state's official webpage, Vietnam's COVID-19 death numbers – or lack thereof – were not taken at face value. Because Western-based reporters believed that the Vietnamese government was not being transparent, they relied on informal modes of audit as means of verification. For organizations, audits are not simply technical tasks used to assess and verify performance. Rather, audits have the "capacity to operationalize and realize accountability" (Power, 1997, p. 134). Power argues that audit practices implicate a relationship of distrust between regulating bodies and practitioners. The distrust lies in an underlying belief that practitioners are unable to self-regulate the quality of their services.

This perspective of an audit is useful for interpreting the relationship between English-language news reporting and Vietnam's COVID-19 response. The case study here demonstrates an audit that comes from a place of distrust and where belief is conditional on outside factors. To be able to claim the truth of numbers and to be believed as transparent depends on the use of first-hand experience of US- or UK-based personnel and the absence of bodies. In other words, the type of personnel undertaking the observation matters.

To be clear, Vietnam has not developed a reputation for transparency where political dissent, state-run media, or other human rights violations are concerned.[1] It is likely because of this history that even those living in Vietnam questioned the numbers at the beginning of the pandemic. After Vietnam announced the first case of COVID-19 in January 2020, rumours that the government was hiding information began to spread among the Vietnamese public. Many on social media expressed doubt and feared that the extent of the coronavirus spread was being repressed by the state-run news media.

On the morning of 25 February 2020, the Ministry of Health's communication arm and People's Committee of Ho Chi Minh City chaired a broadcast session to assert that the government's top priority was open communication and transparency (Quang & Thu, 2020). A month earlier, the Chinese government had been chastised in local social media and international media for censuring several physicians who started to warn the public about the disease. It was incumbent on the Vietnamese state to highlight that not hiding information was a fundamental strategy for prevention in the case of the coronavirus outbreak. A primary justification was that coronavirus information was based on data produced in the country's Emergency Operations Centres, which have direct connection with the United States' Centers for Disease Control and Prevention (CDC).

Indeed, the CDC and Vietnam have had a long history. According to the CDC website, the organization has been working with the government of Vietnam since 1998 to help build capacity for public health and provide technical assistance to address disease outbreaks, laboratory, and surveillance systems. In 2014, both reaffirmed relations under the Global Health Security Agenda (GHSA) – the goal of which is to "achieve the vision of a world safe and secure from global health threats posed by infectious diseases" (GHSA, 2020). Under this agenda, Vietnam and the CDC have built four Emergency Operations Centres, beginning in February 2015, to act as the "nerve centers for epidemic intelligence" (CDC, 2019). State officials reiterated this association with the CDC and added that all the information was also being shared globally.

In *Dan Trí*, a Vietnamese daily newspaper, officials reiterated the close collaboration with WHO experts, who attend all the epidemic-related meetings of the EOC, in an article published on 29 February. "The epidemic is in Vietnam, but will be monitored by the globe. Even if we wanted to hide the epidemic, it would not be possible to" (Vân Sơn, 2020). The article notes that the Vietnamese government even applied strategies that were one step more stringent than the WHO recommendations for handling the epidemic; for example, closing the border with China, restricting travel, requiring medical declarations of all travellers entering Vietnam, and proactive communication about the risks of infection.

Visibility is also a source of truth and transparency in local reporting. In one such interview, a Vietnamese physician, Dr. Truong Huu Khanh, who directs the Infectious Disease department at Children's Hospital 1 in Ho Chi Minh City, explains: "I was asked by a lot of people, including my colleagues, about COVID-19 and whether the health sector or an agency larger than the health sector has hidden or concealed information or not. I affirm, it is impossible to hide the epidemic, when there is a case that requires isolation, all you need to do is look to know because the people involved must wear protective gear" (Vân Sơn, 2020). The article ends with the author making justifications for transparency based on the science of COVID-19:

> If there were infected people in the community, then they would exhibit respiratory symptoms. Among 100 people with the disease, there will be a certain number that experience it more seriously and will have to go to the hospital for an examination. However, in reality, the level of those with respiratory problems is quite low at the hospitals, down by 50% ... I am trying to explain for people clearly there is no one hiding and it is not even possible to hide the situation, but day after day more people ask if the government is hiding it or not, or if the health care is hiding it or not. (Vân Sơn, 2020)

As the epidemic worsened across the globe, with high concentrations of cases in Europe and the United States from April through June 2020, journalists from US- and UK-based media organizations published more features on Vietnam's response to the coronavirus. With low rates of infection and a zero death count, most described the many strategies employed by the state to curb the spread of the virus. The narrative of incredulity and the need to make the absence of dead bodies visible as a form of audit continued. What is notable about this round of reporting is how it centres Western expertise rather than Vietnamese expertise as discussed in early Vietnamese-language reporting. In doubting the numbers, reporters in English-language media relied on first-hand experiences from sources who were affiliated with Western institutions and their team's own investigative reporting.

Reports on Vietnam's success began circulating in early April 2020. Many of them began with the premise that always contextualized Vietnam's pandemic mitigation vis-à-vis other Western countries that were struggling with higher COVID-19 cases and deaths. Doubts were raised alongside praise for the country's strategies. A *Washington Post* article published in April about the lessons Vietnam might offer to the US notes, "Questions remain about whether data from an autocratic nation such as Vietnam is reliable, or if the lower level of tests overall mean there are missing cases, but some U.S. health experts have said they trust the figures" (Taylor, 2020). An NPR article published on 16 April begins with a lede referencing the conundrum of zero deaths. After explaining many of the country's strategies, the writer ends by citing confirmation from John MacArthur, CDC country representative in neighbouring Thailand: "Some may still be skeptical of Vietnam's relatively low COVID-19 case numbers. The CDC's MacArthur is not. 'Our team up in Hanoi is working very, very closely with their Ministry of Health counterparts,' he says. 'The communications I've had with my Vietnam team is that at this point in time, [they] don't have any indication that those numbers are false" (Sullivan, 2020).

An article in Reuters detailing how Vietnam prevented coronavirus outbreaks through aggressive testing describes the country's early and decisive strategies. "The steps are easy to describe but difficult to implement, yet they've been very successful at implementing them over and over again," said Matthew Moore, a Hanoi-based official from the CDC, who had been liaising with Vietnam's government on the outbreak since early January. He added that the CDC has "great confidence in the Vietnamese government's response to the crisis" (Vu et al., 2020). Indeed, the value of zero deaths is then reconfirmed by another doctor, Guy Thwaites, who heads the Oxford University Clinical Research Unit in Ho Chi Minh City. "If there was ongoing and unreported or unappreciated community transmission, we would have seen the patients in our hospital. We have not," he said (Vu et al., 2020).

A CNN article published in May 2020 titled "How Vietnam managed to keep its coronavirus death toll at zero" had an opening lede remarking on the unlikelihood of Vietnam's ability to manage the coronavirus with no deaths and low case numbers (328 confirmed at the time). "To skeptics, Vietnam's official numbers may seem too good to be true," the CNN report noted. Thwaites is interviewed again and discusses his experiences in the hospitals to reconfirm that the Vietnamese government is not hiding COVID-19 deaths. "But Guy Thwaites, an infectious disease doctor who works in one of the main hospitals designated by the Vietnamese government to treat COVID-19 patients, said the numbers matched the reality on the ground. 'I go to the wards every day, I know the cases, I know there has been no death,' said Thwaites, who also heads the Oxford University Clinical Research Unit in Ho Chi Minh City. 'If you had unreported or uncontrolled community transmission, then we'll be seeing cases in our hospital, people coming in with chest infections perhaps not diagnosed – that has never happened,' he said" (Gan, 2020).

The audit of Vietnam's zero death numbers does not stop at interviewing people who were employed in the wards. Reporters with Reuters went as far as interviewing people in the funeral home business to find COVID-19 bodies. Vu et al. (2020) note how "Managers of 13 funeral homes in Hanoi contacted by Reuters said they had seen no uptick in deaths. One said requests for funerals had gone down during the country's lockdown, now lifted, because of the reduction in traffic accidents, one of the biggest killers in Vietnam."

When new cases mysteriously appeared in Da Nang on 29 July 2020, after 99 days with no community transmissions, the government quickly handled the outbreak by locking down movement in the popular tourist city, testing all residents and visitors present within the two-week period, and halting domestic travel to the new epicentre. This second wave ultimately led to the country's first reported COVID-19 death on 31 July 2020. Since the Da Nang outbreak, there have been 35 reported deaths due to COVID-19. For some political analysts, the reporting of the deaths provides evidence that the previous zero death record was indeed true. Dr. Huong Le Thu, a senior analyst at the Australian Strategic Policy Institute, tells the BBC: "The new deaths reported show[s] that there is transparency in reporting COVID-19 in Vietnam and that previous 'no deaths' should have not been questioned in the first place" (Jha, 2020).

**Distrust in Numbers**

On 9 June, Steve Hanke, a professor in the Department of Environmental Health and Engineering at Johns Hopkins University, posted onto Twitter, "These countries are the 'rotten apples' of #coronavirus data. These countries either do not report #covid data or are reporting highly suspicious data" (Hanke, 2020a). His post was accompanied by a screenshot of what appears to be a slide deck from the Johns Hopkins coronavirus dashboard. The title of the slide is "Coronavirus Deaths per Million (population): The Rotten Apples." The slide lists Vietnam with the caption "No Data Reported" and China, Türkiye, Yemen, Syria, Egypt, Venezuela, and India with the caption "Government Reporting Highly Suspect Data."

Vietnam still had no coronavirus deaths in June. Despite extensive documentation of Vietnam's coronavirus management across journalistic and NGO-related media at this point, Hanke's tweet evinced how doubt continued to plague Vietnam's zero death count. Although Hanke is only one person with such dismissive views (albeit with a social media following of over 300,000), the doubts and the manner in which he expresses them articulate with larger colonial patterns that deem successes in the Global South as suspect, especially in situations where countries in the Global North have failed. Even in the context of mounting contrary evidence, he is resolute in his accusations of falsified data and delivers these accusations with representations of Vietnam that are ethnocentric and exoticizing. These strategies work to secure cultural structures of dominance.

His social media post sparked controversy. Almost immediately, Hanke's tweet received 3,800 retweets and over 1,700 responses from experts, academics, journalists, and others from various nationalities vetting his interpretation of COVID-19 data. Even though at the time Vietnam had zero deaths and 334 cases in the country, Hanke nonetheless interpreted no data recorded as a sign that the Vietnamese government was hiding COVID-19's true numbers. Many who commented thus accused Hanke of politically motivated bias because of his affiliation with the libertarian, right-leaning CATO Institute.

Many of the responses also pointed out that Vietnam had already received plaudits for its openness and transparency in sharing coronavirus data and including international institutions in meetings with the Ministry of Health. For example, one Twitter user wrote, "Vietnam is a Rotten apple, good heavens no, then probably the Oxford-Harvard teams in HCM, the CSIS, the US CDC branch in Vietnam, World Economic Forum, Reuters, the WHO, *South China Morning Post*, Worldometer itself, *Yougov*, Poll research, they must be dumb to trust us" (Do, 2020). It is important here to note that again external verification is cited to create confidence in Vietnam's COVID-19 numbers, further demonstrating that even those who may identify with the Global South can uphold colonial hierarchies.

The reaction to Hanke's tweet culminated in a letter addressed to the president, provost, vice provost, and dean of the Whiting School of Engineering at Johns Hopkins University, one signed by 285 public health researchers, experts, and concerned citizens across the world (Rotten Apple Media, 2020a). The signatories criticized Hanke for spreading false information, citing that his claims were not based on evidence. In addition, Hanke's actions were dangerous and detracted from public health communication aimed at building public trust. They also point out Hanke's poor methodology: "We, as part of the audience, became very concerned since COVID-19 data for Vietnam was clearly available on Worldometers (the same website that Professor Hanke cited) at the same time as he tweeted … It initially seemed that Professor Hanke mistook Vietnam's widely verified zero deaths from COVID-19 as no reported data" (Rotten Apple Media, 2020a). Responses to Hanke's tweet were not only confined to social media accounts. Several Vietnam-based online newspapers reported on Hanke and the letter from concerned scholars and citizens about the false information put forth in his tweet (Thanh Niên, 2020; Tuổi Trẻ, 2020).

On 16 June, Hanke responded to his critics, but did not retract his tweet: "Contrary to an image I posted last week, #Vietnam turns out to have a 'perfect' record in its fight against coronavirus. Official Vietnamese data indicate that Vietnam has not suffered any coronavirus fatalities – zero (0) deaths" (Hanke, 2020b). While Hanke acknowledged that he was incorrect, his use of scare quotes around the word "perfect" implied that he still did not trust Vietnam's COVID-19 numbers. This did not sit well with readers, as reported in *Thanh Nien*'s online newspaper, quoting a sentiment of "is that how you apologize?" (Thanh Niên, 2020).

The signatories of the letter received a response from Andrew Douglas, the vice provost for faculty affairs at Johns Hopkins University, on 19 June 2020. Douglas discussed the university's mission to uphold academic freedom, but also recognized the need for academic responsibility. The vice provost said that he would investigate the matter further, following university policy and principles. On the same day, Hanke's original tweet was deleted from his account (Rotten Apple Media, 2020b).

The story did not end with this development. Hanke continued to tweet about Vietnam's coronavirus situation by sharing news stories that represented Vietnam as exoticized and untrustworthy. On 21 July 2020, he attempted to raise suspicions about Vietnam's COVID-19 numbers in another tweet: "According to my @JohnsHopkins colleagues, #Vietnam reports zero COVID deaths. This is miraculous and amazing. After all, 55% of Vietnam's field rats, intended to be eaten in restaurants, are infected with #Coronavirus" (Hanke, 2020c). He provides a link to a *New York Times* article that describes a study on the prevalence of coronavirus infections in field rats from three provinces of Vietnam (Gorman, 2020). The accusation of the article is that the risk of coronavirus transmission is higher the closer one moves from one end of the food supply chain to the dinner table and this can lead to conditions where zoonosis can occur.

There are many problems with the use of this evidence to doubt Vietnam's COVID-19 numbers. Although the *New York Times* article was published on 19 June 2020, the data upon which it is based was collected in 2013 and 2014 – well before the emergence of SARS-CoV-2, which causes COVID-19 in humans. The study also examines six different coronaviruses, none of which has any clear connection to SARS-CoV-2. In addition to such unsupported science, there is an insidious subtext to Hanke's use of field rats to raise doubts about Vietnam's COVID-19 numbers. It is well-covered ground that judging a culture's dining practices and presenting them as strange is an ethnocentric strategy for creating difference and hierarchy among cultures. The strategy has often been used to demean the "East" and set up the "West" as the norm and standard by which all other cultures are judged.[2]

Hanke's tweet was not alone in portraying Vietnamese consumption practices surrounding COVID-19 in an exoticized manner. In April, a news story about black cats being turned into a paste and ingested as treatment for COVID-19 in Vietnam was published across several English-language online news outlets and tabloids. These claims were later debunked in an in-depth investigation by fact-checking website Politifact (Kertscher, 2020). The widespread circulation of the story in several outlets is emblematic of colonial practices for representing and exoticizing the other for the purposes of cultural domination by the "West" over the "East."

## Conclusion

In this chapter, I examined the politics of numbers surrounding Vietnam's response to COVID-19 as represented in English- and Vietnamese-language media. I noted the widespread sentiments of incredulity in response to the news of

Vietnam's zero deaths. In a world where transparency signals integrity, I asked: How is such recognition of transparency conferred, and how is this process marked by legacies of colonialism? In other words, who gets to claim the "indelible truth" of numbers and who does not?

The case study of Vietnam's coronavirus numbers contributes to discussions about the politics of knowledge, how information is counted as a fact by those interpreting, and also existing inequalities in the hierarchy of global power. Power is shown here by which countries are afforded the benefit of the doubt and which require audits to generate confidence in their numbers. Reporting on Vietnam's COVID-19 numbers implicitly reaffirms the view that Western knowledge is legitimating knowledge. While praising Vietnam's swift actions, many stories that circulate about Vietnam's COVID-19 numbers also serve to secure Western-based knowledge as expertise. Ultimately, transparency is not a given for countries that are not aligned with Western ideals. Instead, they are often marked by accusations of corruption, false reporting numbers, and lack of transparency.

There is an unfortunate irony, however. Despite reference to Western-based institutions as sources of expertise and verification, it is in fact Western countries which have largely failed in their coronavirus response, according to the indicators of deaths, cases, and hospitalizations (see also anthropologist Lincoln, 2020). As of 25 November 2020, Vietnam's total coronavirus cases during the pre-vaccination period sits at 1,321 with 1,153 recovered and 35 deaths. This demonstrates that no matter the infrastructure of the health system, in the case of a global pandemic of a highly infectious disease, it is the decisions of leadership that shape the impact of suffering in each respective country.

*Acknowledgments*: Thank you to Kat, Jason, Kiwi, Kristie, Tram, Hung, Doi, Dat, Dat, Nhi, Thi, Martha Lincoln, and my host family in the Mekong for helping me get beyond the headlines when no one could travel in or out of Vietnam. As always, thank you to Steve Eulenberg for being a supportive partner. Between writing this piece and its publication, Quinn came into our lives and has brought us endless joy.

NOTES

1. For example, see the 2016 marine life disaster in which the government initially denied to the public that Taiwanese corporation Formosa had dumped 300 tons of toxic chemicals, killing over 100 tons of fish and decimating the fishing and tourist industry in central Vietnam.
2. Coronaviruses are indiscriminate in what mammals or birds they infect. Coronaviruses can be found in wild game or "non-exotic" or farmed meats such as chicken, cows, and swine. Given that the risk of viral spread comes from the crowded industrial style of raising livestock, many predict that the next pandemic may originate from US farms.

REFERENCES

Abimbola, S., & Pai, M. (2020). Will global health survive its decolonisation? *Lancet*, *396*(10263), 1627–8. https://doi.org/10.1016/S0140-6736(20)32417-X.

Adams, V. (2016). *Metrics: What counts in global health*. Duke University Press.

Ballestero, A. (2018, September). Transparency. In Callan, H. (Ed)., *The International Encyclopedia of Anthropology* (pp. 1–4). John Wiley & Sons.

Beech, H., & Dean, A. (2020, July 17). No One Knows What Thailand Is Doing Right, but So Far, It's Working. *New York Times*. https://www.nytimes.com/2020/07/16/world/asia/coronavirus-thailand-photos.html.

Büyüm, A.M., Kenney, C., Koris, A., Mkumba, L., & Raveendran, Y. (2020, August). Decolonising global health: If not now, when? *BMJ Global Health*, 5:e003394. https://doi.org/10.1136/bmjgh-2020-003394.

CDC (Centers for Disease Control and Prevention). (2019). Vietnam: Connecting for stronger emergency response. US Centers for Disease Control. https://www.cdc.gov/globalhealth/security/stories/vietnam_emergency_response.html.

Dao, A., & Nichter, M. (2016). The social life of health insurance in low- to middle-income countries: An anthropological research agenda. *Medical Anthropology Quarterly*, *30*(1), 122–43. https://doi.org/10.1111/maq.12191.

Davis, S.L.M. (2020). *The uncounted: Politics of data in global health*. Cambridge University Press.

Do, B. [@dishwasher1910]. (2020, June 10). Vietnam is a rotten apple, good heavens no, then probably the Oxford-Havard teams in HCM, the CSIS, the US [Tweet]. Twitter. Retrieved 10 June 2020, from https://twitter.com/Dishwasher1910/status/1270663424351956992.

Drabek, Z., & Payne, W. (2002). The impact of transparency on foreign direct investment. *Journal of Economic Integration*, *17*(4), 777–810. https://doi.org/10.11130/jei.2002.17.4.777.

Exemplars of Global Health. (2020, July 23). *COVID-19 exemplars: Vietnam's story of success against COVID*. YouTube. Retrieved on 23 July 2020, from https://www.youtube.com/watch?v=uaeiBbHbl3g.

FT Visual & Data Journalism Team. (2020, October 18). COVID-19: The global crisis – in data. *Financial Times*. Retrieved on 1 November 2020, from https://ig.ft.com/coronavirus-global-data/.

Gan, N. (2020, May 30). How Vietnam managed to keep its coronavirus death toll at zero. CNN. https://www.cnn.com/2020/05/29/asia/coronavirus-vietnam-intl-hnk/index.html.

GHSA. (2020). Global Health Security Agenda. Retrieved 2 October 2020, from https://ghsagenda.org/.

Glass, R.I. (2020, July/August). Decolonizing and democratizing global health are difficult, but vital goals: Opinion by Fogarty director Dr. Roger I. Glass. *Fogarty International Center*, *19*(4), 1–12. https://www.fic.nih.gov/News/GlobalHealthMatters/july-august-2020/Pages/roger-glass-decolonizing-global-health.aspx.

Gorman, J. (2020, June 19). Wildlife trade spreads coronaviruses as animals get to market. *New York Times.* https://www.nytimes.com/2020/06/19/science/coronavirus-rats-vietnam.html.

Green, A. (2019, May 21). The activists trying to "decolonize" global health. *Devex.* https://www.devex.com/news/sponsored/the-activists-trying-to-decolonize-global-health-94904.

Guinto, R. (2019, February 11). #DecolonizeGlobalHealth: Rewriting the narrative of global health. *International Health Policies.* https://www.internationalhealthpolicies.org/blogs/decolonizeglobalhealth-rewriting-the-narrative-of-global-health/.

Hanke, S. [@steve_hanke]. (2020a, June 9). These countries are the "rotten apples" of #coronavirus data. These countries either do not report #covid data or [Tweet]. Twitter. Retrieved 10 June 2020 (Tweet deleted on 19 June 2020, by Hanke).

Hanke, S. [@steve_hanke]. (2020b, June 16). *Contrary to an image I posted last week, #Vietnam turns out to have a "perfect" record in its fight* [Tweet]. Twitter. Retrieved 29 July 2020 from https://twitter.com/steve_hanke/status/1272876513147400193.

Hanke, S. [@steve_hanke]. (2020c, July 21). *According to my @JohnsHopkins colleagues, #Vietnam reports zero COVID deaths. This is miraculous and amazing. After all, 55%* [Tweet]. Twitter. Retrieved 29 July 2020, from https://twitter.com/steve_hanke/status/1285726250426945536.

Hetherington, K. (2011). *Guerrilla auditors: The politics of transparency in neoliberal Paraguay.* Duke University Press.

Hirsch, A. (2020, May 21). Why are Africa's coronavirus successes being overlooked? *Guardian.* http://www.theguardian.com/commentisfree/2020/may/21/africa-coronavirus-successes-innovation-europe-us.

Holshue, M.L., DeBolt, C., Lindquist, S., Lofy, K.H., Wiesman, J., Bruce, H., … & Pillai, S.K. (2020, January 31). First case of 2019 novel coronavirus in the United States. *New England Journal of Medicine.* https://doi.org/10.1056/NEJMoa2001191.

Jackson, S. (2020, June 21). COVID-19 in Vietnam – the fear, the tears, the pride and the debt. *Medium.* https://medium.com/@stevejacksonHN/covid-19-in-vietnam-the-fear-the-tears-the-pride-and-the-debt-f0932a71012d.

Jha, P. (2020, August 8). Coronavirus Vietnam: The mysterious resurgence of COVID-19. BBC News. https://www.bbc.com/news/world-asia-53690711.

Kertscher, T. (2020, May 8). Evidence lacking that cats eaten as COVID-19 cure. *Politifact.* https://www.politifact.com/factchecks/2020/may/08/south-west-news-service/Evidence-lacking-that-cats-eaten-as-COVID-19-cure/.

Konanykhin, A. (2018, October 10). How transparency can help the global economy to grow. World Economic Forum. https://www.weforum.org/agenda/2018/10/how-transparency-can-help-grow-the-global-economy/.

Lincoln, M. (2020, September 15). Study the role of hubris in nations' COVID-19 response. *Nature.* https://www.nature.com/articles/d41586-020-02596-8?utm_source=twt_nnc&utm_medium=social&utm_campaign=naturenews.

Merry, S.E. (2011). Measuring the world: Indicators, human rights, and global governance: With CA comment by John M. Conley. *Current Anthropology, 52*(S3), S83–S95. https://doi.org/10.1086/657241.

Merry, S.E., Davis, K.E., & Kingsbury, B. (2015). *The quiet power of indicators: measuring governance, corruption, and rule of law*. Cambridge University Press.
Nelson, D.M. (2010). Reckoning the after/math of war in Guatemala. *Anthropological Theory*, *10*(1–2), 87–95. https://doi.org/10.1177/1463499610365374.
Nguyen, H. (2020, April 15). Vietnam specifies cash fines on fake news. *VnExpress International*. https://e.vnexpress.net/news/news/vietnam-specifies-cash-fines-on-fake-news-4085431.html.
PAHO TV. (2020, July 24). *Exemplars in global health COVID-19 webinar*. YouTube. Retrieved on 25 July 2020, from https://www.youtube.com/watch?v=s8e6MczjqWA.
Pailey, R.N. (2020). De-centering the "white gaze" of development. *Development and Change*, *51*(3), 729–45. https://doi.org/10.1111/dech.12550.
Pollack, T., Thwaites, G., Rabaa, M., Choisy, M., van Doorn, R., Duong, H.L., Dang, Q.T., Tran, D.Q., Phung, C.D., Ngu, D.N., Tran, A.T., La, N.Q., Nguyen, C.K., Dang, D.A., Tran, N.D., Sang, M.L., & Thai, P.Q. (2020, March 5). Emerging COVID-19 success story: Vietnam's commitment to containment. *Our World in Data*. Retrieved 30 November 2020, from https://ourworldindata.org/covid-exemplar-vietnam.
Porter, T.M. (1996). *Trust in numbers: The pursuit of objectivity in science and public life*. Princeton University Press.
Power, M. (1997). *The audit society: Rituals of verification*. Oxford University Press.
Quang, H., & Thu, H. (2020, February 25). Bộ y tế: Việt Nam không giấu thông tin về dịch COVID-19 [Ministry of Health: Vietnam does not hide information about the epidemic COVID-19]. *ZingNews.vn*. https://zingnews.vn/zingnews-post1051408.html.
Rotten Apple Media. (2020a, June 16). Vietnam not COVID-19 "rotten apple": 285 sign open letter to Johns Hopkins University. *Medium*. https://medium.com/@vietnamnotcovid19rottenapple/vietnam-not-covid-19-rotten-apple-285-sign-open-letter-to-johns-hopkins-university-9ffb551b85d7.
Rotten Apple Media. (2020b, June 19). UPDATE: "Rotten apple" tweet finally deleted by Johns Hopkins professor. *Medium*. Retrieved 21 September 2020, from https://medium.com/@vietnamnotcovid19rottenapple/update-rotten-apple-tweet-finally-deleted-by-johns-hopkins-professor-478df6ba3b9.
Said, E.W. (1978). *Orientalism*. Knopf Doubleday Publishing Group.
Smith, N. (2020, August 1). Vietnam miracle escape from COVID may be down to "natural immunity." *Telegraph*. https://www.telegraph.co.uk/news/2020/08/01/vietnam-miracle-escape-covid-may-natural-immunity/.
Strathern, M. (2000). The tyranny of transparency. *British Educational Research Journal*, *26*(3), 309–21. https://doi.org/10.1080/713651562.
Sullivan, M. (2020, April 16). In Vietnam, there have been fewer than 300 COVID-19 cases and no deaths. Here's why. NPR.Org. https://www.npr.org/sections/coronavirus-live-updates/2020/04/16/835748673/in-vietnam-there-have-been-fewer-than-300-covid-19-cases-and-no-deaths-heres-why.
Tatarski, M. (2020, June 19). The think tank erasure of Vietnam's success. *Medium*. https://medium.com/@vietnamnotcovid19rottenapple/the-think-tank-erasure-of-vietnams-success-3d3fdc0d1dad.

Taylor, A. (2020, April 30). Analysis | Vietnam offers tough lessons for U.S. on coronavirus. *Washington Post.* https://www.washingtonpost.com/world/2020/04/30/vietnam-offers-tough-lessons-us-coronavirus/.

*Thanh Niên.* (2020, June 18). Nghi ngờ kết quả Việt Nam chống dịch, giáo sư mỹ phải đính chính [Doubting Vietnam's anti-epidemic results, the American professor had to correct it]. https://thanhnien.vn/the-gioi/nghi-ngo-ket-qua-viet-nam-chong-dich-giao-su-my-phai-dinh-chinh-1239068.html.

*Tuổi Trẻ.* (2020, June 17). Nói đi rồi nói lại về kết quả chống dịch của Việt Nam, giáo sư nổi tiếng Steve Hanke gây bất bình [Speaking and talking again about Vietnam's anti-epidemic results, renowned professor Steve Hanke causes grievances]. https://tuoitre.vn/noi-di-roi-noi-lai-ve-ket-qua-chong-dich-cua-viet-nam-giao-su-noi-tieng-steve-hanke-gay-bat-binh-20200617161958419.htm.

Vân Sơn. (2020, February 29). Việt Nam có âm thầm giấu dịch COVID hay không? [Is Vietnam secretly hiding the COVID epidemic?]. *Báo điện tử Dân Trí.* https://dantri.com.vn/suc-khoe/viet-nam-co-am-tham-giau-dich-covid-hay-khong-20200225171439082.htm.

Vu, K., Nguyen, P., & Pearson, J. (2020, April 29). After aggressive mass testing, Vietnam says it contains coronavirus outbreak. Reuters https://www.reuters.com/article/us-health-coronavirus-vietnam-fight-insi/after-aggressive-mass-testing-vietnam-says-it-contains-coronavirus-outbreak-idUSKBN22B34H.

Ward, A. (2020, May 5). Vietnam, Slovenia, and 3 other overlooked coronavirus success stories. *Vox.* https://www.vox.com/2020/5/5/21247837/coronavirus-vietnam-slovenia-jordan-iceland-greece.

Zhao, J.H., Kim, S.H., & Du, J. (2003). The impact of corruption and transparency on foreign direct investment: An empirical analysis. *MIR: Management International Review, 43*(1), 41–62. https://www.jstor.org/stable/40835633?seq=1&cid=pdf-reference#references_tab_contents.

# A Global Address

# 6  An "Unseen Enemy" and the "Shadow Pandemic": The COVID-19 Pandemic and Domestic Violence

BY SHWETA ADUR AND ANJANA NARAYAN

> While Lele was holding her 11-month-old daughter, her husband began to beat her with a high chair … Eventually, she says, one of her legs lost feeling and she fell to the ground, still holding the baby in her arms … A photograph she took after the incident shows the high chair lying on the floor in pieces, two of its metal legs snapped off – evidence of the force with which her husband wielded it against her. (Taub, 2020)

Earlier this year, 2020, just as the world was coming to terms with the horror of living in unprecedented pandemic times, Lele, a 26-year-old mother, was confronting an additional terror of her own in Anhui Province in China. Elsewhere, an advocate at the National Domestic Violence Hotline updated the organization's logbook with the following statement,

> I spoke to a female caller in California that is self-quarantining for protection from COVID-19 due to having asthma … Her partner strangled her tonight. While talking to her, it sounded like she has some really serious injuries. She is scared to go to the ER due to fear around catching COVID-19 … (Godin, 2020)

The two women, separated by the vast Pacific Ocean, were united in their fate as they confronted the precarity of life under the twin terrors of a pandemic and a violent partner, both deadly and unpredictable. As the leaders of the world strategized to go to "war" with the "unseen enemy" that is COVID-19, many others – particularly, women and children – battled the ubiquitous public health crises of domestic violence. Citing the horrifying surge in violence against women inside their own homes, weaponized by the very strategies that were being used to combat the virus, Phumzile Mlambo-Ngcuka, executive director of UN Women, cautioned the world to not ignore the "shadow pandemic" that endangered women's and children's lives (Mlambo-Ngcuka, 2020; see also UN Women, 2020).

Domestic violence is a broad term that encompasses a range of violations that occur in the "home" and includes intimate partner violence (IPV), elder abuse, and child abuse. This chapter focuses on domestic violence in the context of IPV as it is experienced by women and perpetrated by men who are their current or former partners. Though we do not discount that IPV happens to men, as well as within same-sex relationships, our focus on women within heterosexual relationships highlights the gendered aspect of IPV, whereby the majority of perpetrators tend to be males and majority of victim-survivors are women.

Drawing on feminist scholarship and the current media reporting of the coronavirus pandemic, this chapter reveals a somewhat understudied yet gendered consequence of the pandemic: the alarming surge of domestic violence. Our chapter begins by unpacking the gendered underpinnings of IPV from a feminist standpoint. The section underscores the importance of examining wars, natural disasters, and catastrophes through a gendered lens but also underscores that this shift is comparatively recent, albeit long overdue. This section is followed by a brief note about the pandemic itself that then scaffolds to the next section, which describes the links between the pandemic and a concurrent surge in IPV. We close the chapter by assessing what has been done and what could be done more efficaciously in the future.

**Theoretical Underpinnings**

Since the latter part of the twentieth century, the interdisciplinary field of intimate partner violence has witnessed complex shifts in approaches and theoretical models. Yet, it is the feminist framework and feminist organizing that has predominantly driven collective action to raise awareness as well as influence laws, policies, and services for survivors of IPV.

Moving beyond psychological frameworks that focus on individual-level variables to understand domestic violence, early feminist approaches in the 1960s and 1970s fundamentally shifted the framing of violence within the context of gender dynamics wherein they noted that the majority of perpetrators are men while its victims are women and children. Central to the feminist model is that intimate partner violence is deeply entrenched in patriarchal social structures and ideologies that have historically and culturally allowed women to be oppressed by their male partners (Dobash & Dobash, 1979; Walker, 1979). A more recent turn in feminist scholarship that centres masculinity studies demonstrates the centrality of socialization that normalizes tropes of toxic masculinity and its impact on heightening violence against women. Feminist activists and scholars underscore that violence in intimate relationships is about power and control, and the rate of everyday and severe violence including death is much higher for women than for men. These gender-based explanations were extremely important on two related grounds. *First*, they offered a powerful critique to psychological interventions that viewed intimate partner violence solely through the lens of abnormal individual disorders and family dysfunction. *Second*, they led to the recognition that IPV is not a private family matter. Rather, it is a public problem that requires state- and policy-level intervention (MacKinnon, 1989). As a result, in the United States over the last five decades, the women's movement made great strides in terms of institutionalizing federal legislation and funding, criminalizing IPV, and establishing an expansive network of domestic violence shelters, hotlines, and other services throughout the country (Fagan, 1990; Shepard, 2005).

However, by the mid-1990s, second wave feminist scholars' predominantly gender-based explanation of intimate partner violence was being challenged for its myopic focus on the experiences of white middle-class heterosexual women. Drawing on experiences of women marginalized by colour, poverty, sexuality, immigrant status, etc., a growing number of feminist scholars began calling a for a more structural and intersectional approach (Crenshaw, 1991; Renzetti, 1998; Bhattacharjee, 1997; hooks, 1984; West, 2004). Increasingly, critics began to highlight that criminalizing domestic violence and greater state intervention are embedded within a system that has perpetrated institutional violence against communities of colour who are often subject to much higher rates of harassment, police violence, and incarceration (Mama, 1989; Davis, 2000; Bohmer et al., 2002; Baskin, 2003; Smith, 2005). One of the first scholars to criticize the failures of the criminal justice system when it comes to intimate partner violence and marginalized women was Kimberlé Crenshaw (1991). In her groundbreaking article "Mapping the Margins: Intersectionality, Identity, Politics, and Violence against Women of Color," Crenshaw notes:

> Women of color are often reluctant to call the police, a hesitancy likely due to a general unwillingness among people of color to subject their private lives to the scrutiny and control of police, frequently hostile. There is also a more generalized community ethic against public intervention, the product of a desire to create a private world from the diverse assaults on the public lives of racially subordinated people. The home is not simply a man's castle in the patriarchal sense, but function[s] as a safe haven from the indignities of life in a racist society. (1991, p. 1257)

Overall, activists and scholars increasingly called for a reconceptualization of dominant feminist frameworks and practices to incorporate an intersectional lens to address intimate partner violence. Transcending simplistic binaries of public and private, scholars such as hooks (1984), Abraham (1995, 2000), Dasgupta (1998), Purkayastha (2000), Josephson (2002), Menjivar and Salcido (2002), Sokoloff and Dupont (2005), and Price (2012) argue that violence in the home is often the consequence of structural oppression and social injustice faced by marginalized groups in the public realm. As Sokoloff and Dupont emphasize, there is a need to challenge "the primacy of gender as an explanatory model of domestic violence" and "examine how other forms of inequality and oppression, such as racism, ethnocentrism, class privilege, and heterosexism, intersect with gender oppression. This approach calls for public policies that address these structural root causes of domestic violence" (Sokoloff & Dupont, 2005, p. 39). In response to these critiques, anti-violence activism in marginalized communities is now moving beyond an over-reliance on criminal justice interventions and experimenting with more progressive and community-based strategies to address IPV (Presser & Gaarder, 2000).

By the early 1990s, there was also an expansion in transnational feminist organizing and the articulation of intimate partner violence as a human rights issue (Thomas & Beasley, 1993; Roth, 1994; Coomaraswamy, 2000; Bunch & Frost, 2000; Morgaine, 2007, 2011; Ertürk & Purkayastha, 2012; Hall, 2015; McQuigg, 2015; Purkayastha, 2018). The conceptualization of intimate partner violence as a specific form of gender-based human rights violation was the result of decades of grassroots feminist activism and advocacy at various United Nations conferences (Third and Fourth World Conferences on Women in Vienna in 1993 and Beijing in 1995) that culminated in the adoption of the 1993 UN Declaration of the Elimination of Violence against Women and the appointment of a United Nations Special Rapporteur on Violence against Women in 1994. These international legal instruments became tools to pressure states to provide equal and adequate protections for victims of domestic violence under the law and hold countries accountable for perpetuating human rights violations. Furthermore, scholars (see Obreja, 2019) argue that the human rights language allows for a more structural and intersectional analysis of IPV, "as it expands the focus beyond a relatively narrow civil rights approach, which requires only punishment of the offender, to a broader social justice one combining civil and political rights with social, economic and cultural ones" (Merry & Shimmin, 2011, p. 116).

More recently, feminist researchers (Enarson, 2012; First et al., 2017; Jenkins & Phillips, 2008) have studied the intersection of gender-based violence and crises such as natural disasters, epidemics, and pandemics. They highlight that women are often more vulnerable during times of unrest and crisis as they are often exacerbated by entrenched structural inequalities that shape their lives. Studies on large-scale catastrophes and public health crises, such the 1997 Red River Valley Flood (Enarson & Scanlon, 1999), hurricanes Katrina and Rita (Jenkins & Phillips, 2008; Peek & Fothergill, 2008; Ross-Sheriff, 2007), the Great Recession (Schneider et al., 2016), the Ebola epidemic (O'Brien & Tolosa, 2016), and the HIV/AIDs pandemic (Pellowski et al., 2013; Mukherjee, 2007; Njiru & Purkayastha, 2017), all document an increase in the rate and severity of intimate partner violence. Taken together, these studies highlight several interrelated scenarios.

*First*, while the link between economic instability and IPV is well documented in the literature, these studies show that unemployment and job losses during a crisis can dramatically increase the risk of interpersonal violence. *Second*, the disruption of social support networks such as friends and family, as well as curtailed access to public support systems, leads to increased isolation. Access to legal aid like restraining orders, crisis counselling, medical assistance, shelters, and other forms of crisis support is often limited during a crisis because of reduced manpower and high demand for services. *Finally*, longstanding structural inequities such as poverty, inadequate housing, and lack of affordable child care and health care, in conjunction with racism and homophobia, make it particularly challenging for women of colour and LGBTQ survivors of IPV to access the resources they need during a crisis (First et al., 2017). By and large, these barriers isolate survivors and significantly reduce their already very limited options for leaving violent relationships during a disaster and crisis. Moving forward, feminist scholars have called for the critique of gender-neutral approaches to addressing disasters and health crises and integration of an intersectional lens to planning policies related to crisis response and recovery. Our following discussion points to the urgency of this shift.

**An "Unseen Enemy": The COVID-19 Pandemic**

In December 2019, a new strain of coronavirus – SARS-CoV-2 – was identified in Wuhan, China. The exponential speed of transmission, followed by acute respiratory distress and high rates of fatality among its patients, quickly made it a significant source of concern. In a matter of weeks, an epidemic that had originated in central China became a global outbreak, crossing international boundaries and affecting people on a worldwide scale. Soon after, in March 2020, the World Health Organization (WHO) officially described the spread of COVID-19 as a pandemic, an admission that the devastation was global and that the outbreak had become uncontrollable. Historically, only a handful of epidemics have attained the notoriety of a pandemic. According to the Johns Hopkins Coronavirus Resource Center (2020), as of 15 December 2020, the number of people infected with COVID-19 had reached 72,869,705 worldwide, and the global death toll at the time of this writing in late 2020 was at 1,621,635. The United States had the highest number of cases, followed by India and Brazil.

While this is not the first pandemic and neither will it be the last, for most of us who have not lived through the 1918 Influenza Pandemic, the COVID-19 pandemic is unparalleled in terms of its scale as well as in social and economic disruption. Even in its earliest months, the contagion proved indomitable as it forced countries to hastily retreat into lockdown modes. By April 2020, more than 3.9 billion people, approximately half of the global population, in more than 90 countries had been asked to shelter in place. Schools, offices, retail places, and non-essential services were all closed and curfews were enforced. People panicked as they hoarded food and cleaning supplies, creating artificial shortages of essential items. Consumer supply chains creaked and heaved under the crushing weight of an unexpected increase in demand. At one point, the media reported fights breaking out in supermarkets over supplies as basic as toilet paper. Though the financial consequence of this unprecedented shutdown has yet to be accounted for, all signs point to its long-lasting and deleterious impact. Except for a handful of giants that have profited from the pandemic (e.g., delivery services, information and communications technology (ICT), medical equipment producers), a significant number of businesses have perished. As unemployment rates shot up, household savings crumbled, leaving despair and distress in their wake. For example, "half of New York City households (50%) say they suffered job or wage loss, and among these households, 73% say they are having serious financial problems. Nearly half (47%) of those with employment changes say they've used up all or most of their savings, with 8% more saying they didn't have any savings at the start of the outbreak" (Neel, 2020).

At its peak, at least 90 countries issued shelter-in-place orders, and many closed their borders and instituted significant travel restrictions. The widescale disruptions, perhaps, resembled the most cynical of dystopian novels. Yet as we live through these extraordinary times, scholars and scientists across disciplinary boundaries have simultaneously sought to find order in the chaos. They have sought to understand not only the medico-biological aspects of the virus but also its social ramifications. In this regard, the sociological scholarship, though still nascent, has been definitively influential. Social scientists have warned against considering the virus an "equal opportunity" offender; a virus does not proliferate in a vacuum. Instead, the way it spreads, how it is experienced, and who it impacts is more shaped by intractable politics and structural inequality. Whereas a virus, by itself, does not discriminate and does not care for borders and boundaries, it does cleverly map onto pre-existing socioeconomic inequalities to deliver vastly disproportionate outcomes.

Though arguably COVID-19 has not spared the most powerful dignitaries of our times – princes, politicians, presidents, and entertainers alike – their privileged access to superior health care implicitly allows them to fight the virus differently from a common person. As we describe briefly below, the impact of the pandemic is stratified by race, class, gender, sexuality, nationality, and other such inequalities and therefore must be studied as such. Foregrounding the fact that social conditions are at the bottom of disease disparities is essential to sensitize policy solutions and recommendations. For example, while "social distancing" has been recommended as a means to control the spread of the virus, for millions of people living in refugee camps, prisons, detention centres, and overcrowded housing, "social distancing" is hardly an option. For the homeless populations, "sheltering in place" is not a possibility. Looking at the disproportionate impact on low-income communities of colour, Laster Pirtle (2020) asserts that "racial capitalism" – a system wherein capital accumulation and racial exploitation are co-constitutive – has endangered poor communities of colour more, since they are over-represented in insecure jobs with low wages and higher risks of exposure. Moreover, jobs such as the grocery worker, restaurant worker, cleaner, and transit driver are also often seasonal, part-time, and/or constitutive of the gig economy. As a result, *"telework and social distancing are privileges that evade low-wage workers, many of whom are Black Americans and Latinos"* [emphasis added] (Ray & Rojas, 2020). In Los Angeles, for example, "more than half of Black households report serious financial problems (52%), compared with 37% of whites. And in Los Angeles, 71% of Latinos – almost twice the percentage of whites – report serious financial problems" (Neel, 2020).

Similarly, feminists and gender scholars have examined gendered disease disparities and experiences. In this category, a significant majority have grappled with the gender gap in COVID-19 death rates, i.e., the proclivity of men succumbing to the virus at rates higher than those of women. An influential stream has reviewed the impact quarantining has had on families' lives and relationships, especially the strains of work-life balance and child care. A sliver of these studies has systematically looked into marital relationships and intimate partnerships, and only a handful examine the pandemic's impact on the darker sides of intimate partnerships – particularly relationship abuse and violence. Our work builds on and contributes to this needed genre of scholarship. In the following paragraphs, we discuss the implications of the pandemic; specifically, violence against women within their own homes in both the short and long term.

## A "Shadow Pandemic": The Pandemic and the Horrifying Surge in Intimate Partner Violence

Though "shelter in place" orders have been announced to contain the outbreak, for millions of women and children the order comes with its own dangers. Mlambo-Ngcuka (2020), the executive director of UN Women, warned that "with 90 countries in lockdown, four billion people are now sheltering at home from the global contagion of COVID-19 … We see a shadow pandemic growing, of violence against women." Although domestic violence, as one of the leading causes for human rights violations, has preexisted the pandemic, the latter has provided opportune conditions to intensify the violence. While data is limited, given the recentness of the pandemic coupled with the fact that IPV is also notoriously under-reported, a review of what is available reveals disturbing trends: the Canadian Broadcasting Corporation (CBC) reported

20–30% more calls for help in Canada (Patel, 2020), with some areas seeing calls increase by 400%. France saw a 30% hike in calls to its first responders. In Australia, search engines such as Google were seeing the highest magnitude of searches for domestic violence help in the past five years. The USA is no different in this regard. Boserup et al. (2020) found that the Portland Police Bureau of Oregon recorded a 22% increase in arrests related to domestic violence after the stay-at-home orders were announced. Similar trends were found in other states. In Texas, the San Antonio Police Department reported an 18% increase in calls in March 2020 compared to March 2019. Similarly, in Alabama the Jefferson County Sheriff's office reported a 27% increase in domestic violence calls in March 2020 compared to March 2019.

As we have mentioned earlier, feminist scholarship unequivocally establishes that wars, disasters, and pandemics result in the concurrent rise of violence against women. The surge is both *conditional* and *opportunistic* – in other words, pandemics often present "conditions" that are primed for the escalation of violence and abuse as well as "opportunities" to do so with impunity. The current pandemic is no different; its public health mandates for social distancing and sheltering in place provide both the *conditions* for escalating violence and *opportunities* for perpetrators to evade scrutiny. Confinement exacerbates strains and conflicts in a relationship. Research shows that victims of abuse are often forced to socially isolate at home by their abusers through a myriad of threats such as physical violence and mental manipulation, with the pandemic exacerbating the entrapment by quarantining perpetrators at home for longer stretches of time with little else to do for distraction. This essentially translates into increasing the number of hours they are present to terrorize their victims. The victims, on the other hand, are at the mercy of their perpetrators more than ever before as social distancing and isolation rob them of protective factors against violence.

The Centers for Disease Control (CDC) has revealed that social support and connectedness are important protective factors against violence. Physical confinement is coupled with emotional duress as the stresses of finances, precarious livelihoods, fears, and job losses become precipitous factors that further exacerbate relationships. The restructuring of household routine as well as added caregiving responsibilities on account of school closures further leads to volatility in relationships. The concurrent uptick in gun and alcohol sales in the US, both linked to the hysteria and fear caused by the pandemic, adds another layer of precarity to women's lives behind closed doors. To put it more bluntly, "the number one way victims die from domestic violence is by a gun" (Flock, 2020), and over-consumption of alcohol is a positive predictor of relationship violence.

While, on the one hand, quarantining and its associative stresses complicate relationships and intensify IPV on interpersonal levels, on the other hand

Image 6.1. Thousands gathered in Tel Aviv's Charles Clore Park on 1 June 2020, to protest against domestic violence. Israel's Welfare and Social Services Ministry published figures showing a 112% increase in complaints about domestic violence since the beginning of the pandemic (Alamy/Eddie Gerald).

the institutions that are meant to support victims of IPV are in dire straits on account of the pandemic. Historically, healthcare workers have been the first to identify, detect, and provide referrals in IPV cases, as "victims tend to have more health problems and frequently look for assistance from health care services in the context of acute traumatic injuries, physical, sexual and psychological sequels of abuse, and many acute and chronic pathologies" (Moreira & Pinto da Costa, 2020). However, with the pandemic, healthcare workers are stretched thin as the systems are collapsing under record levels of COVID-19 hospitalizations. As of December 2020, available data from the US Department of Health and Human Services confirmed that at least 200 hospitals were at full capacity, while in one-third of all hospitals more than 90% of all ICU beds were occupied, and COVID-19 patients accounted for 46% of all staffed ICU beds (Yan et al., 2020).

First responders are also under excessive strain. For example, in April 2020, when New York City was the hotspot for the virus, 20% of uniformed members of the New York Police Department (NYPD) were not working due to illness, translating to more than 6,974 members (Holcombe, 2020). The fewer available personnel also meant that action against IPV became a low priority. Unsurprisingly, some victims who did report their aggressors received little help from the authorities. Feng Yuan, a co-founder of the Equality-Chinese anti-domestic violence advocacy group, recounted to a *New York Times* reporter how one of her clients was told that the responders will "come to your place after the crisis" (Taub, 2020). In the United States, victims calling domestic violence hotlines reported that they had "first tried calling law enforcement and were told that the police were not currently making arrests for misdemeanors" (Flock, 2020). Additionally, victims of IPV may also refrain from using social and protective services for fear of exposure to the virus and/or concern about overwhelming the responders and frontline workers.

Domestic violence shelters have not been spared in this downward spiral. The majority of shelters operate on the basis of donations/grants and are collective living facilities. In April 2020, the National Coalition against Domestic Violence (NCADV, 2020) documented the ways in which various anti-violence organizations had restructured their operations because of safety considerations. Some shelters had to limit the number of new intakes while others operated under reduced capacity as staff took ill or worked remotely and/or with staggered schedules. Some others reported that they were forced to postpone all community support groups, presentations, and group counselling and were limited to providing telehealth consults. A few other shelters announced that they were no longer accepting donations of used clothing and relevant items and that group-home beds were being moved six feet apart, essentially further restricting capacity (NCADV, 2020). These safety planning measures, though crucial, limit the viability and efficacy of domestic violence shelters, removing more lifelines from the victims. For example, while shelters moved their counselling services to a safer virtual platform, the move presented significant risk for victims of IPV who live under constant surveillance by their tormentors. They are thus less likely to use these services lest they are discovered by their abusive partners. Altogether, the pandemic has led to a situation where *individual*-level risk factors have increased tremendously, and protective *societal* factors have diminished, leaving IPV victims increasingly more vulnerable to violence and abuse.

More recently, coronavirus vaccines have been greenlighted for use and there is hope that we shall see the proverbial light at the end of the tunnel. However, when life begins to limp back to normalcy, the impact that the lockdowns have had on escalating IPV will echo for a very long time to come, and possibly across generations. From the macro-structural standpoint, the global cost of violence, which had previously been estimated at approximately USD 1.5 trillion, is expected to rise as violence increases and continues (UN Women, 2020). Social services for victims are likely to experience budget cuts under the impending financial setback. At the individual level, the world economic recession that is predictably looming in the aftermath of the pandemic will only limit the financial independence that is often critical to escape situations of IPV (Godin, 2020). Moreover, reverberations of the trauma inflicted in the short run will impact not only women but also children who have witnessed episodes of abuse without the respite of having schools and friends to offer comfort. Existing research on domestic violence has shown that trauma begets trauma: in homes where there is IPV, there is also a strong possibility for child neglect and/or abuse (Forke et al., 2019). Children who witness IPV often experience intense psychological stress as well as behavioural and cognitive impairments (Ireland & Smith, 2009; Bair-Merritt et al., 2006; Martin, 2002) and some may also go on to emulate the violence they witness in their own personal relationships (Bensley et al., 2003; Ireland & Smith, 2009; Ernst et al., 2006; Forke et al., 2018), creating an ongoing trans-generational cycle of trauma and violence.

**Some Consideration for the Future and Concluding Remarks**

Our chapter described the toll that the pandemic and its accompanying safety measures have had on victims-survivors of IPV. However, what implications does it have for feminist scholarship and mobilization? What do we learn from this historical period? And how do we take lessons from this moment to guide ourselves towards a better future? Overall, in a prescient move, leaders of the world were quick to sound the alarm regarding the surge of IPV. The executive director of UN Women, Phumzile Mlambo-Ngcuka, asserted that "*the increase in violence against women must be dealt [with] urgently with measures embedded in economic support and stimulus packages that meet the gravity and scale of the challenge and reflect the needs of women who face multiple forms of discrimination*" [emphasis added] (Mlambo-Ngcuka, 2020; see also UN Women, 2020). She also called for all governments to make the prevention and redress of violence against women a key part of their national response plans for COVID-19 mitigation. In step with the declaration, governments have responded by incorporating considerations for women's safety in their stimulus plans. For example, the US CARES package grants $45 million to helping victims and survivors of violence. The Canadian government announced a US$82 billion COVID-19 aid package on 18 March 2020, which includes earmarked funding increases of US$50 million for gender-based violence shelters and sexual assault centres. We find this shift to be a welcome change, as in the past women's lives and experiences have often been relegated to less importance or a "private" concern. In that sense, it almost appears as if the decades of feminist efforts and activism have paid off. Although steps have been taken in the right direction, however, the optimism for top-down strategies must also be tempered, since there is much more ground to cover in terms of generating public policies favourable towards women's dignity and welfare.

Continuing decades of engagement with the issue of violence in intimate relationships, feminist groups are yet again at the forefront of reframing the response to the COVID-19 pandemic. The heightened risk and severity in rates of intimate partner violence in the context of COVID-19 have called attention to the need for a more "intersectionally informed analysis" to critically examine current frameworks, policies, and interventions (Hankivsky, 2020). Scholars argue that more support is needed for a progressive research agenda to inform policy on intimate partner violence during national and international crises. For instance, given the differential impact of the pandemic on different groups of people, especially women and communities of colour, scientists are calling for a reconceptualization of the way data is collected. There are now efforts to ensure that data is significantly disaggregated to include multiple indicators such as income, citizenship, location, education, race, ethnicity, ability, age, etc. (Cleveland, 2020; Berkhout & Richardson, 2020; Lokot & Avakyan, 2020; Eaves & Al-Hindi, 2020). Furthermore, they highlight that effective and meaningful interventions will only happen if mainstream research addresses "how experiences of and responses to COVID-19 are influenced by globalisation, capitalism, urbanisation, war, conflict, climate change, racism, and xenophobia" (Hankivsky, 2020). Thus, there is a need to contextualize individual experiences of IPV during the pandemic within a larger socio-historical context to enhance our policy and programmatic responses.

Additionally, an intersectional approach calls for the need to move beyond top-down strategies that rely primarily on criminal justice responses to IPV. Recent developments noted by Qazi (2020) demonstrate that grassroots and community-based organizations have emerged as the most powerful advocates for survivors, often filling in the vacuum of services left by an overwhelmed social service infrastructure during the pandemic. Therefore, it is crucial that the perspectives of groups and organizations working at the frontlines are included during policy debates so that community needs and resources for women who are at risk of violence during a crisis such as our COVID-19 pandemic are prioritized (Berkhout & Richardson, 2020). Finally, scholars such as Kulkarni (2019) and Bowleg (2020) emphasize that the voices of survivors, especially from structurally marginalized communities, who are directly impacted by the inequities resulting from the pandemic have to be heard and made visible. The pandemic presents an opportunity to rethink and reconfigure existing systems by prioritizing women's lived experiences at every stage of decision making.

REFERENCES

Abraham, M. (1995). Ethnicity, gender, and marital violence: South Asian women's organizations in the United States. *Gender & Society*, 9(4), 450–68. https://doi.org/10.1177/089124395009004004.

Abraham, M. (2000). Fighting back: Abused women's strategies of resistance.
In Abraham, M. (Ed.), *Speaking the unspeakable: Marital violence among South Asian immigrants in the United States* (pp. 132–53). Rutgers University Press.

Bair-Merritt, M.H., Blackstone, M., & Feudtner C. (2006). Physical health outcomes of childhood exposure to IPV: A systematic review. *Pediatrics, 117*(2), 278–90. https://doi.org/10.1542/peds.2005-1473.

Baskin, C. (2003). From victims to leaders: Activism against violence towards women. In Anderson, K., & Lawrence, B. (Eds.), *Strong women stories: Native vision and community survival* (pp. 213–27). Sumach Press.

Bensley, L., van Eenwyk, J. & Simmons, K.W. (2003). Childhood family violence history and women's risk for IPV and poor health. *American Journal of Preventive Medicine, 25*(1), 38–44. https://doi.org/10.1016/S0749-3797(03)00094-1.

Berkhout, S.G., & Richardson, L. (2020). Identity, politics, and the pandemic: Why is COVID-19 a disaster for feminism(s)?. *History and Philosophy of the Life Sciences, 42*(49). https://doi.org/10.1007/s40656-020-00346-7.

Bhattacharjee, A. (1997). The public/private mirage: Mapping homes and undomesticating violence work in the South Asian immigrant community. In Alexander, M.J., and Mohanty, C.T. (Eds.), *Feminist genealogies, colonial legacies, democratic futures* (pp. 203–29). Routledge.

Bohmer, C., Brandt, J., Bronson, D., & Hartnett, H. (2002). Domestic violence law reforms: Reactions from the trenches. *Journal of Sociology and Social Welfare, 29*(3), 71–87. https://scholarworks.wmich.edu/jssw/vol29/iss3/5.

Boserup, B., McKenney, M., & Elkbuli, A. (2020). Alarming trends in US domestic violence during the COVID-19 pandemic. *American Journal of Emergency Medicine, 38*(12), 2753–5. https://doi.org/10.1016/j.ajem.2020.04.077.

Bowleg, L. (2020). We're not all in this together: On COVID-19, intersectionality, and structural inequality. *American Journal of Public Health, 110*, 917_917. https://doi.org/10.2105/AJPH.2020.305766.

Bunch, C., & Frost, S. (2000). Women's human rights: An introduction. In Kramarae, C., & Spender, D. (Eds.), *Routledge international encyclopedia of women: Global women's issues and knowledge*, Volume 2. Routledge.

Cleveland, N. (2020, March 23). An intersectional approach to a pandemic? Gender data, disaggregation, and COVID-19. *data2x*. Retrieved 14 December 2020, from https://data2x.org/an-intersectional-approach-to-a-pandemic-gender-data-disaggregation-and-covid-19/.

Coomaraswamy, R. (2000, February 29). Report of the special rapporteur on violence against women, its causes, and consequences, on trafficking in women, women's migration and violence against women, submitted in accordance with Commission on Human Rights Resolution 1997/44. United Nations Doc. E/CN.4/2000/68.

Crenshaw, K. (1991). Mapping the margins: Intersectionality, identity politics, and violence against women of color. *Stanford Law Review, 43*(6), 1241–99. https://doi.org/10.2307/1229039.

Dasgupta, S.D. (1998). Women's realities: Defining violence against women by migration, race, and class. In Bergen, R.K. (Ed.), *Issues in intimate violence* (pp. 209–19). Sage.

Davis, A. (2000). The color of violence against women. *ColorLines Magazine*, 10. http://colorlines.com/archives/2000/10/the_color_of_violence_against_women.html.

Dobash, R.E., & Dobash, R. (1979). *Violence against wives – A case against the patriarchy*. Free Press.

Eaves L., & Al-Hindi, K.F. (2020). Intersectional geographies and COVID-19. *Dialogues in Human Geography*, *10*(2), 132–6. https://doi.org/10.1177/2043820620935247.

Enarson, E. (2012). *Women confronting natural disaster: From vulnerability to resilience*. Lynne Rienner.

Enarson, E., & Scanlon, J. (1999). Gender patterns in a flood evacuation: A case study of couples in Canada's Red River Valley. *Applied Behavioral Science Review*, *7*(2), 103–25. https://genderandsecurity.org/projects-resources/research/gender-patterns-flood-evacuation-case-study-canadas-red-river-valley. https://doi.org/10.1016/S1068-8595(00)80013-6.

Ernst, A.A., Weiss, S.J., & Enright-Smith, S. (2006). Child witnesses and victims in homes with adult IPV. *Academy Emergency Medicine*, *13*(6), 696–9. https://doi.org/10.1197/j.aem.2005.12.020.

Ertürk, Y., & Purkayastha, B. (2012). Linking research, policy and action: A look at the work of the special rapporteur on violence against women. *Current Sociology*, *60*(2), 142–60. https://doi.org/10.1177/0011392111429216.

Fagan, J. (1990). Natural experiments. In Kempf, K. (Ed.), *Measurement issues in criminology* (pp. 108–37). Springer-Verlag.

First, J.M., First, N.L., and Houston, B.J. (2017). Intimate partner violence and disasters: A framework for empowering women experiencing violence in disaster settings. *Affilia: Journal of Women and Social Work*, *32*(3), 390–403. https://doi.org/10.1177/0886109917706338.

Flock, E. (2020, March 31). He forced her to wash her hands over and over until they were raw and bleeding: For victims of domestic violence, stay-at-home orders are a worst-case scenario. *Cosmopolitan*. Retrieved 14 December 2020, from https://www.cosmopolitan.com/lifestyle/a31932146/domestic-violence-abuse-coronavirus-resources/.

Forke, C., Catalozzi, M., Localio, A.R., Grisso, J.A. Weibe, D.J., & Fein, J.A. (2019). Intergenerational effects of witnessing domestic violence: Health of the witnesses and their children. *Preventive Medicine Reports 15*. https://doi.org/10.1016/j.pmedr.2019.100942.

Forke, C.M., Myers, R.K., Fein, J.A., Catallozzi, M., Localio, A.R., Wiebe, D.J., & Grisso, J.A. (2018). Witnessing IPV as a child: How boys and girls model their parents' behaviors in adolescence. *Child Abuse and Neglect*, *84*, 241–52. https://doi.org/10.1016/j.chiabu.2018.07.031.

Godin, M. (2020, March 18). As cities around the world go on lockdown, victims of domestic violence look for a way out. *Time*. https://time.com/5803887/coronavirus-domestic-violence-victims/.

Hall, R.J. (2015). Feminist strategies to end violence against women. In Baksh, R., & Harcourt, W. (Eds.), *The Oxford handbook of transnational feminist movements* (pp. 1–25). Oxford University Press.

Hankivsky, O. (2020, May 6). Using intersectionality to understand who is most at risk of COVID-19. *Pursuit*. https://pursuit.unimelb.edu.au/articles/using-intersectionality-to-understand-who-is-most-at-risk-of-covid-19.

Holcombe, M. (2020, April 7). 12 NYPD members have died from suspected cases of coronavirus. CNN. https://www.cnn.com/2020/04/07/us/nypd-coronavirus-out-sick/index.html.

hooks, b. (1984). *Feminist theory: From margin to center*. South End Press.

Ireland, T.O., & Smith, C.A. (2009). Living in partner-violent families: Developmental links to antisocial behavior and relationship violence. *Journal of Youth and Adolescence*, *38*(3), 323–39. https://doi.org/10.1007/s10964-008-9347-y.

Jenkins, P., & Phillips, B. (2008). Battered women, catastrophe, and the context of safety after Hurricane Katrina. *NWSA Journal*, *20*(3), 49–68. https://www.muse.jhu.edu/article/256898.

Johns Hopkins Coronavirus Resource Center. (2020). *COVID-19 dashboard*. https://coronavirus.jhu.edu/map.html.

Josephson, J. (2002). The intersectionality of domestic violence and welfare in the lives of poor women. *Journal of Poverty*, *6*(1), 1–20. https://doi.org/10.1300/J134v06n01_01.

Kulkarni, S. (2019). Intersectional trauma-informed intimate partner violence (IPV) services: Narrowing the gap between IPV service delivery and survivor needs. *Journal of Family Violence*, *34*, 55–64. https://doi.org/10.1007/s10896-018-0001-5.

Lokot, M., & Avakyan, Y. (2020). Intersectionality as a lens to the COVID-19 pandemic: Implications for sexual and reproductive health in development and humanitarian contexts. *Sexual and Reproductive Health Matters*, *28*(1), 40–3. https://doi.org/10.1080/26410397.2020.1764748.

MacKinnon, C. (1989). *Toward a feminist theory of the state*. Harvard University Press.

Mama, A. (1989). Violence against black women: Gender, race, and state responses. *Feminist Review*, *32*(1), 30–48. https://doi.org/10.1057/fr.1989.18.

Martin, S.G. (2002). Children exposed to domestic violence: Psychological considerations for health care practitioners. *Holistic Nursing Practice*, *16*(3), 7–15. https://doi.org/10.1097/00004650-200204000-00005.

McQuigg, R. (2015). Domestic violence as a human rights issue: Rumor v. Italy. *European Journal of International Law*, *26*(4), 1009–25. https://doi.org/10.1093/ejil/chv057.

Menjivar, C., & Salcido, O. (2002). Immigrant women and domestic violence: Common experiences in different countries. *Gender & Society*, *16*(6), 898–920. https://doi.org/10.1177/089124302237894.

Merry, S., & Shimmin, J. (2011). The curious resistance to seeing domestic violence as a human rights violation in the United States. In Hertel, S., & Libal, K. (Eds.),

*Human rights in the United States: Beyond exceptionalism* (pp. 113–31). Cambridge University Press.

Mlambo-Ngcuka, P. (2020, April 6). Violence against women and girls: The shadow pandemic statement by Phumzile Mlambo-Ngcuka, Executive Director of UN Women. https://www.unwomen.org/en/news/stories/2020/4/statement-ed-phumzile-violence-against-women-during-pandemic.

Moreira, D.N., & Pinto da Costa, M. (2020). The impact of the COVID-19 pandemic in the precipitation of intimate partner violence. *International Journal of Law and Psychiatry*, *71*(01606). https://doi.org/10.1016/j.ijlp.2020.101606.

Morgaine, K. (2007). Domestic violence and human rights: Local challenges to a universal framework. *Journal of Sociology and Social Welfare*, *34*(1), 109–29. https://scholarworks.wmich.edu/jssw/vol34/iss1/7.

Morgaine, K. (2011). How would that help our work?: Domestic violence and human rights on the ground. *Violence against Women*, *17*(1), 6–27. https://doi.org/10.1177/1077801209347749.

Mukherjee, J.S. (2007). Structural violence, poverty, and the AIDS pandemic. *Development*, *50*, 115–21. https://doi.org/10.1057/palgrave.development.1100376.

National Coalition against Domestic Violence (NCADV). (2020, March 13). What domestic violence organizations need to know about coronavirus blog post. Retrieved 14 December 2020, from https://ncadv.org/blog/posts/what-dv-orgs-need-to-know-coronavirus.

Neel, J. (2020, September 9). NPR poll: Financial pain from coronavirus pandemic "much, much worse" than expected. NPR. Retrieved 14 December 2020, from https://www.npr.org/sections/health-shots/2020/09/09/909669760/npr-poll-financial-pain-from-coronavirus-pandemic-much-much-worse-than-expected.

Njiru, R., & Purkayastha, B. (2017). "As a woman I cannot just leave the house": Gendered spaces and HIV vulnerability in marriages in Kenya. *Journal of Gender Studies*, *27*(8), 957–68. https://doi.org/10.1080/09589236.2017.1377064.

O'Brien, M., & Tolosa, X. (2016). The effect of the 2014 West Africa Ebola virus disease epidemic on multi-level violence against women. *International Journal of Human Rights in Healthcare*, *9*(3), 151–60. https://doi.org/10.1108/IJHRH-09-2015-0027.

Obreja, L.D. (2019). Human rights law and IPV: Towards an intersectional development of due diligence obligations. *Nordic Journal of Human Rights*, *37*(1), 63–80. https://doi.org/10.1080/18918131.2019.1589209.

Patel, R. (2020, April 27). Minister says COVID 19 is empowering domestic abusers as rates rise in parts of Canada. CBC News. Retrieved 14 December 2020, from https://www.cbc.ca/news/politics/domestic-violence-rates-rising-due-to-covid19-1.5545851.

Peek, L., & Fothergill, A. (2008). Displacement, gender, and the challenges of parenting after Hurricane Katrina. *NWSA Journal*, *20*(3), 69–105. https://doi.org/10.1353/nwsa.0.0044.

Pellowski, J.A., Kalichman, S.C., Matthews, K.A., & Adler, N. (2013). Pandemic of the poor: Social disadvantage and the US HIV epidemic. *American Psychologist*, *68*(4), 197–209. https://doi.org/10.1037/a0032694.

Pirtle, L. (2020). Racial capitalism: A fundamental cause of novel coronavirus (COVID-19) pandemic inequities in the United States. *Health Education & Behavior*, *47*(4), 504–8. https://doi.org/10.1177/1090198120922942.

Presser, L., & Gaarder, E. (2000). Can restorative justice reduce battering? Some preliminary considerations. *Social Justice*, *27*(1), 175–94. https://www.jstor.org/stable/29767197.

Price, J. (2012). *Structural violence: Hidden brutality in the lives of women*. State University of New York Press.

Purkayastha, B. (2018). Gender and human rights. In Risman, B., Froyum, C., & Scarborough, W. (Eds.), *Handbook of the sociology of gender* (pp. 523–35). Springer.

Purkayastha, B. (2000). Liminal lives: South Asian youth and domestic violence. *Journal of Social Distress and Homelessness*, *9*, 201–19. https://doi.org/10.1023/A:1009408018107.

Qazi, M. (2020). Women collectives articulate grassroots response to shadow pandemic. *Daily Sabah*. Retrieved 14 December 2020, from https://www.dailysabah.com/opinion/op-ed/women-collectives-articulate-grassroots-response-to-shadow-pandemic.

Ray, R., & Rojas, F. (2020, April 16). Inequality during the coronavirus pandemic. *Contexts*. Retrieved 14 December 2020, from https://contexts.org/blog/inequality-during-the-coronavirus-pandemic/.

Renzetti, Claire. M. (1998). Violence and abuse in lesbian relationships: Theoretical and empirical issues. In Bergen, R.K. (Ed.), *Issues in intimate violence* (pp. 117–27). Sage.

Ross-Sheriff, F. (2007). Women and disasters: Reflections on the anniversary of Katrina and Rita. *Affilia: Journal of Women and Social Work*, *22*(1), 5–8. https://doi.org/10.1177/0886109906295813.

Roth, Kenneth. (1994). Domestic violence as an international human rights issue. In Cook, R. (Ed.), *Human rights of women: National and international perspectives* (pp. 326–39). University of Pennsylvania Press.

Schneider, D., Harknett, K., & McLanahan, S. (2016). Intimate partner violence in the Great Recession. *Demography*, *53*, 471–505. https://doi.org/10.1007/s13524-016-0462-1.

Shepard, M. (2005). Twenty years of progress in addressing domestic violence: An agenda for the next 10. *Journal of Interpersonal Violence*, *20*(4), 436–41. https://doi.org/10.1177/0886260504267879.

Smith, A. (2005). *Conquest: Sexual violence and American Indian genocide*. South End Press.

Sokoloff, N.J., & Dupont, I. (2005). Domestic violence at the intersection of race, class, and gender: Challenges and contributions to understanding violence against marginalized women in diverse communities. *Violence against Women*, *11*(1), 38–64. https://doi.org/10.1177/1077801204271476.

Taub, A. (2020, April 6). A new COVID-19 crisis: Domestic abuse rises worldwide. 2020. *New York Times*. Retrieved 14 December 2020, from https://www.nytimes.com/2020/04/06/world/coronavirus-domestic-violence.html.

Thomas, D.Q., & Beasley, M.E. (1993). Domestic violence as a human rights issue. *Human Rights Quarterly*, *15*(1), 36–62. https://doi.org/10.2307/762650.

UN Women. (2020). *The shadow pandemic: Violence against women during COVID-19.* Retrieved 14 December 2020, https://www.unwomen.org/en/news/in-focus/in-focus-gender-equality-in-covid-19-response/violence-against-women-during-covid-19.

Walker, L.E. (1979). *Battered women: A psychosociological study of domestic violence.* Van Nostrand Reinhold.

West, C.M. (2004). Black women and intimate partner violence: New directions for research. *Journal of Interpersonal Violence*, *19*(2), 1487–93. https://doi.org/10.1177/0886260504269700.

Yan, H., Maxouris, C., & McPhillips, D. (2020, December 10). 200 hospitals have been at full capacity, and 1/3 of all US hospitals are almost out of ICU space. CNN Health. https://www.cnn.com/2020/12/10/health/us-coronavirus-thursday/index.html.

# The United States

# 7 An Investigation into the Economic, Social, and Psychological Dimensions of COVID-19

BY KEVIN MCCAFFREE AND ANONDAH SAIDE[*]

## Introduction

When the Black Death was decimating Europe, wealthy political and military elites escaped to the countryside to avoid the dense urban environments that seemed hardest hit by disease (McKinley, 2020). Peasants, labourers, and other commoners, so dependent on markets in the city, disproportionately stayed behind, and disproportionately died as a result. How applicable are these observations to present-day pandemic responses? More specifically, to what extent does economic, social, and psychological vulnerability continue to play a role in people's responses to pandemics (even ones quite a bit less severe than the bubonic plague)? In this chapter we turn our attention towards these questions and others, using data on the COVID-19 pandemic as a case study.

### Impact of COVID-19 on Well-Being

The COVID-19 virus is many times deadlier than seasonal influenza and spreads more easily (Cates et al., 2020). Worldwide, as of this writing in December 2020, nearly 82 million people have been officially diagnosed with COVID-19 and 2 million people have died (*New York Times*, 2020b). Governments around the world have responded with various measures, from face-mask mandates to six-foot social distancing guidelines, to restrictions on travel, to the closing of schools and businesses (Centers for Disease Control and Prevention (CDC), 2020; *New York Times*, 2020a). The research tells us that healthcare workers perceive a greater risk related to the spread of COVID-19 than does the general public and that, unsurprisingly, healthcare workers are also more knowledgeable about the pandemic (Simione & Gnagnarella, 2020).

---

[*] Both authors contributed equally to the manuscript and are listed alphabetically.

The medical and public health concerns related to the pandemic should be familiar to us. The individual social, physical, economic, and psychological toll of adjusting to a global pandemic, however, has gone relatively less discussed.

Some concerns related to how COVID-19 might interact with economic vulnerabilities spurred examinations during the beginning of the pandemic. In April of 2020, polling data revealed that fewer than half (47%) of Americans had the savings/funds necessary to cover economic expenses for three months. This number, though dire, hid a starker reality: only 33% of people with a high school degree (or less) and only 23% of those with lower income reported having the funds to cover three months out of work (Pew Research Center, 2020b). By August 2020, the initial magnitude of the economic damage of the pandemic became more evident as a *quarter* of all adults in the United States and a *third* of those with lower income reported having been laid off or lost a job (Pew Research Center, 2020c). Nearly half (44%) of those in the lower income bracket said they had to use money from their savings or retirement in order to pay their bills. Even a third (33%) of Americans reporting middle-range income noted how they had to tap into their savings and retirement to get through the economic implications of the pandemic (Pew Research Center, 2020c).

Other pre-existing economic vulnerabilities that were being negatively affected by the pandemic had to do not with *amount* of income but with *how* people earn a living. Working-class jobs have long been more physical, more hands-on, than the bourgeois white-collar professions. According to the Bureau of Labor Statistics (2020), fewer than 30% of people employed as office workers, service workers, installation and repair workers, construction workers, or in production or transportation are able to work remotely. In fact, analysts suggest that only 37% of *all* jobs can be "plausibly performed at home" (Dingel & Neiman, 2020, p. 4). Moreover, this is an upper-bound estimate. Indeed, in 2019, only 24% of people had worked from home some or all of the time (Bureau of Labor Statistics, 2019).

The economic impact of COVID-19 will, of course, be felt worldwide. The Gross Domestic Product (GDP) of many countries may drop 3–6%, and of some 10%, perhaps even 15%, depending on which projected scenarios manifest themselves (e.g., how long the portions of that country's economy are shut down; see also Fernandes, 2020). As with service and production workers in the United States who have been mauled by business closings, those countries that specialize in tourism, service, or production will see the most dramatic losses.

In addition to economic changes, people's everyday routines have also changed and this has plausibly impacted their physical, psychological, and social well-being. After the government-imposed lockdown in Austria, for example, researchers analysed data from 1.2 million cell phones and found that people were making fewer, longer, calls to others (Reisch et al., 2020). This was not just due to having fewer conversations at work – even on the weekend people were having longer calls with fewer people. Rather, it was consistent with other data showing that people's communication networks began to shrink with the onset of the pandemic. This may have been very positive; perhaps people were just reconnecting with concerned loved ones.

Other developments, however, are distinctly less positive. The data indicate that, during lockdowns, people sat more, worked out less, ate in a less healthy fashion, and felt uncomfortable attending parties, eating at restaurants, or going to vote (Ammar et al., 2020; Pew Research Center, 2020a). One innovative study even compared samples of people in pandemic lockdown and samples of first-time prisoners to see if there were any similarities in experiences of confinement (Dhami et al., 2020). Results revealed that those in lockdown reported being less physically active than first-time prisoners and, worse, they reported feeling more hopeless.

People have suffered psychologically even though data suggest suicide rates have not risen (John et al., 2020; Rück et al., 2020). Studies consistently show that around 20% of people report significant distress associated with COVID-19, mostly anxiety or depression, and symptomologies appear to be worsening as the lockdowns persist (Kelly, 2020). An online survey with over six thousand respondents from 59 countries found that approximately 24% of people reported symptoms consistent with moderate to severe depression (Alzueta et al., 2020). Researchers were also able to directly link worse mental health with changes made to one's lifestyle in response to quarantine restrictions such as working remotely and being separated from family and friends. Finally, they found that those respondents scoring worse on mental health measures also reported more conflict at home and more difficulty paying bills – illustrating that social, psychological, and economic vulnerabilities are interrelated.

More studies on the social and psychological implications of COVID-19 are emerging every day. For example, another study found that in response to COVID-19 restrictions, students began studying together less and their anxiety, depression, loneliness, and concern for health and family increased (Elmer et al., 2020). People even experienced a slowing-down of their time perception, itself correlated with greater stress and lower satisfaction (Ogden, 2020).

As a result, some struggled with or did not bother to follow public health guidelines. Data with samples recruited from social media sites (see also Leary et al., 2020) shows that the more people feared missing out on social events, the less stringently they reported following social distancing guidelines. Studies also show that men, perhaps predictably, were more likely than women to think that wearing a face mask was a sign of weakness; men were also found to be less likely to take health precautions (Capraro & Barcelo, 2020; Clark et al., 2020).

Over the duration of the pandemic during the pre-vaccination period, we saw a number of instances of large-scale disruptions to social distancing behaviour contributing to the spread of COVID-19. The Sturgis Motorcycle rally in South Dakota lasted 10 days and drew hundreds of thousands of people, while protests against police brutality inspired by the George Floyd incident drew similar numbers in cities across numerous states (Valentine et al., 2020; Walker, 2020). Churches around the country became "hot spots" for COVID-19 outbreaks, students from large universities contributed to a spike in cases after Spring Break, and, of course, birthday parties and other gatherings led to outbreaks (Mangrum & Niekamp, 2020; Plater, 2020; Stump, 2020). In each case, COVID-19 spread to further numbers of people. Even the quite forgivable desire to stay with others contributed to the spread: work by Kelley and colleagues (2020) found that staying overnight at someone's house (or having someone else stay overnight at your house) doubled people's risk of becoming infected.

Trust in others is also important to people's well-being (Helliwell & Wang, 2010). Trust in others has classically been related to social support and community cohesion (Sampson et al., 1999). However, recent work has begun to show why trust in others might influence people's tendency to follow pandemic-related guidelines. Alessandri et al. (2020), for example, find that generalized social trust moderates how people's personality traits interact with their tendency to morally disengage and, thus, whether they follow pandemic guidelines.

These and other studies are also beginning to examine trust in institutions, particularly trust in government. One study using a sample from New Zealand found that people reported greater trust in a variety of institutions, as well as a stronger national identity, 18 days into a lockdown compared to a pre-lockdown sample (Sibley et al., 2020). However, though people post-lockdown reported greater levels of institutional trust in authorities like scientists, politicians, and police, they also reported greater degrees of mental distress. These sorts of results – increasing mental distress alongside increasing institutional trust – may be part of a pattern across democratic institutions (Bækgaard et al., 2020).

However, further work is needed to explore institutional trust outside of government trust, and, moreover, work is needed to investigate if the trust dynamic mentioned above holds in highly polarized democracies. Trust in news media, for example, may be a promising avenue. One study of over 1,100 adults across the United States at the beginning of the pandemic found that people who reported greater trust in public health institutions such as the CDC, lower trust in President Trump, and avoidance of Fox News scored lower on measures of "misinformation" (e.g., knowledge about what helps prevent the spread, or about what symptoms are) (Dhanani & Franz, 2020). Another study found a fairly small but consistent effect indicating that Fox News viewers engaged in fewer preventive behaviours than did CNN viewers (Zhao et al., 2020).

**COVID-19 and Partisan Attitudes**

People the world over tend to form groups defined by complex symbolic understandings of events, other groups of people, and political policies. Once people develop a *political identity*, believing themselves to be supporters of a particular constellation of policies, any incoming information from their environment gets filtered through a set of biases which promote their existing political-tribal affiliations. Information that appears congruent with existing expectations or political identities is accepted or, perhaps, investigated lightly; while information that violates expectations or political identities is rejected, or at least, investigated more critically.

In order to apply an identity framework (see also Clark et al., 2019) to the study of COVID-19 spread, Freira and colleagues (2020) employed a "tribal partisanship" model to interpret why people might give optimistic or pessimistic estimates of COVID-19 deaths in their country. What seemed to best predict people's attitudes about COVID-19 were not any particular facts but, rather, their political attitudes. Other research (e.g., Collins et al., 2020) has found that measures of people's political identity predict threat perception, emotional suffering, and level of support for pandemic-related restrictions.

COVID-19 struck the United States during a presidential election year and a period of rising political polarization, particularly amongst partisans (Doherty, 2014; Hare & Poole, 2014). Given that Republicans controlled the executive branch of the White House when the pandemic struck, Republicans (perhaps especially Christian nationalists; see also Perry et al., 2020) were quicker than Democrats to think of reports of epidemic severity as nothing more than baseless critiques of President Donald Trump's successful pandemic leadership. Democrats, on the other hand, being the party out of power and responding to an unpredictable President Trump, were quicker than Republicans to regard Trump's leadership, which exacerbated the public health emergency, as a catastrophe worthy of the highest condemnation. For example, researchers have found that 69% of Republicans, but only 33% of Democrats, say they are very confident or extremely confident in the federal government's ability to address COVID-19 (Pickup et al., 2020). Interestingly, Republicans and Democrats have not differed in their rates of compliance with pandemic guidelines (e.g., not going into work or working from home more than usual, keeping a distance of at least two arms' lengths, approximately six feet, from others as much as possible). Other work shows political liberals to be more willing than conservatives to control and punish others in order to address their COVID-19 related concerns (Graso et al., 2020). These authors attribute their findings to liberals' greater on average willingness to avoid harm and conservatives' greater willingness to express personal liberties.

Compared to Democrats, Republicans have been found to be less trusting of scientists' ability to understand COVID-19 (Evans & Hargittai, 2020). Political tribalism was also a strong predictor of whether school districts reconvened face to face. Republican districts were more likely to return to face-to-face instruction while Democrat-run districts were more likely to stay fully online (Hartney & Finger, 2020). In fact, the COVID case rate was not significantly related to the likelihood of a district staying online when measures of partisanship were in the model. Another interesting study found that the strongest correlate of state-enforced mask laws was the political affiliation of the state governor (Adolph et al., 2020). Consider that in a sample taken at the outset of the pandemic, 51% of Democrats (and those leaning Democrat) reported that their personal life had been "changed in a major way" by COVID-19, compared to only 38% of Republicans (and those leaning Republican) (Pew Research Center, 2020a).

Finally, one study of over 5,500 Americans found that Democrats were more worried than Republicans about getting infected, being hospitalized, or dying from COVID-19 (de Bruin et al., 2020). Political partisanship, the researchers found, was a much stronger predictor of COVID-19-related policy attitudes than even participants' perceived risk of infection or dying. They also found, perhaps unsurprisingly, that Democrats were more likely than Republicans to report getting their news from CNN or MSNBC, whereas Republicans reported higher viewing rates for Fox News.

## Current Study

### Research Objective and Hypotheses

In the present study, we hope to extend some of the aforementioned and existing research on social vulnerabilities and well-being, trust, and political identity. We start from the premise that aspects of people's existing well-being (financial, social, or psychological) impacted their willingness and ability to worry about COVID-19 infection. In addition to this, we looked at the extent to which trust in news media mediates well-being and concern about COVID-19. Lastly, we will assess these relationships according to people's political affiliations. Figure 7.1 depicts the theoretical relationships that informed our research questions and hypotheses examined herein.

Our first research question was: "Will concern about the spread of COVID-19 be related to trust in news media (i.e., an institution tasked with disseminating COVID-19 information to the public) and related to indices of individuals' well-being?" We hypothesized the following patterns:

a  Individuals with higher *psychological* well-being scores, who trust news media more, will report greater concern about COVID-19 spread. That is, trust in institutions will mediate the relationship between psychological well-being and concern about the pandemic.
b  Individuals with higher *financial* well-being scores, who trust news media more, will report greater concern about COVID-19 spread. That is, trust in institutions will mediate the relationship between financial well-being and concern about the pandemic.

The Economic, Social, and Psychological Dimensions of COVID-19   183

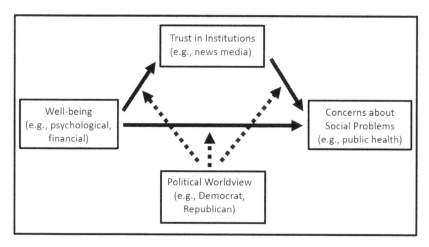

Figure 7.1. Visual depiction of conceptual relations.

c Individuals with higher *social* well-being scores, who trust news media more, will report greater concern about COVID-19 spread. That is, trust in institutions will mediate the relationship between social well-being and concern about the pandemic.

Our second research question was exploratory and related to political identity. In particular, we wondered, "To what extent do the hypothesized patterns discussed above hold for both *self-reported* Democrats and *self-reported* Republicans?" We wanted to explore the possibility that political affiliation moderates the relationships discussed above, given what is known about partisan attitudes concerning the pandemic. In particular, conservative political pundits have expressed concern on conservative news outlets (e.g., Fox News) about the negative effect on the economy of implementing certain pandemic-related restrictions (e.g., closing or limiting business). Republicans also tend to be more behaviourally religious and opposed to restrictions on personal liberties than Democrats are (e.g., Perry et al, 2020; Samore et al., 2020), which may manifest as decreased concern for something (i.e., the pandemic) that would inhibit them from engaging in their religious practices. As a result, in addition to exploring political identity as a moderating factor, we hypothesized that:

a Compared to Republicans, Democrats will express greater concern with coronavirus spread relative to potential costs to the economy (i.e., unemployment rate).

b   Compared to Republicans, Democrats will report greater trust in (presumably pandemic-focused) news media.
c   Compared to Republicans, Democrats will be less likely to report that the pandemic has inhibited their ability to engage in religious practices.

Given the literature reviewed above, this study is not entirely exploratory, but it *is* exploratory. The social and political response to COVID-19 is unprecedented, and previously understood relationships between well-being, trust in media, and public health concerns might not hold under present conditions. This is the primary reason why the literature review above has been largely confined to work since the onset of the COVID-19 pandemic. Nevertheless, data regarding the relationships investigated here will be useful for researchers and theorists alike, given the urgency of the topic and the well-established relevance of the variables in play.

*Methods*

Qualtrics survey software was used to create and distribute a 15-minute survey measuring political attitudes and indicators of social, psychological, and economic well-being, as well as attitudes towards the COVID-19 pandemic. Participants were recruited with the intention of having a nationally representative sample that reflects the United State adult population in terms of educational attainment, gender, and household income. Qualtrics panel services was used to recruit participants.

*Participants*

Our original sample included 1,401 adults from the US. However, 323 adults were excluded from the analyses because those individuals did not identify with one of the two major political parties analysed here (self-reported Democrats and self-reported Republicans). The final sample used for this study included 587 Democrats ($M = 45.47$, $SD = 17.47$, 54% Female) and 491 Republicans ($M = 48.64$, $SD = 16.13$, 49% Female).

*Measures*

**Demographic Characteristics:** *Household Income* was measured with the question, "What was your household income last year (that is, 2019)?" Answers ranged from "$0–$24,999" [1] to "200,000 or more" [7]. *Age* was measured with the question, "What is your age in years?" The answers provided ranged from 18 to 94. *Gender* was measured with the question, "What is your gender?" Males were coded as "0," females were coded as "1."

**Well-Being:** We measured well-being across three domains. *Economic Well-Being* was measured with the question, "In the past month how worried have you been about being able to pay your bills? (e.g., medical, child-care, housing)?" Answers ranged from "not at all worried" [1] to "extremely worried" [7]. Answers were then reverse-coded such that higher scores indicated greater economic well-being. *Psychological Well-Being* was measured with the question, "Overall, how satisfied are you with life as a whole these days?" Answers ranged from "not at all" [1] to "completely" [10] (VanderWeele et al., 2020). Higher scores on this variable indicate greater psychological well-being. *Social Well-Being* was measured with the six-item revised UCLA Loneliness Scale (Wongpakaran et al., 2020). This scale assesses the degree to which individuals feel socially disconnected and isolated (e.g., lacks companionship, feels alone). On a five-point Likert scale, answers ranged from "strongly disagree" [0] to "strongly agree" [5]. Scores were reverse-coded and then averaged for an overall score where higher scores indicated better social well-being (i.e., greater feelings of social connectedness) (Cronbach > .9 for Democrats and Republicans).

**Attitudes about the Pandemic:** *Pandemic Concern* was measured with the question, "In your opinion, how concerned should people in your community be about the spread of coronavirus (COVID-19)?" Answers ranged from "not at all concerned" [0] to "very concerned" [3]. *Religious Restriction* was measured with the question, "COVID-19 has restricted my ability to engage in the religious practices I want to engage in." On a five-point Likert scale, answers ranged from "strongly disagree" [-3] to "strongly agree" [3]. *Prioritizing Pandemic* was measured with the question, "If you had to choose, which issue would you prioritize?" Answer options were placed on a continuum from "reducing the unemployment rate" [-3] to "reducing the spread of coronavirus" [3]. The higher the score on this question, the greater reported prioritization of reducing the spread of coronavirus in lieu of reducing the unemployment rate.

**Trust in Media:** Participants were asked two questions related to trust in media. The first asked, "How often do you read/watch/listen to the news?" Answers ranged from "never/rarely" [1] to "multiple times per day" [7]. The second question asked, "How trustworthy would you say news media is in general?" with answers ranging from "not at all trustworthy" [1] to "very trustworthy" [7]. Answers to these to two questions were multiplied to create a new variable, *Trust in Media*, which represents their overall trust in news media contextualized by the degree to which they reference news media for information.

## Results

### Correlational Analyses

**Republicans:** The descriptive and correlational statistics discussed below among Republicans are presented in Table 7.1. There was a moderate significant relationship between Pandemic Concern and Trust in Media. There was a small significant relationship between all three well-being indices

Table 7.1. Correlations, means, and standard deviations: self-reported Republicans.

| Variable | 1 | 2 | 3 | 4 | 5 | 6 | 7 | 8 | 9 | 10 | n | M | SD |
|---|---|---|---|---|---|---|---|---|---|---|---|---|---|
| 1. Pandemic Concern | — | | | | | | | | | | 491 | 2.150 | 0.955 |
| 2. Age in Years | -.049 | — | | | | | | | | | 491 | 48.640 | 16.134 |
| 3. Household Income | .225** | -.274** | — | | | | | | | | 491 | 3.540 | 1.911 |
| 4. Gender[a] | -.178** | .164** | -.386** | — | | | | | | | 491 | — | — |
| 5. Trust in Media | .431* | -.314** | .414** | -.383** | — | | | | | | 491 | 25.424 | 16.299 |
| 6. Psychological Well-being | .180** | .040 | .323** | -.195** | .407** | — | | | | | 491 | 7.450 | 2.309 |
| 7. Social Well-being | -.132** | .274** | -.099* | .135** | -.328** | .048 | — | | | | 491 | 3.013 | 1.625 |
| 8. Economic Well-being | -.229** | .378** | -.039 | .108* | -.340* | .021 | .457** | — | | | 491 | 4.018 | 2.299 |
| 9. Prioritizing Pandemic | .562** | -.094 | .161** | -.134** | .445** | .215** | -.276** | -.219** | — | | 491 | 1.050 | 2.008 |
| 10. Religious Restriction | .225** | -.106* | .222** | -.194** | .310** | .205** | -.231** | -.212** | .199** | — | 491 | 0.550 | 2.211 |

[a] Effect sizes with Gender are point bi-serial correlation coefficients. All other effects sizes are Pearson's r correlation coefficients.
† < .10. * p < .05. ** p < .01.

(Social, Psychological, and Economic) and Pandemic Concern. There was a small-to-moderate significant relationship between all three well-being indices and Trust in Media. Overall, Republicans who reported greater life satisfaction, greater trust in news media, less concern about paying bills, and less social disconnectedness also reported greater concern about the spread of COVID-19. As a result of these analyses, all three well-being indices were included in the planned mediation analyses among Republicans.

**Democrats:** The descriptive and correlational statistics discussed below among Democrats are presented in Table 7.2. There was a small significant relationship between Pandemic Concern and Trust in Media. None of the well-being indices were significantly related to Pandemic Concern. There was a small significant relationship between Trust in Media and Psychological Well-Being. Trust in Media was not related to Social Well-Being or Economic Well-Being. Overall, Democrats who reported greater life satisfaction and greater trust in news media also reported greater concern about the spread of COVID-19. As a result of these analyses, Social Well-Being and Economic Well-Being were not included in the planned mediation analyses among Democrats.

*Mediation Analyses*

The following mediation analyses were conducted using SPSS Amos Version 25.

**Republicans:** Economic Well-Being was a significant predictor of Trust in Media, = -.349, $SE$ = .043, $p < .01$, and Trust in Media was a significant predictor of Pandemic Concern, = .399, $SE$ = .044, $p < .01$. Economic Well-Being was a significant direct predictor of Pandemic Concern, = -.093, $SE$ = .046, $p$ = .041. The indirect effect of Economic Well-Being on Pandemic Concern was tested using a bootstrap estimation approach with 2,000 samples. The indirect effect was significant, consistent with partial mediation, = -.136, $SE$ = .024, $p < .01$, 95% CI [-.192, -.096]. That is, Republicans who reported lower concern about paying their bills (i.e., higher economic well-being) also reported trusting media less and were also less concerned about COVID-19. See Figure 7.2 for a visual depiction of these relationships.

Psychological Well-Being was a significant predictor of Trust in Media, = .407, $SE$ = .038, $p < .01$, and Trust in Media was a significant predictor of Pandemic Concern, = .428, $SE$ = .041, $p < .01$. Psychological Well-Being was not a significant direct predictor of Pandemic Concern, = .006, $SE$ = .044, $p$ = .888. The indirect effect of Psychological Well-Being on Pandemic Concern was tested using a bootstrap estimation approach with 2,000 samples. The indirect effect was significant, consistent with full mediation, = .174, $SE$ = .024, $p < .01$, 95% CI [.134, .229]. That is, Republicans who reported greater life satisfaction were more likely to express concern about the pandemic when they also reported greater trust in media. See Figure 7.2 for a visual depiction of these relationships.

Table 7.2. Correlations, means, and standard deviations: self-reported Democrats.

| Variable | 1 | 2 | 3 | 4 | 5 | 6 | 7 | 8 | 9 | 10 | n | M | SD |
|---|---|---|---|---|---|---|---|---|---|---|---|---|---|
| 1. Pandemic Concern | — | | | | | | | | | | 587 | 2.560 | 0.718 |
| 2. Age in Years | .108** | — | | | | | | | | | 587 | 45.370 | 17.471 |
| 3. Household Income | .011 | −.049 | — | | | | | | | | 587 | 3.070 | 1.716 |
| 4. Gender[a] | .111** | −.035 | −.296** | — | | | | | | | 584 | — | — |
| 5. Trust in Media | .258** | .263** | .204** | −.120** | — | | | | | | 587 | 29.254 | 12.930 |
| 6. Psychological Well-being | .002 | .067 | .223** | −.121** | .284** | — | | | | | 587 | 6.760 | 2.375 |
| 7. Social Well-being | .068† | .230** | .066 | .023 | .036 | .298** | — | | | | 587 | 2.994 | 1.452 |
| 8. Economic Well-being | .059 | .382** | .079† | −.026 | .045 | .177** | .365** | — | | | 587 | 3.961 | 2.167 |
| 9. Prioritizing Pandemic | .305** | .213** | −.057 | .094* | .233** | .124** | .050 | .112** | — | | 587 | 1.750 | 1.572 |
| 10. Religious Restriction | .026 | −.121** | .231** | −.152** | .164** | .241** | −.059 | −.147** | −.040 | — | 587 | −0.030 | 2.264 |

[a] Effect sizes with Gender are point bi-serial correlation coefficients. All other effects sizes are Pearson's r correlation coefficients.
† < .10. * p < .05. ** p < .01.

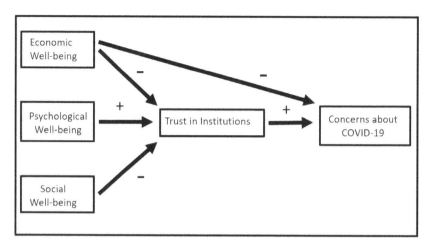

Figure 7.2. Visual depiction of the pared-down significant mediation effects among Republicans.

Social Well-Being was a significant predictor of Trust in Media, = -.328, $SE = .044$, $p < .01$, and Trust in Media was a significant predictor of Pandemic Concern, = .434, $SE = .043$, $p < .01$. Social Well-Being was not a significant direct predictor of Pandemic Concern, = .011, $SE = .043$, $p = .786$. The indirect effect of Social Well-Being on Pandemic Concern was tested using a bootstrap estimation approach with 2,000 samples. The indirect effect was significant, consistent with full mediation, = -.141, SE = .025, p < .01, 95% CI [-.197, -.099]. That is, Republicans who reportedly felt less social isolation (i.e., higher social well-being) expressed more concern about the pandemic when they also reported greater trust in news media. See Figure 7.2 for a visual depiction of these relationships.

**Democrats:** Psychological Well-Being was a significant predictor of Trust in Media, = .284, $SE = .040$, $p < .01$, and Trust in Media was a significant predictor of Pandemic Concern, = .279, $SE = .040$, $p < .01$. Psychological Well-Being was not a significant direct predictor of Pandemic Concern, = -.077, $SE = .040$, $p = .061$. The indirect effect of Psychological Well-Being on Pandemic Concern was tested using a bootstrap estimation approach with 2,000 samples. The indirect effect was significant, consistent with full mediation, = .079, SE = .017, p < .01, 95% CI [.049, .115]. That is, Democrats reporting higher levels of life satisfaction, and who also reported higher levels of trust in news media, were more likely than other Democrats to express concern about the pandemic. Figure 7.3 visualizes these relationships.

Figure 7.3. Visual depiction of the pared-down significant mediation effect among Democrats.

*Political Affiliation and Concern about the Pandemic*

In order to contextualize the differences in moderated mediation for Republicans and Democrats, we ran a series of Independent-Samples t-tests. Please revisit Table 7.1 and Table 7.2 for the means and standard deviations that correspond to the variables and groups discussed below. Democrats were more likely than Republicans to report that people in their community should be "very concerned" about the spread of coronavirus, $t(895.9) = 7.866,^{1}$ $p < .001$. The difference between Democrats and Republicans was small-to-moderate ($d = .493$). Democrats reported greater trust in news media, compared to their Republican counterparts, $t(926.86) = 4.215^{1}$, $p < .001$. The difference between Democrats and Republicans was small ($d = .263$). Democrats reported prioritizing "reducing the spread of coronavirus" over "reducing the unemployment rate," more than did Republicans, who still prioritized reducing the spread of coronavirus, but not to the same extent, $t(919.32) = 6.217^{1}$, $p < .001$. The difference between Democrats and Republicans was small-to-moderate ($d = .388$). Republicans were more likely than Democrats to report that the pandemic had restricted their ability to engage in the religious practices they want to engage in, $t(1076) = -4.195$, $p < .001$. The difference between Republicans and Democrats was small ($d = -.257$).

In summary, relative to their Republican counterparts, Democrats in this sample expressed greater concern about the spread of coronavirus generally, and in light of the unemployment rate specifically. Democrats also reported greater trust in news media and a neutral perspective on whether the pandemic has restricted their ability to engage in their religious practices.

## Discussion

In this study we measured three types of well-being (social, psychological, financial), as well as attitudes related to the pandemic, among 587 Democrats and 491 Republicans. We assessed whether well-being, concern about the pandemic, and trust in media were related to each other. Further, we assessed whether those relationships were the same for members of both major political parties. Thus, we tested the hypothesized relations between trust, well-being, and pandemic concern separately for our sample of self-reported Republicans and self-reported Democrats.

Our study revealed differences between Democrats and Republicans in trust and concern about COVID-19. It also revealed that trust in news media was the strongest predictor of COVID-19 concern and that trust mediated many relations between well-being and COVID-19 concern. The data from this study also suggests that economically and socially vulnerable Republicans are more concerned about the pandemic than "elite" Republicans. In other words, this study provided support for existing work, but also pointed us in some new and interesting directions discussed below.

### Democrats Are More Concerned about COVID-19 and More Trusting of News Media

First, the hypothesized mean level differences between Democrats and Republicans were supported, but the differences were small. Democrats reported greater concern with reducing the spread of the coronavirus than did Republicans. This difference held even in our tradeoff question, that is, Democrats reported greater concern for reducing the spread of COVID-19, even at the expense of potentially increasing the unemployment rate. Importantly, members of both political parties expressed concern about reducing the spread of coronavirus; the difference between these groups was therefore a difference in magnitude, not kind.

Democrats also reported slightly greater trust in news media, though, notably, the correlation between trust in news media and coronavirus concern was stronger for Republicans than it was for Democrats (despite being significant for both). However, recall that Republicans in our sample reported both lower average trust in media and lower average pandemic concern compared to Democrats. Unfortunately, this analysis could not determine with certainty that the news coverage people watched was focused on COVID-19, though given news media saturation about outbreaks, vaccine developments, rising case totals, the president's high-profile COVID-19 diagnosis, and the like, it is not surprising that we found consistent relationships between trust in news media and concern about COVID-19.

### Trust in News Media Was Strongly Related to Pandemic Concern

In general, we found that among our sample of Republicans, concern about the COVID-19 pandemic was most strongly, and positively, correlated with trust in media, household income, and a sense that one's ability to practise religion was being restricted. Also, among Republicans, we found that a willingness to prioritize the pandemic over the economy was correlated positively, and strongly, with trust in media and negatively with social connectedness. Trust in news media covaried the most with pandemic concern; the greater the trust in news media, the greater the concern with the spread of coronavirus. However, social well-being and financial well-being were both negatively related to trust in news media, indicating that Republicans who were more financially and socially secure were less trusting of news media. These findings add to the existing research literature in clarifying the importance of media in influencing people's (reported) perceptions. These findings also suggest a new avenue, in particular, that support for quarantine policies, such as those that could lead to unemployment, might be resisted to a greater extent by more socially connected or financially secure Republicans.

Among our sample of Democrats, we found that concern about the COVID-19 pandemic was most strongly, and positively, correlated with trust in news media, gender (i.e., being female), and age (i.e., being older). Interestingly, gender was also predictive of concern amongst Republicans, but for them it was men who reported being more concerned with COVID-19's spread in the community. With regard to our question about prioritizing reducing the spread of COVID-19 at the expense of potential unemployment, age emerged as a particularly strong effect for Democrats, relative to Republicans. Among Democrats, those participants who reported being older, female, and having greater psychological well-being were more willing to prioritize reducing the spread of coronavirus over reducing unemployment. Similar to Republicans, though, trust in news media correlated the strongest with concern over the spread of coronavirus.

*Economically and Socially Vulnerable Republicans Are More Concerned*

Our mediation tests revealed a different set of patterns for Republicans and Democrats. First, we did indeed find that psychological well-being (i.e., satisfaction with life) was related to pandemic concern for both Democrats and Republicans, though it was fully mediated by greater trust in, and viewership of, news media. In other words, individuals who reported greater life satisfaction also reported greater concern about the spread of coronavirus when they reported greater trust in news media. However, this was the only well-being measure that was significant among Democrats. Among Democrats, financial well-being and social well-being were not significantly related, directly, or indirectly, to COVID-19 concern in our sample.

Where things become interesting and unexpected is among Republicans. Republicans revealed a more complex relationship regarding the trust and well-being variables. Unlike the sample of Democrats, both economic well-being and social well-being were directly related to trust in news media and indirectly related to concern about the spread of COVID-19 among Republicans.

These findings were somewhat unexpected: among Republicans, those who were more financially secure and socially connected trusted media less and expressed lower concern about the spread of COVID-19. How does one interpret this? Well, from a tribal partisanship perspective, we may speculate that Republicans would be more motivated than Democrats, on average, to see COVID-19 as minor, insignificant, or well controlled. Our findings suggest that this relationship might be especially strong for high-status (i.e., financially comfortable and socially well connected) Republicans. These individuals may be especially motivated to see COVID-19 as insignificant or well controlled because they, and much news media, equate a critique of COVID-19's effects with critiques of President Trump (and thus, of them). Possibly as a result of that, such individuals have lower trust in news media and less concern for the spread of COVID-19.

*Study Limitations*

Perhaps the biggest strength of this study is that we have examined self-reported Democrats and self-reported Republicans separately, instead of merely treating "political affiliation" as an independent variable in a regression model. This has allowed us to peer into some of the relevant financial, psychological, social, and institutional trust (in this case, news media) variables which might plausibly be related to COVID-19 concern. Yet, as with any study, there are limitations. For one, we do not have a measure of affiliation strength. We do not doubt that some of those reporting themselves to be "Democrats" are more committed to their political affiliation than others. We can say the same about Republicans. It is possible that identity salience or affiliative strength could add important texture to our results here.

Another limitation is that we cannot know how much COVID-19 news coverage individuals in our sample have been exposed to. This is especially important given our target variable of trust in news media, which is measured as a composite variable including frequency of viewership and trustworthiness of news media "in general." While satisfyingly internally valid (i.e., we are confident that people who watch more news and have more general trust in news can be said to have greater trust in news media), we cannot be certain that our trust in media variable is externally valid without knowing something about the content of the coverage for the motivation behind consumption of news media. Equally important, we cannot be sure if our respondents spent their time watching Fox News, CNN, or neither; all we have is an aggregate measure of news-watching frequency and trust in news media in general. Our measure is therefore cautiously broad and global, with all of the attendant strengths (generalizability and applicability) and weaknesses (specificity, context).

Other limitations might be pointed out – we could have, for example, examined other political affiliations outside of the mainstream, or other institutions in which people put their trust (i.e., medicine or government). We focused here on Democrats and Republicans because of the current discourse that points out their political polarization, and because much existing work uses these categories. As for examining other institutional trust measures, we are pursuing this in follow-up papers. This is also, of course, an area quite open for other investigators.

## Conclusion

Where does our study leave the discussion of social, psychological, and financial well-being as it relates to pandemic concern? Our sample can only tell us some things about Americans, and we focused particularly on differences within political groups. Yet, the cautious extrapolations that can be drawn from our findings are nevertheless suggestive not only that Democrats and Republicans are generally concerned about COVID-19, but also that the more socially well connected and financially secure Republicans (i.e., the least socially and financially vulnerable) are most likely to report lower concern about the spread of COVID-19 in their communities. By socially well connected and financially secure, we do not mean celebrities or very wealthy CEOs. Our sample was of normal, everyday Americans. Thus, what our results suggest is that there is perhaps a more subtle sociological status effect driving attitudes towards COVID-19 amongst Republicans. Republicans who, just relative to their peers, feel less socially isolated and concerned about paying their bills are those similar Republicans who are less likely to take pandemic-related regulation seriously.

Taken together, we can carefully extrapolate from the study the following: First, that when Republicans become more similar to Democrats in their COVID-related attitudes, the less financially or socially secure they feel. Second, most partisans, across parties, are concerned with COVID-19. Third, the "tribal partisanship" model discussed above may need to take account of within-group status hierarchy effects. Republicans are not only defining themselves in opposition to (perceived) Democrat positions, but they are also likely magnifying group-model positions differentially *within* their own group. Future work might assess the degree to which a common group belief (e.g., "Donald Trump handled the pandemic well and COVID-19 is not a concern for our society") might be particularly amplified by sociologically high-status in-group members.

Finally, and obviously, the COVID-19 pandemic, damaging as it is to life and to economies, is nothing like the lethal Black Death and no doubt other cases of plague throughout history. In rich democracies today, people have access to health care that others in the rest of the world, and certainly those in 14th-century Europe, did not. Plagues around the world are not and have never been equally damaging or equally treatable. But we can be reasonably certain that there are variables that matter now, mattered then, and will matter to those studying pandemics in the future. These variables include people's social connectedness, their ability to support themselves financially, and their life satisfaction. Where they get their news about the wider world will also, always, be important (even if our measure here is, and it is, imperfect). Though our analysis is tuned to the present politicization of public health, we encourage others to operationalize and analyse these perennially important variables in future work.

NOTE

1  For these Independent-Samples t-tests, the Levene's Test for Equality of Variances was significant. As a result, we reported the t statistic for "equal variances not assumed."

REFERENCES

Adolph, C., Amano, K., Bang-Jensen, B., Fullman, N., Magistro, B., Reinke, G., & Wilkerson, J. (2020, August 31). Governor partisanship explains the adoption of statewide mandates to wear face coverings. *medRxiv Preprint Server for Health Sciences*. https://www.medrxiv.org/content/10.1101/2020.08.31.20185371v1.full.pdf.

Alessandri, G., Filosa, L., Tisak, M.S., Crocetti, E., Crea, G., & Avanzi, L. (2020, August 27). Moral disengagement and generalized social trust as mediators and moderators of rule-respecting behaviors during the COVID-19 outbreak. *Frontiers in Psychology*, *11*, 2102. https://doi.org/10.3389/fpsyg.2020.02102.

Alzueta, E., Perrin, P., Baker, F.C., Caffarra, S., Ramos-Usuga, D., Yuksel, D., & Arango-Lasprilla, J.C. (2020, October 31). How the COVID-19 pandemic has changed our lives: A study of psychological correlates across 59 countries. *Journal of Clinical Psychology*. https://doi.org/10.1002/jclp.23082.

Ammar, A., Brach, M., Trabelsi, K., Chtourou, H., Boukhris, O., Masmoudi, L., Bouaziz, B., Bentlage, E., How, D., Ahmed, M., Müller, P., Müller, N., Aloui, A., Hammouda, O., Paineirias-Domingos, L.L., Braakman-Jansen, A., Wrede, C., Bastoni, S., ... Hoekelmann, A. (2020, May 8). Effects of COVID-19 home confinement on physical activity and eating behaviour: Preliminary results of the ECLB-COVID19 International Online-Survey. *medRxiv Preprint Server for Health Sciences*. https://doi.org/10.1101/2020.05.04.20072447.

Bækgaard, M., Christensen, J., Madsen, J.K., & Mikkelsen, K.S. (2020, July 12). Rallying around the flag in times of COVID-19: Societal lockdown and trust in democratic institutions. *Journal of Behavioral Public Administration*, *3*(2), 1–12. https://doi.org/10.30636/jbpa.32.172.

Bureau of Labor Statistics. 2019. American time use survey summary. Retrieved 27 December 2020, from https://www.bls.gov/news.release/atus.nr0.htm.

Bureau of Labor Statistics. 2020. Ability to work from home: Evidence from two surveys and implications for the labor market in the COVID-19 pandemic. Retrieved 27 December 2020, from https://www.bls.gov/opub/mlr/2020/article/ability-to-work-from-home.htm.

Capraro, V., & Barcelo, H. (2020). The effect of messaging and gender on intentions to wear a face covering to slow down COVID-19 transmission. *arXiv preprint*. https://arxiv.org/pdf/2005.05467.pdf.

Cates, J., Lucero-Obusan, C., Dahl, R.M., Schirmer, P., Garg, S., Oda, G., Hall, A.J., Langley, G., Havers, F.P., Holodniy, M., & Cardemil, C.V. (2020, October 23). Risk for in-hospital complications associated with COVID-19 and influenza – Veterans Health Administration, United States, October 1, 2018–May 31, 2020. *Morbidity and Mortality Weekly Report*, *69*(42), 1528–34. https://doi.org/10.15585/mmwr.mm6942e3.

Centers for Disease Control and Prevention (CDC). (2020). COVID-19 travel recommendations by destination. Retrieved 27 December 2020, from https://www.cdc.gov/coronavirus/2019-ncov/travelers/map-and-travel-notices.html.

Clark, C.J., Davila, A., Regis, M., & Kraus, S. (2020). Predictors of COVID-19 voluntary compliance behaviors: An international investigation. *Global Transitions*, *2*(2020), 76–82. https://doi.org/10.1016/j.glt.2020.06.003.

Clark, C.J., Liu, B.S., Winegard, B.M., & Ditto, P.H. (2019, August). Tribalism is human nature. *Current Directions in Psychological Science*, *28*(6), 587–92. https://doi.org/10.1177/0963721419862289.

Collins, R.N., Mandel, D.R., & Schywiola, S.S. (2020, September 17). Political identity over personal impact: Early US reactions to the COVID-19 pandemic. *PsyArXiv Preprints*. https://psyarxiv.com/jeq6y/.

de Bruin, W.B., Saw, H.W., & Goldman, D.P. 2020. Political polarization in US residents' COVID-19 risk perceptions, policy preferences, and protective behaviors. *Journal of Risk and Uncertainty*, *61*(2), 177–94. https://doi.org/10.1007/s11166-020-09336-3.

Dhami, M.K., Weiss-Cohen, L., & Ayton, P. (2020, November 19). Are people experiencing the "pains of imprisonment" during the COVID-19 lockdown?. *Frontiers in Psychology*, *11*(578430). https://doi.org/10.3389/fpsyg.2020.578430.

Dhanani, L.Y., & Franz, B. 2020. The role of news consumption and trust in public health leadership in shaping COVID-19 knowledge and prejudice. *Frontiers in Psychology*, *11*(560828). https://doi.org/10.3389/fpsyg.2020.560828.

Dingel, J.I., & Neiman, B. (2020, April). How many jobs can be done at home? National Bureau of Economic Research. Retrieved 27 December 2020, from https://www.nber.org/papers/w26948.

Doherty, C. (2014, June 12). 7 things to know about polarization in America. Pew Research Center. https://www.pewresearch.org/fact-tank/2014/06/12/7-things-to-know-about-polarization-in-america/.

Elmer, T., Mepham, K., & Stadtfeld, C. (2020, July 23). Students under lockdown: Comparisons of students' social networks and mental health before and during the COVID-19 crisis in Switzerland. *PsyArXiv Preprints*. https://psyarxiv.com/ua6tq/.

Evans, J.H., & Hargittai, E. (2020). Who doesn't trust Fauci? The public's belief in the expertise and shared values of scientists in the COVID-19 pandemic. *Socius: Sociological Research for a Dynamic World*, *6*, 1–13. https://doi.org/10.1177/2378023120947337.

Fernandes, N. (2020, March 23). Economic effects of coronavirus outbreak (COVID-19) on the world economy. IESE Business School Working Paper No. WP-1240-E.

Freira, L., Sartorio, M., Boruchowicz, C., Boo, F.L., & Navajas, J. (2020, October 28). The irrational interplay between partisanship, beliefs about the severity of the COVID-19 pandemic, and support for policy interventions. *PsyArXiv Preprints*. https://psyarxiv.com/4cgfw/.

Graso, M., Chen, F.X., & Reynolds, T. (2020, March). Moralization of COVID-19 health response: Asymmetry in tolerance for human costs. *Journal of Experimental Social Psychology*, *93*(2021), 1–12. https://doi.org/10.1016/j.jesp.2020.104084.

Hare, C., & Poole, K.T. (2014). The polarization of contemporary American politics. *Polity*, *46*(3), 411–29. https://doi.org/10.1057/pol.2014.10.

Hartney, M.T., & Finger, L.K. (2020). Politics, markets, and pandemics: Public education's response to COVID-19. EdWorkingPaper No. 20-304. Retrieved 27 December 2020, from https://edworkingpapers.com/sites/default/files/ai20-304.pdf.

Helliwell, J.F., & Wang, S. (2010, April 2010). Trust and well-being. *International Journal of Wellbeing*, *1*(1), 42–78.

John, A., Pirkis, J., Gunnell, D., Appleby, L., & Morrissey, J. (2020, November 12). Trends in suicide during the COVID-19 pandemic. *BMJ*, *2020*:371:m4352. https://www.bmj.com/content/371/bmj.m4352. https://doi.org/10.1136/bmj.m4352.

Kelley, J., Evans, M.D.R., & Kelley, S. (2020, October 27). Let's spend the night together: A challenge for medically optimal coronavirus social distancing policies. *SocArXiv Papers*. https://osf.io/preprints/socarxiv/tsqyb/.

Kelly, B.D. (2020, December 7). Quarantine, restrictions and mental health in the COVID-19 pandemic. *QJM: An International Journal of Medicine*, *114*(2), 93–4. https://doi.org/10.1093/qjmed/hcaa322.

Leary, A., Dvorak, R., & De Leon, A., Peterson, R., & Troop-Gordon, W. (2020, May 13). COVID-19 social distancing. *PsyArXiv Preprints*. https://psyarxiv.com/mszw2.

Mangrum, D., & Niekamp, P. (2020, May 21). College student contribution to local COVID-19 spread: Evidence from university spring break timing. *Journal of Urban Economics*. https://papers.ssrn.com/sol3/Papers.cfm?abstract_id=3606811. https://doi.org/10.2139/ssrn.3606811.

McKinley, K. (2020, April 16). How the rich reacted to the bubonic plague has eerie similarities to today's pandemic. *The Conversation*. https://theconversation.com/how-the-rich-reacted-to-the-bubonic-plague-has-eerie-similarities-to-todays-pandemic-135925.

*New York Times*. (2020a). Coronavirus restrictions and mask mandates for all 50 states. Retrieved 28 December 2020, from https://www.nytimes.com/interactive/2020/us/states-reopen-map-coronavirus.html.

*New York Times*. (2020b). Coronavirus world map: Tracking the global outbreak. Retrieved 28 December 2020, from https://www.nytimes.com/interactive/2020/world/coronavirus-maps.html.

Ogden, R.S. (2020, July 6). The passage of time during the UK COVID-19 lockdown. *Plos One*, *15*(7), 1–16. https://doi.org/10.1371/journal.pone.0235871.

Perry, S.L., Whitehead, A.L., & Grubbs, J.B. (2020). Culture wars and COVID-19 conduct: Christian nationalism, religiosity, and Americans' behavior during the coronavirus pandemic. *Journal for the Scientific Study of Religion*, *59*(3), 405–16. https://doi.org/10.1111/jssr.12677.

Pew Research Center. (2020a, March 30). Most Americans say coronavirus outbreak has impacted their lives. https://www.pewsocialtrends.org/2020/03/30/most-americans-say-coronavirus-outbreak-has-impacted-their-lives/.

Pew Research Center. (2020b, April 21). About half of lower-income Americans report household job or wage loss due to COVID-19. Retrieved 27 December 2020, from https://www.pewsocialtrends.org/2020/04/21/about-half-of-lower-income-americans-report-household-job-or-wage-loss-due-to-covid-19/.

Pew Research Center. (2020c, September 24). Economic fallout from COVID-19 continues to hit lower-income Americans the hardest. https://www.pewsocialtrends.org/2020/09/24/economic-fallout-from-covid-19-continues-to-hit-lower-income-americans-the-hardest/.

Pickup, M., Stecula, D., & van der Linden, C. (2020). Novel coronavirus, old partisanship: COVID-19 attitudes and behaviors in the United States and Canada. *Canadian Journal of Political Science*, *57*, 357–64. https://doi.org/10.1017/S0008423920000463.

Plater, R. (2020, December 3). Indoor church services are COVID-19 hot spots: Here's why. *Healthline*. https://www.healthline.com/health-news/indoor-church-services-are-covid-19-hot-spots-heres-why.

Reisch, T., Heiler, G., Hurt, J., Klimek, P., Hanbury, A., & Thurner, S. (2020, October 20). Behavioral gender differences are reinforced during the COVID-19 crisis. *ArXiv* https://arxiv.org/abs/2010.10470.

Rück, C., Mataix-Cols, D., Malki, K., Adler, M., Flygare, O., Runeson, B., & Sidorchuk, A. (2020, December 11). Will the COVID-19 pandemic lead to a tsunami of suicides? A Swedish nationwide analysis of historical and 2020 data. *MedRxiv Preprint Server for Health Sciences*. https://doi.org/10.1101/2020.12.10.20244699.

Samore, T., Fessler, D.M.T., Sparks, A.M., & Holbrook, C. (2020). Of pathogens and party lines: Social conservatism positively associates with COVID-19 precautions among Democrats but not Republicans. *PsyArxiv Preprints*. https://psyarxiv.com/9zsvb/download?format=pdf.

Sampson, R.J., Morenoff, J.D., & Earls, F. (1999). Beyond social capital: Spatial dynamics of collective efficacy for children. *American Sociological Review*, *64*(5), 633–60. https://doi.org/10.2307/2657367.

Sibley, C.G., Greaves, L.M., Satherley, N., Wilson, M.S., Overall, N.C., Lee, C.H.J., Milojev, P., Bulbulia, J., Osborne, D., Milfont, T.L., Houkamau, C.A., Duck, I.M., Vickers-Jones, R., Barlow, F.K. (2020). Effects of the COVID-19 pandemic and nationwide lockdown on trust, attitudes toward government, and well-being. *American Psychologist*, *75*(5), 618–30. https://doi.org/10.1037/amp0000662.

Simione, L., & Gnagnarella, C. (2020). Differences between health workers and general population in risk perception, behaviors, and psychological distress related to COVID-19 spread in Italy. *PsyArxiv Preprints*. Retrieved 11 November 2020, from https://psyarxiv.com/84d2c.

Stump, S. (2020, June 25). 18 family members diagnosed with coronavirus after surprise party in Texas. *Today*. https://www.today.com/health/18-family-members-diagnosed-coronavirus-after-surprise-birthday-party-t185221.

Valentine, R., Valentine, D., & Valentine, J.L. (2020). Relationship of George Floyd protests to increases in COVID-19 cases using event study methodology. *Journal of Public Health*, *42*(4), 696–7. https://doi.org/10.1093/pubmed/fdaa127.

VanderWeele, T.J., Trudel-Fitzgerald, C., Allin, P., Farrelly, C., Fletcher, G., Frederick, D.E., Hall, J., Helliwell, J.F., Kim, E.S., Lauinger, W.A., Lee, M.T., Lyubomirsky, S., Margolis, S., McNeely, E., Messer, N. Tay, L., Visawanth, V., Węziak-Białowolska, D., Kubzansky, L D. (2020). Current recommendations on the selection of measures for well-being. *Preventive Medicine*, *133*, 1–6. https://doi.org/10.1016/j.ypmed.2020.106004.

Walker, M. (2020, August 17). Sturgis Motorcycle Rally rolls to a close as virus tracking remains complex. *New York Times*. https://www.nytimes.com/2020/08/17/us/sturgis-motorcycle-rally-coronavirus.html.

Wongpakaran, N., Wongpakaran, T., Pinyopornpanish, M., Simcharoen, S., Suradom, C., Varnado, P., Kuntawong, P. 2020. Development and validation of a 6-item Revised UCLA Loneliness Scale (RULS-6) using Rasch analysis. *British Journal of Health Psychology*, *25*(2), 233–56. https://doi.org/10.1111/bjhp.12404.

Zhao, E., Wu, Q., Crimmins, E.M., & Ailshire, J.A. (2020). Media trust and infection mitigating behaviours during the COVID-19 pandemic in the USA. *BMJ Global Health*, *5*(10), 1–10. https://doi.org/10.1136/bmjgh-2020-003323.

# 8 Unsettling Contact: The Collapse of Emotional Distance at a COVID-19 Medical Frontline

BY JUNBIN TAN AND PHU TRAN

Ms. A, a patient in her 30s, heaved deep breaths while lying in a half-prone position. The ward's nurse, busy monitoring another patient's vitals, turned to Dr. Tran, the anesthesiologist on duty, as he neared Ms. A's bed. Ms. A's oxygen saturation level had fallen dangerously low, and this prompted the nurse to alert Dr. Tran. Ms. A did not exhibit the loud wheezing that most doctors and nurses had associated in the earliest stages of the COVID-19 pandemic with the sickest patients. Now, in April 2020, after weeks caring for COVID-19 patients, medical workers realized that the COVID-19 patients' illness trajectories were far from straightforward. Thorough examinations were still needed to ascertain how each patient fared but were extremely difficult in times shaped by the pandemic where health personnel and resources were stretched thin.

After examining Ms. A's vitals, Dr. Tran concurred with the nurse's request for her to be intubated. After he explained the process to Ms. A and obtained her verbal consent (difficult as she was oxygen-starved and exhibited delirium), she had to be sedated before the intubation. Dr. Tran then proceeded to insert plastic tubing into her mouth. Opening her mouth and bending forward so that he was mere inches from her, Dr. Tran employed a laryngoscope to peer into her throat before manoeuvring the tube past her vocal cords and into the deeper parts of her airway. Such proximity to Ms. A placed him at greater risk of being infected even if he had donned his personal protective equipment (PPE). One's sense of personal safety, however, was a lesser preoccupation when faced with patients who were battling death. Dr. Tran listened through a stethoscope to ensure that the tube was placed correctly before giving a signal for it to be linked to a ventilator.

Macy, a respiratory therapist, was waiting at Ms. A's bedside with a portable ventilator – a machine that uses pressure to deliver oxygen-saturated air into patients' respiratory systems. Macy connected the breathing tube to the ventilator, tried turning the ventilator on, and discovered that it had malfunctioned. Did Ms. A receive the tube the wrong way, or was the ventilator faulty?

And if the ventilator was faulty, what could they do to fix it? A minute passed and a second minute came to an end as they struggled to render the ventilator operative. During this period, Ms. A's oxygen saturation kept on dipping. The ventilator, they decided, had to be placed aside. They hurriedly took Ms. A off the ventilator and placed her on an anti-valve. Macy hooked up the anti-valve and tried "breathing" Ms. A by hand to increase her oxygen saturation. Her vitals had fallen critically low. Both did whatever else they could, but the time that had elapsed proved to be lethal.

"That girl, she can't handle this whole time without oxygen, and she died. So, they die," Dr. Tran said over Zoom as he narrated this incident, which had occurred the previous week at the time of this writing in late fall 2020. Macy, he said, was extremely shocked by the sudden death, "Before, you see a healthy person, you talk to them for 10 minutes," Dr. Tran continued, "And now, they're dead. And it's probably because of something you did. Macy thinks she's responsible. The respiratory therapist brings the ventilators around and takes care of them." There was a long, uncomfortable pause, and Dr. Tran continued: "And Macy cried. We took 10 or 15 minutes to tell her it's not her fault. The machines weren't in good condition. The room wasn't one that had a good enough power supply. These were beyond one's control. And we did the best we could. Macy must be overwhelmed." "And you too," I responded as best I could, as I tried to grasp their feelings of being at the COVID-19 medical frontline.

When the pandemic first peaked in New York City in April and May 2020, medical workers braced themselves as they attended to wave after wave of patients. All of New York City's hospitals were converted for COVID-19 treatment. Only the most severely ill patients with respiratory and associated complications were admitted. Many required intensive care. They depended on life-saving technologies that were now in short supply. Ms. A's demise and Macy's emotional response occurred within this context. This incident, Dr. Tran noted, was one of many that occurred in April and May, and it was unusual, as medical workers were typically in control of their emotions. However, medical workers wrestled with treating patients amidst uncertainties about how long the pandemic would last. Indeed, when this chapter was submitted in November 2020, New York City was facing a resurgence of COVID-19 infections, rising numbers of patients and deaths, and impending shortages in human personnel and medical supplies, along with increasing risks of residents being infected. This chapter addresses medical workers' emotionally unsettling encounters and breakdowns, which we analyse against a medical institution that was torn asunder during the spring 2020 emergence of COVID-19 in many parts of the United States. In such a context, we hope to frame how health professionals and workers, medical supplies, technologies, and normative expectations floundered during the pandemic.

What do these unsettling encounters indicate about changing social relations, of differing valences of closeness and distance, and engagements and vulnerabilities among medical workers and between them and their patients? And how did they respond to them? Additionally, we also ask: Under what institutional conditions and processes did these deeply emotional encounters develop? To address these questions, we turn to Emmanuel Levinas's *Totality and Infinity* (1969) and *Otherwise Than Being* (1981) and Maurice Merleau-Ponty's *The Phenomenology of Perception* (2012) and *The World of Perception* (2004) so as to reflect on the relevance of their writings as much as build upon their frameworks.

We tend to think of medical institutions as structured, so much so that works on emotions in hospitals typically highlight their institutionalized character and the emotional labour involved in maintaining these norms – the works of Arlie Hochschild (2003) are instructive here. Levinas's works, conversely, provide a language for us to deliberate on the "openness" of the Self (Macy, in this case) when the Self is confronted with the suffering and death of an "other" (Ms. A). Merleau-Ponty allows us to analyse the affectivity of our "world," or the environment in which we live (the hospital amidst the pandemic), on our embodied selves, on our perceptions and interactions. Unsettling contact and emotional breakdown in such contexts, while disruptive to medical workers' professional responsibilities, are also moments for bridging social distances. In this manner, medical workers are able to express their felt responsibilities to the lives around them (see also Levinas 1969, 1981). Such contexts thus offer moments of apprehending and cognizing "broader" disruptions to health, life, and familiar environments (see also Merleau-Ponty, 2012).

Research for this chapter proceeded through a series of Zoom conversations and interviews between your author, an anthropologist, and coauthor and anesthesiologist Dr. Tran, based at New York-Presbyterian Queens Hospital in Flushing, New York, during the period between May and July 2020. During our Zoom sessions, we discussed Dr. Tran's experiences and observations at the hospital. The discussions were conducted on his rest days during the spring 2020 height of the pandemic and in subsequent months when the pandemic eased, and patient numbers edged lower for the summer season before its later fall resurgence. While Dr. Tran works and lives in New York City, I had returned to Singapore to be with family when infection numbers had started to increase for the first time in New York City. This project began with my concern about what was happening in New York City after watching news on the outbreak, the lives that it claimed, and the medical, community, and state efforts that surfaced to mitigate the spread of infections. Our conversations led us to realize that an account could be written to document and reflect upon encounters at the medical frontline. Dr. Tran's experiences and observations are corroborated with media sources, news articles, and other scholars' narrations, which we both employed to flesh out encounters at such medical frontlines.

The insights and analyses in our essay, while largely anthropological, would have been impossible were it not for Dr. Tran's experiences and observations. Compared to conventional anthropological writings based on ethnographic field work, this chapter relies mostly on one interlocutor's narration and perspective and could therefore be said to be partly autobiographical and partly interlocutionary or interview based. As fieldwork is neither possible nor desirable during the pandemic, we had, like other socially oriented researchers, resorted to working through media platforms, the affordances and negative implications of which are discussed in detail elsewhere (Saxena & Johnson, 2020). Researching from afar requires, what Mead and Métraux (1953, p. 11) called greater "reconstructive imagination" on the part of the anthropologist, which is "the ability to recreate in the absence of either the once experienced or of the never fully experienced living scene," since in-person fieldwork is not possible under pandemic restrictions. As such, despite its departure from anthropology's disciplinary norms, our chapter demonstrates how collaboration with interlocutors – and recognition of their effort through coauthorship, as Dunia et al. (2020) remarked in their critique of "facilitating" and "contracting" researcher relationships – further enhanced by video and conferencing platforms (e.g., Zoom), makes anthropological research possible in pandemic times and amidst distancing.

**The Hospital amidst the Pandemic**

Ms. A was among the tens of thousands of people hospitalized, and among the 600 to 700 people who passed away on a single day in early April 2020. The number of New Yorkers diagnosed with COVID-19 had exceeded 200,000 individuals by mid-April (*New York Times*, 2020; Rothfield et al., 2020). What happened to Ms. A and the medical workers who treated her, while astounding, was hardly exceptional, and is testament to the harshness of pandemic times: "It would be unique if it wasn't for COVID, but from March to May, this took place several times. We got a chain of emails on these cases," noted Dr. Tran.

New York City's hospitals, among the best equipped in the United States and the world, found themselves ill equipped for the pandemic. New York Presbyterian Queens, where Dr. Tran practises, like other New York City hospitals, converted all wards and available spaces for COVID-19 treatment. However, not all rooms had "negative pressure" – a ventilation system that allows air to flow in but not out, and which prevents room-to-room contamination. Divisions were created through curtains positioned between beds, while at some hospitals tarps were hung from ceilings. Operating theatres were made into intensive care unit (ICU) wards, and anesthesia machines were deployed as ventilators for patients with breathing difficulties. Even the cafeteria was cleared out to make space for 20 additional beds (Kaye, 2020). More beds, Dr. Tran said, would have been added if there had been more power outlets, which ventilators require, apart from those located at the café's perimeter (see also Choi & Velasquez, 2020). The demand for ventilators outran hospitals' supplies (Kliff et al., 2020). The state obtained portable ventilators from the federal stockpile, but some were found to be faulty and in poor condition (Dodge, 2020), and, as we saw in our opening scenario, led to severe consequences.

Doctors and nurses rely on telemetry units to monitor patients' oxygen levels and other vital signs. Vital signs would be projected onto a screen placed at a prominent location, which sounds an alarm when patients are in a critical state. Medical workers could perform their tasks and still be notified of medical emergencies. While rooms were modified, beds arranged, and portable ventilators brought in, telemetry units, which were required especially when staff was stretched thin, could not be added. The patients who were hospitalized were usually very sick, and the lack of telemetry units meant that they could not be adequately monitored. Some had succumbed to cardiac arrests with no chance of being resuscitated. New York City's medical infrastructure thus fell short during the pandemic, which limited medical workers' ability to provide care.

Changes of hospital space were matched with the mobilization of more medical workers, especially nurses, for the care of COVID-19 patients. Although they experienced greater work volumes, anesthesiologists and medical therapists were not given new roles, since focusing on intubating patients and caring for their respiratory condition were critical processes for COVID-19 treatment. Intubation cases increased three- to fourfold, to as many as 10 to 12 intubations per day, many of which were performed for severely ill patients. Dr. Tran notes:

> We rush between patients. As I'm doing an intubation, they would call me for another one. As an anesthesiology consultant, I don't usually have to attend to normal wards, but during COVID, every ward had sick people. We had a floor of not-too-sick patients, but one of those 20 patients might get really sick and need intubation. I was running up and down between different floors.

New York City, it was reported, was suffering from a shortage of respiratory specialists (Campanile, 2020; Choi & Velasquez, 2020). The shortage of staff, compounded by resource scarcity, had severe implications: "If there is a small number of COVID patients in the hospital, every intubation is going to be perfect. But when practitioners are tired and resources stretched thin, nurses and therapists cannot pay as much attention to the intubation process. Equipment may fail or be forgotten. More patients may die," Dr. Tran emphasized.

To address staff shortages, hospitals employed doctors and nurses from other parts of the United States who flew to New York City to take up required positions, many of whom were paid *per diem*, more than what they usually earn. Medical workers also left their families and sought employment at the New York City frontline because state policies that prohibited elective surgeries and converted hospitals for COVID-19 treatment left them without work and without pay (Ansari & Prang, 2020). While larger hospitals in bigger cities, such as Columbia University Hospital and New York Presbyterian, not only have financial reserves that can tide them over but also employ additional medical workers and provide bonuses to employees, the financial situations of smaller hospitals differ significantly. When elective surgery – for many hospitals, the main source of revenue – is curtailed, medical workers suffer pay cuts or do not get paid (Sanger-Katz, 2020; Abelson, 2020). Apart from medical workers who voluntarily travelled to New York City, these factors, relating to how the medical system interacts with state policies, also contributed to the movement of medical workers to cover the shortfall.

The movement of medical workers from other departments and wards, and the employment of medical workers from elsewhere, might have resolved the numbers issue, but problems relating to skills, training, and experience surfaced (Campanile, 2020; Chen, 2020). Nurses who worked in non-ICU wards, for example, did not have as much experience caring for ICU patients, and were unfamiliar with pre-intubation procedures. Dr. Tran described situations where nurses from other wards or hospitals did not know how to care for ICU patients, or were unsure of how much sedation to give, and anesthesiologists, respiratory therapists, and ICU nurses had to step in, thus increasing their workload.

In addition to hospital size, technological and resource-related difficulties, and workforce shortages and changing job scopes, the pandemic also disrupted medical norms and practices. Lapses in "best practices" usually followed from other kinds of disruptions at hospitals that greatly limited how medical workers cared for patients. Because converted wards were not installed with telemetry units, the hospital was facing medical personnel shortages, and some nurses were unfamiliar with running ICU units, patients ended up not being as closely monitored as they would be in an ideal situation. For example, some nurses did not notice when patients succumbed to viral infections and cardiac arrests. Moreover, while patients would usually, in non-pandemic times, receive the Advanced Cardiac Life Support (ACLS) sequence, which involves intubation, administration of medication, chest compressions, and defibrillation (delivery of electric shocks to the heart), many who suffered cardiac arrests during the pandemic went unnoticed. Many died before medical workers had the opportunity to resuscitate them. Moreover, some medical workers hesitated to deliver chest compressions, and others modified existing resuscitation procedures in order to prevent the spread of the virus from sick patients into the environment and onto themselves, their colleagues, and other patients. Their concerns were especially warranted since many converted wards were not installed with "negative pressure" isolation devices:

> When we give chest compressions, we're also compressing the lungs. COVID is going to enter the environment. Then, the issue becomes: what do we do? And who's responsible for doing it? Should the anesthesiologist come and place the airway [intubate]? Should the critical care doctor who's resuscitating the patient who has cardiac arrest place the airway? Or should you not do this at all, and just do compressions, place the oxygen mask on patients, and hope they survive? The confusion went on for weeks. Eventually, the solution was that the ICU specialist would call us when they think that a patient has a good chance of survival, and we will place the airway. We don't do that for every cardiac arrest. Some ICU specialists choose not to do compressions. They just give electricity and medication, which was not how we used to do things.

"Best practices" were suspended, and exceptional measures were taken in exceptional times. Suspension of institutionalized norms regarding treatment and health care was not, however, limited to hospitals. Members of the community were fearful of going to hospitals, even when they showed symptoms of COVID-19. Between March and May 2020, paramedics found people dead in their homes much more frequently than usual (Cox, 2020). Paramedics did not perform cardiopulmonary resuscitation (CPR) on people who were dying at home during that period, for fear of contracting COVID-19 (Callimachi, 2020; Herbert, 2020). The dead were sent to morgues at hospitals, where many were held for long periods – often in refrigerator trucks – since crematoriums were running at full capacity owing to the extraordinary death rate (Rose, 2020).

Hospitals, Wolf and Hall (2018) argued, "organize people around specific emergency infrastructures and communication routines" (p. 487) such that "techne [oriented to emergencies] matters in the discussion of how pandemic scenarios are performed and translated into [techno-social] infrastructures" (p. 490). While medical infrastructure had responded to disruptions, as administrators and medical workers made use of and modified facilities, technologies, and practices, such infrastructure and practices also posed limits to the degree to which changes could be made. Treatment and caregiving during pandemic times depended on modifications, but were also limited by technological infrastructures (e.g., lack of telemetry units and "negative pressure" wards) and a shortage of frontline professionals. Strains in different parts of the system developed in relation to one another, as did conditions of possibilities.

This overview of how hospitals experienced disruption and adapted to it provides the context for our discussion of medical frontliners' encounters with patients. We had discussed Macy's response to Ms. A's death on more than one occasion in our Zoom conversations. This encounter, we felt, served as an anchor for analysing the poignant emotions that medical workers felt during the pandemic, the individualization of responsibility, and new forms of emotional attachment that developed, all of which unsettled normative expectations in ways that accommodated movement between emotional detachment and emotional warmth in the delivery of health care.

## Beyond Normative Expectations: Intersubjective and Situated Encounters

Amidst our conversations on emotions, from frustration and grief to guilt and remorse, that medical frontliners might have experienced, Dr. Tran mentioned time and again the emotional detachment that medical professionals had to observe. Arlie Hochschild (2003, p. 53) terms these normative standards "feeling rules," which institutions prescribe and which "set limits to the emotional possibilities" of people living or working in a community (see also Hochschild, 1979). Feeling rules operate by governing the conditions that evoke or limit feeling, such as when surgical drapes cover patients' bodies and transform individuals into aseptic "operating fields" (Hirschauer, 1991). Such rules can alternatively and directly refer to emotive states, for instance, when medical workers are socialized to suppress sadness and grief. Medical workers, "in treating bodies, [must] also treat feelings about bodies" (Hochschild, 2003, p. 151), if they are to perform their tasks without being hampered by patients' passing. Macy's emotional breakdown and her colleagues' attempts to comfort her (and Dr. Tran's acknowledgment and disapproval) gesture to the presence of these norms on structuring emotions, albeit at their moment of rupture.

Hochschild (2003), building upon Goffman's (1978) *The Presentation of Self in Everyday Life*, emphasizes the contours of emotional expression, which Crewe, Warr, Bennett, and Smith (2014, p. 59) pointedly describe as "the rules of emotional display." We are able, through this approach, to study strategies of "masking" and "fronting" that interlocutors harness to suppress their emotions, control others' emotions to ease interactions, or advance their agendas (Ellingson, 2004; Portillo et al., 2013; Reissman, 1990). We can also analyse emotional governance or regulation and the emotional work that medical workers perform (James, 1992; von Scheve, 2012). These "feeling rules" are integrated into a hospital's "emergency infrastructure" (Wolf & Hall, 2018) in the social and cultural sense. However, these normative expectations are inadequate for the purposes of our inquiry on the embodied experiences and abrupt emotions of helplessness and sorrow, and of grief and remorse, that Macy and medical workers experienced.

Hochschild's work appears more relevant if we read her as conceiving "feeling rules" not as purely a discursive matter, but as mediated by expectations developed over time (2003, p. 35; see also Gross & Barrett, 2011). For medical workers, deaths come as no surprise when patients are older, suffer existing life-threatening conditions, or are victims of fatal accidents. But the pandemic situation was different:

> Macy was definitely used to seeing patients "expire." But the COVID situation presents new difficulties. It's expected that when old and really sick patients enter the ICU, they might die. But patients [like Ms. A] are more or less healthy people a week ago, don't have any medical problems, suddenly they have cold-like symptoms, they can't breathe, and it's only one problem. Usually that one problem does not kill them. Now this one problem suddenly kills them. You might be talking to them before the procedure, and then they suddenly die on you. That's also unexpected. And these [things] happen at much larger numbers.

To the extent that Hochschild acknowledges changing situations, she also assumes the continuity of experiences and the adherence to "feeling rules," which can be modified but are ultimately not abandoned. This relative stability of emotional expression, it is clear, does not resonate with the emotional intensity of Macy's response to Ms. A's death, a response where affective ambivalence and a sense of responsibility towards the other ruptures into being. Feeling rules, in this sense, form the backdrop of our analyses, but our primary concern rests upon their breakdown.

The helplessness of being caught up in the suffering or death of the other has been elaborated by Emmanuel Levinas in *Totality and Infinity* (1969) and *Otherwise Than Being* (1981), both of which conceive a dialectical opposition between vulnerability and empathy. Acts of empathy, Levinas argues, begin from the self and involve actively imposing one's frames of thought and feeling onto the other as one desires to grasp the other's feelings. Levinas argues that despite one's desire to empathize, it is never possible for one to fully access or comprehend the other's perspective. In contrast, vulnerability, as opposed to empathy, is the process by which one becomes mired in, affected by, and implicated in the other's suffering, pain, or death. Vulnerability, Levinas argues, attests to "the passivity of exposure":

> [The] exposure to affection, sensibility, a passivity more passive still than any passivity, an irrecuperable time, an unassemblable diachrony of patience, an exposedness always to be exposed the more, an exposure to expressing, and thus to saying, thus to giving. (Levinas, 1981, p. 50)

Vulnerability is, to use Louis Althusser's (2006) term, a condition of being "interpellated" or hailed into an experience by the other. It involves the "decentering of subjectivity," as Boublil (2018) notes, or perhaps even more radically, the fragmentation of the self, of "being torn up from oneself in the core of one's unity," which engenders an intersubjective state of "one-penetrated-by-the-other" (Levinas, 1981, p. 50). The condition of being affected by the other, Bernasconi (2001, p. xii) clarified in a foreword to *Existence and Existents* (Levinas, 2001), suggests that we conceive "the body as an event and not as a substantive" and thus attend to the "arising of the human being" produced via its engagement with the other through the transformation of the event into an existent. Rather than a quality of one's action, Levinas attends to the self being acted upon and the process through the self is remade through its exposure to the other.

Medical workers' experiences of being perturbed by the fragility of what had been healthy lives and the unexpected frequency of deaths must also be understood as a condition of the self's situatedness in fractured social spaces. The distinction between interiority and exteriority collapses not only between the self (medical worker) and the other (particular patient), but also between the self and the broader environment in which the healthcare worker is situated: the hospital, New York City, and the general atmosphere that characterizes a pandemic. Here, Maurice Merleau-Ponty (2012) offers us some guidance through his theorizing of how individuals are always already shaped, at a pre-reflexive level, by individuals and by the world around us. The "world," which Merleau-Ponty conceives of as physical, social, and subjective, is *a priori* to every action, practice, and reflection that individuals engage in, such that embodied persons not only act but are always already acted upon. The world also includes the social norms and language that we use to make sense of our perceptions, and that which we discussed as "feeling rules" (Hochschild, 2003), but not only such. As we shall see, Merleau-Ponty also suggests that we adopt a more perceptive and reflexive approach, compared to Hochschild's (2003) analyses of institutionalized rules and expectations.

While Merleau-Ponty (2012) believes, like Levinas (1969, 1981) in the self's ontological openness, he holds a more positive attitude towards knowing the other, which Levinas relegates to the margins when centring vulnerability. Knowing, Merleau-Ponty argues, can occur through the encounter of the sensing body and the sensed, the self and the environment, and the condition of "touching" and "being touched" (p. 93); that is, he acknowledges the actor as acting *and* acted upon. In his analysis of anger in a later text, *The World of Perception* (2004, p. 83), Merleau-Ponty writes that one person's sensing of another's anger takes place as a movement "from the outside" to "the body that is the very presence of these [affective] possibilities":

> [Seen from the outside, it appears that it is really] in this room and in this part of the room, that the anger breaks forth. It is in the space between him and me that it unfolds … When I recall being angry at Paul, it does not strike me that this anger was in my mind or among my thoughts but rather, that it lay entirely between me

who was doing the shouting and that odious Paul who just sat there calmly and listened with an ironic air … [But] when I reflect on what anger is and remark that it involves a certain evaluation of another person, that I come to the following conclusion … as soon as I turn back to the real experience of anger, which was the spur to my reflections, I am forced to acknowledge that this anger does not lie beyond my body, directing it from without, but rather that in some inexplicable sense it is bound up with my body. (Merleau-Ponty, 2004, pp. 83–5)

We take two lessons from Merleau-Ponty, in addition to insights that Levinas (1969, 1981) offered: firstly, that the intersubjective encounter, beyond one's "exposure" to an other, also involves one's embeddedness in a broader physical, social, and subjective space. Secondly, perception *eventually* involves an interactive encounter between the self and the other rather than mere exposure, and reflection can be part of that encounter. In the following section, building upon Levinas (1969, 1981) and Merleau-Ponty (2004, 2012), we return to our understanding about Macy's unsettling encounter with Ms. A in the context of the pandemic-stricken hospital and medical system that we have described.

**The Collapse of Emotional Distance**

Death is not new, as Dr. Tran explained time and again. Before the pandemic, it was not infrequent for medical workers to see patients "expire" before their eyes, especially patients who were aged or very sick and/or those who arrived at the hospital with serious injuries. Death is not new, especially for anesthesiologists and respiratory therapists, who are often called on to intubate patients in these conditions. Medication is administered before placing patients on ventilators, a process that paralyses their chest muscles so that ventilators can, with less obstruction, pump oxygen-saturated air into their lungs. While a patient is paralysed, oxygen saturation rapidly falls until it reaches 10%, sometimes 0%, during which time the anesthesiologist performs intubation and connects the patient to a ventilator. The time that elapses before oxygen saturation rises from 0% to 50%, and hopefully towards 80–90%, is when the sickest patients may die.

> Patients sometimes stop breathing during the procedure. They "expire." You feel so helpless when that happens. To save them, you did an intervention that failed. For some [medical workers] who are not used to it, they are shocked when it happens. When a patient dies on them … Patients are in bad condition, so if it happens, it is foreseeable. If the heart is failing and kidneys are failing, you will know that he's probably going to die. It's still emotional but largely foreseeable. And if you have a faulty machine, and they die, that's very unexpected. The machine is supposed to save lives. That person will be saved, if not for COVID. If you're running the machine, you'll feel much worse.

Deaths confront medical workers who see them for their first time and those who witnessed the unusual numbers of untimely deaths during the pandemic. What Dr. Tran described as "helplessness" in the face of death does not intimate apathy, but the deep emotions that accompany one's "passivity" – one is thus confronted, as Levinas (1969, 1981) would note. The sense of passivity, of "feeling bad," is amplified for medical workers, who try to save patients and whose practices trudge the difficult terrain between hope and despair, between furthering possibilities of extending lives and giving up when "you know that the patient's probably going to die." One recalls Macy's hands fumbling as she tried working the anti-valve, hoping that Ms. A would start breathing again. Judith Butler (2012, p. 136), working through questions of our capacity to respond ethically when confronted by images of war, violence, and death, described these images as containing a "structural paranoia" and "an indefinite form of address" to which our responses are inevitably and helplessly bound. Individuals like Macy and other workers caught up in emotions when confronted with the many deaths at the COVID-19 medical frontline are compelled, as Butler writes, "to listen to the voice of someone we never chose to hear [and] see an image that we never elected to see" (p. 136). What death evokes, as she expressed in an account of grief,

> Is the thrall in which our relations with others hold us, in ways we cannot always recount or explain, in ways that often disrupt the self-conscious account of ourselves we might try to provide, in ways that challenge the very notion of ourselves as autonomous and in control. (Butler, 2004, p. 23)

Encounters with untimely deaths compelled medical workers to feel, to be unsettled.

The untimeliness of death and the astounding frequency with which it occurred were a source of emotional discomfort even for medical workers who claimed that they were emotionally prepared. Indeed, when I asked Dr. Tran during our Zoom conversation, "Which kinds of occurrences during the pandemic struck you the most?" Dr. Tran replied:

> When I see young people with no health problems, really skinny, 25 years old or something, needing intubation. They're young and healthy, and they don't usually have problems. It hits home that the disease could affect anybody. COVID-19 could affect you, anybody, to the degree that you need to be warded and you need critical care. So obviously, it's not just 65 years and above in nursing homes, but people like you.

"Events of death," Christensen, Willerslev, and Meinert (2013, p. 3) argue, "have strong potentials of reworking senses of time and questions about existence in time in a way that few other events have." The biggest problem in death, they write, "is that the specifics – the concrete time and sequencing of our deaths – are unknown" (p. 3). Unlike Christensen et al. and their interlocutors, however, medical workers concern themselves with the specifics of timing deaths, a process that is pegged onto their clinical assessments of patients' health. Doctors and nurses deal with sickness and death in the course of their work, and, as evident from the anecdote on "foreseeable deaths" above, are familiar with differentiating pathological bodies from healthy ones – that is, those who they *knew* would die from others who they *knew* would live. The emotional weight that doctors and nurses experienced was not due to their inability to predict deaths in the way Christensen et al. (2013) envisioned, but to circumstances unfolding before their eyes that eluded their time-tested beliefs of when deaths should occur. The collapse of these predictabilities led to the unsettling of common sense, as Mary Douglas (1966) argued for the case of classificatory logics coming under threat. However, the emotive disruptions we account for are exacerbated because, not unlike Douglas's prediction, the disrupted logics are not easily restored. The dead do not breathe again.

Related to the emotional poignancy of untimely deaths was the unusual frequency with which they occured, scenarios which also ran against medical workers' common sense:

> During COVID, the number of patients who were intubated rose three- to fourfold, up to as many as 10 to 12 intubations per day. We see many more dying patients. In residency [in 2019], patients die on me maybe thrice a year, either on my table, or in front of me. Right now, this happens thrice a week, on me, or even more than that. One night, I had eight intubations. Three just died, one of them just when I was putting the tube in.

"What was it like at that moment?" I asked, to which Dr. Tran replied, "We do all that we can. But many of them eventually die a day later. We can't save everyone. We couldn't too, in the past, but now COVID created further problems in manpower and resources that we can't do anything about." The huge number of patients warded and who required critical care at ICUs placed an immense strain on New York City's medical system, as we discussed earlier (see also Chen, 2020; Rothfield et al., 2020). These constraints, which medical workers struggled with while trying to treat COVID-19 patients, were also part of what they had to confront emotionally. "Doing all that we can," to paraphrase Dr. Tran's reply, gestures in the pandemic context to the struggle against the limits that technological infrastructure and a shortage of staff posed, and falls short of what medical workers could have done under pre-pandemic conditions. In these situations, as Chen (2020) reported, it was not uncommon for nurses to quit in tears: "They were fed up. They couldn't take it any longer," she wrote.

The experiences of Macy, Dr. Tran, and other medical professionals could be understood through Levinas's (1969, 1981) analytic of exposure, vulnerability, and passivity, but it also appears that the emotional weight that medical workers felt was not entirely pre-reflexive. While affected by what they saw and experienced, some medical workers also reacted emotionally out of a sense of responsibility and complicity:

> It wasn't entirely because of the virus. It's also equipment malfunction, which doesn't usually happen. The hospital is well equipped. We usually have time to set everything up. So, it's just astonishing to her that this [Dr. Tran and Macy failing to start the ventilator] could happen. This is America. You're not supposed to die because you don't have enough oxygen or inadequate power supply. Well, we are responsible for the airway. But we don't have the proper equipment, so she wasn't able to properly do her job, and now she felt responsible for it.

Returning again to the opening scene, Macy, Dr. Tran said, felt remorseful because respiratory therapists were charged with the responsibility of ensuring that ventilators and other equipment were in working condition. The hospital's shortage of ventilators and its reliance on ventilators that were sometimes faulty (Dodge, 2020; Kliff et al., 2020), coupled with frontline staff shortages and anesthesiologists' and respiratory therapists' hectic schedules, suggest that Macy probably did not have the opportunity to verify that the ventilator was in proper working condition. The sense of responsibility and complicity in Ms. A's death seemed greater when accompanied by the knowledge that use of these technologies is supposed to save lives, and did save lives in the past. Sentiments of incredulity, such as "in America, no one would conceive of lives being lost due to inadequate power, poor oxygen supply, or faulty equipment," only further intensified the sense of culpability. Factors beyond medical workers' control, such as fatigue and faulty ventilators, became invisibilized relative to the human hands that operated them. This meant that medical workers felt more palpably responsible for operating such equipment, and for potentially contributing to the death rates.

One's greater tendency to view these responsibilities as one's own is perhaps exacerbated by the "thinning out" of frontline professionals, which meant medical workers were more directly responsible for the individual patients they attended to:

> [In non-pandemic times] there are few cases where we feel directly responsible for someone's life. As with the paramedics, most of us act as part of a team. Someone gives an order, and they do it, when someone dies, it's foreseeable and the blame is spread equally between everyone involved. And sometimes, there are people to talk it over. But this time, Macy felt directly responsible for that, because there were few other people present and able to help.

The thinning out of the medical teams and homogenization of medical procedures, as well as treating COVID-19 patients with breathing complications and organ failure resulting from being on a ventilator for long durations, compelled medical workers to work in smaller teams, meaning that there were fewer people to assist them with procedures such as intubation and ensuring that faulty machines are repaired (Chen, 2020). The staffing changes, while necessary, may also contribute to increased perceptions of responsibility towards patients, and correspondingly, the greater sense of remorse about these patients' deaths.

For anesthesiologists and respiratory therapists, COVID-19 also brought about changes in the kinds of interactions that they were familiar with. Before the pandemic, they cared for a more even distribution of patients, which ranges from those with non-life-threatening conditions and others who undergo surgeries for emergencies. Most of these patients, Dr. Tran said, "see us once in their lives, or once in a decade, unless they have multiple related surgeries." In the pandemic, however, the situation differed:

> Not all COVID patients who are hospitalized are ventilated. But if they are [and that's when Dr. Tran and his respiratory therapists colleagues are called to their beds], that's probably the last thing they see. You will know that they are in a terrible condition, with respiratory system or multiple organs failing. So, it's a lot of intense interaction during a short term. For the few people who are awake, and they may not last long, I will have given them confidence that I can do it, that I can save them. Hopefully, they can recover and get extubated (removed from their breathing tube). Even if they can't understand me, see me, or talk to me, I still have to treat them with respect and with the knowledge that this might be their final moment of, you know, being able to function normally.

For these and the preceding reasons, the sudden introduction of unexpected situations led medical workers to become emotionally overwhelmed. While they were familiar with death, it is precisely the rupture of this familiarity with certain frequencies of deaths and the ability to forecast patients' recovery that further compounds their shock and sorrow. For anesthesiologists and respiratory therapists who directly cared for critically ill COVID-19 patients, some of whom interacted with them before being intubated, the fact that the patients they recognize may not necessarily survive the ordeal (compared to patients undergoing elective surgeries in usual times) added to their emotional pressure.

Our conversations usually drifted from Dr. Tran's observations at the hospital to the pandemic's impacts and state and public responses in New York City, where Dr. Tran lives and where I visited weekly before leaving the United States. We took up these conversations as friends, but I was also personally curious about what happened in the city where he lived. It was much later, however, that we realized that the broader spatiality that is the city of New York, where many medical workers had lived for many years, would also have informed their experiences. That is, not only are they doctors and nurses but they are also members of a community of New Yorkers:

> But most New Yorkers will know somebody who has died or who was very sick with COVID. I know some friends, personal friends who were very sick. They are all fine now, but that's because my friends are young. So that makes it close to home, because you live here, you see it, you experience all of these people who fell sick and died, even if I am not a doctor, if I walk around and see these things. The refrigerator truck outside hospitals and empty streets, for example.

Medical workers' emotionally heavy encounters with patients, and their ability to pull through despite these encounters, are shaped not only by the immediate hospital environment but also by broader circumstances outside the hospital. Many of these, such as donation drives and community efforts (Harris, 2000; Paybarah, 2020), eateries' support for frontline medical workers (Wells, 2000), and the evening rituals that New Yorkers engaged in at 7:00 p.m. as they clapped and cheered for medical and essential workers (Newman, 2020), created a more liveable environment that, in Dr. Tran's view, motivated medical workers as they continued onward with their care for the severely ill.

**Unsettling Contact, Mediating Distance**

Our openness to being affected by individuals around us – as Levinas (1969, 1981) theorized through the concepts "exposure" and "vulnerability" – is apparent in the many examples that were provided herein. These include medical workers' shock and despair at the untimely deaths of their patients and the frequency at which these occurred, the felt responsibility for and perceived complicity in deaths of patients, the blurring of boundaries between self and patient, and the environments of the hospital and the city. Many of these were face-to-face encounters and attest to the emotional poignancy of personal, embodied encounters that Levinas privileged in the theorizing of "face," although it seems, as Butler (2012) offered, that exposure and vulnerability are also possible in less personal, non-embodied encounters. While we cannot fully understand the other (or, rather, we should not claim to do so), being exposed, we argue, has the effect of eroding and, in the case of more extreme emotions like Macy's at Ms. A's death, collapsing the boundaries between self and other, as well as between medical worker and patient. While Dr. Tran's words "It hits home that the disease could affect anybody" cannot be understood most literally as COVID-19's democratizing effects, since its impacts are still mediated by social location (e.g., class, race, and gender) and lived experiences, they convey the sentiment that we are not ontologically different and separable. If anything, we are connected through our human condition, through our compelled responses towards others' suffering and death.

Levinas's (1969, 1981) theorizing on "exposure" and "vulnerability" prompts us to consider how our subjectivities are shaped by encounters with individuals around us. While the intersubjective and interactional emphases that he highlights prove to be useful, medical workers' experiences are situated within the broader material and subjective "worlds," to evoke Merleau-Ponty (2004, 2012), in which they live. These worlds, for medical workers in pandemic times, are not characterized by settled cultural logics and institutionalized practices, but by radically unsettling experiences situated against physical and social structures under strain. The intensity of experience is generated in our encounters and felt in the spaces "between us" and "from the outside" – that is, in medical workers' engagement with social and infrastructural strains, and the negative (but also positive) moods in New York City which some have internalized. The internalization process, which is part of how one's perceptions/experiences and relations to the other develop, led some medical workers such as Macy to feel remorseful about the far from usual deaths and death counts. Being "exposed," to quote Merleau-Ponty (2004, 2012), is thus a condition that fuels the human propensity for reflection. Medical workers tried, as social beings, to grapple with their relations with the other to whom they were exposed.

This emergent sociality between Macy and Ms. A, we must clarify, does *not* mean that Macy is accountable or should feel responsible for Ms. A's death. In fact, Macy returned to work after 30 minutes of "feeling bad." Instead, we explain Macy's attitude towards Ms. A as an accountability *to make sense* of a human situation: the compulsion to grapple with what one is confronted with, which is the suffering and death of the other. Whether Macy eventually felt responsible for Ms. A's death is less important than the compulsion to think about the latter's untimely demise and one's relationality and possible complicity in the matter. Whether other medical workers outwardly displayed their emotions as Macy had done is also less consequential than the fact that many sensed that something radically problematic was occurring and tried to make sense of the emergency on site. Moreover, as Dr. Tran replied when queried whether doctors ever grieve:

> For doctors, there is little space in normal times for grieving. Especially when you are a surgery-related specialist, everything you do involves people who can possibly die. The chances of them living are very limited. We give injections and put devices in the body that reduce pain, and we build relationships with patients, which is also part of critical care. There's obviously a lot of dead and dying people, but we're specialists and are in very high demand, so we can't just take time off to deal with emotions.

The call for responsibility, as Murphy (2018, p. 193) argued in her work on hunger, along with the incorporation of Levinas's and Merleau-Ponty's ideas, must therefore always be preceded by one's response to the call for attention. Only through being attentive can one be inspired to contemplate one's responsibility for the pain, suffering, and death of the other.

We would be mistaken if we prematurely concluded that, since medical workers do not grieve, they simply have no feelings about their patients' demise. To do so would be to conflate hospitals' feeling rules (Hochschild, 2003), that medical workers ought to maintain a distance from the emotions they feel, which in this chapter is argued to be situational and diverse even if all are affected by these rules that form part of the "world" they inhabit, as discussed in Merleau-Ponty's (2004) analyses of language (pp. 86–7). Our medical workers' exposure to various situations during the pandemic compelled them to respond, and also created the space for ethical reflection in ways that Merleau-Ponty (2012) briefly addressed. Emotions, he argued, are inevitably bound to oneself (see also Levinas, 1981), since they involve one's attempts to perceive and cognize one's social relations in the world (Merleau-Ponty, 2004, pp. 83–5). Similarly, the remorse that medical workers experienced gestures to generative spaces of ethical reflection following the collapse of "rules" on appropriate emotional distance between medical workers and patients. The moments of emotional breakdown – exacerbated more frequently during the pandemic – are therefore bound up with attempts to reconcile, as Macy and other medical workers sought to make sense of the unsettling forms of contact they encountered in such exceptional times of crisis.

*Acknowledgments*: The authors thank Princeton-Mellon Initiative in Architecture, Urbanism and Humanities for funding research for this project. We dedicate this chapter to the healthcare workers who risked their lives to save many others, and in memory of the many who lost their lives.

REFERENCES

Abelson, R. (2020, May 9). Hospitals struggle to restart lucrative elective care after coronavirus shutdowns. *New York Times*. https://www.nytimes.com/2020/05/09/health/hospitals-coronavirus-reopening.html.

Althusser, L. (2006). Ideology and ideological state apparatuses (notes towards an investigation). In Gupta, A., & Sharma, S. (Eds.), *The anthropology of the state: A reader* (pp. 86–98). Blackwell Publishing.

Ansari, T., & Prang, A. (2020, August 9). Nurses travel from coronavirus hot spot to hot spot, from New York to Texas. *Wall Street Journal*. https://www.wsj.com/articles/nurses-travel-from-coronavirus-hot-spot-to-hot-spot-from-new-york-to-texas-11596988800.

Bernasconi, R. (2001). Foreword. In Levinas, E. (Ed.), *Existence and existents* (pp. vii–vx). Duquesne University Press.

Boublil, E. 2018. The ethics of vulnerability and the phenomenology of interdependency. *Journal of the British Society for Phenomenology, 49*(3), 183–92. https://doi.org/10.1080/00071773.2018.1434952.

Butler, J. (2004). *Undoing gender*. Routledge.

Butler, J. (2012). Precarious life, vulnerability, and the ethics of cohabitation. *Journal of Speculative Philosophy, 26*(2), 134–51. https://doi.org/10.5325/jspecphil.26.2.0134.

Callimachi, R. (2020, May 19). Paramedics, strained in the hot zone, pull back from CPR. *New York Times*. https://www.nytimes.com/2020/05/10/nyregion/paramedics-cpr-coronavirus.html.

Campanile, C. (2020, March 29). Here's the next big problem after New York gets ventilators. *New York Post*. https://nypost.com/2020/03/29/heres-the-next-big-problem-after-new-york-gets-ventilators/.

Chen, P.W. (2020, May 7). The calculus of coronavirus care. *New York Times*. https://www.nytimes.com/2020/03/20/well/live/coronavirus-covid-doctor-nurse-shortage-staff.html.

Choi, A., & Velasquez, J. (2020, March 20). Trained operators for coronavirus patient ventilators scarce in New York. *The City*. https://www.thecity.nyc/2020/3/20/21257921/trained-operators-for-coronavirus-patient-ventilators-scarce-in-new-york.

Christensen, D.R., Willerslev, R., & Meinert, L. (2013). Introduction. In Christensen, D.R., & Willerslev, R. (Eds.), *Taming time, timing death* (pp. 1–16). Routledge.

Cox, C.E. (2020, July 17). New York City's cardiac arrest spike in COVID-19 tracks with missing STEMIs. *TCTMD*. https://www.tctmd.com/news/nycs-cardiac-arrest-spike-covid-19-tracks-missing-stemis.

Crewe, B., Warr, J., Bennett, P., & Smith, A. (2014). The emotional geography of prison life. *Theoretical Criminology, 18*(1), 56–74. https://doi.org/10.1177/1362480613497778.

Dodge, B. (2020, March 31). New York hospitals received damaged ventilators with missing parts in emergency shipments from a national stockpile. *Business Insider*. https://www.businessinsider.com/new-york-got-damaged-ventilators-from-national-stockpile-2020-3.

Douglas, M. (1966). *Purity and danger: An analysis of concepts of pollution and taboo*. Routledge and Kegan Paul.

Dunia, O.A., Baaz, M.A., Mwambari, D., Parashar, S., Oseema, A., Toppo, M., & Vincent, J.B.M. (2020, June 18). The COVID-19 opportunity: Creating more ethical and sustainable research practices. *Items: Insights from the Social Sciences*. https://items.ssrc.org/covid-19-and-the-social-sciences/social-research-and-insecurity/the-covid-19-opportunity-creating-more-ethical-and-sustainable-research-practices/.

Ellingson, L.L. (2004). *Communicating in the clinic: Negotiating frontstage and backstage teamwork*. Hampton Press.

Goffman, E. (1978). *The presentation of self in everyday life*. Penguin.

Gross, J.J., & Barrett, L.F. (2011). Emotion generation and emotion regulation: One or two depends on your point of view. *Emotion Review*, 3(1), 8–16. https://doi.org/10.1177/1754073910380974.

Harris, A. (2000, April 1). Remembering the neediest during the coronavirus pandemic. *New York Times* https://www.nytimes.com/2020/04/01/reader-center/neediest-cases-covid-19-relief-campaign.html.

Herbert, G. (2020, April 23). NY rescinds new resuscitation guidelines for EMS. *EMS1*. https://www.ems1.com/resuscitation/articles/ny-rescinds-new-resuscitation-guidelines-for-ems-jTislkdO8t4oslEZ/.

Hirschauer, S. (1991). The manufacture of bodies in surgery. *Social Studies of Science*, 21(2), 279–319. https://doi.org/10.1177/030631291021002005.

Hochschild, A.R. (1979). Emotion work, feeling rules, and social structure. *American Journal of Sociology*, 85(3), 551–75. https://doi.org/10.1086/227049.

Hochschild, A.R. (2003). *The managed heart: Commercialization of human feeling*. University of California Press.

James, N. (1992). Care = organization + physical labor + emotional labor. *Sociology of Health and Illness*, 14(4), 488–509. https://doi.org/10.1111/1467-9566.ep10493127.

Kaye, J. (2020, April 17). New York-Presbyterian Queens transforms cafeteria into extra room for patients. *Flushing*. https://qns.com/2020/04/newyork-presbyterian-queens-transforms-cafeteria-into-extra-room-for-patients/.

Kliff, S., Satariano, A., Silver-Greenberg, J., & Kulish, N. (2020, March 18). There aren't enough ventilators to cope with the coronavirus. *New York Times*. https://www.nytimes.com/2020/03/18/business/coronavirus-ventilator-shortage.html.

Levinas, E. (1969). *Totality and infinity: An essay on exteriority*. Duquesne University Press.

Levinas, E. (1981). *Otherwise than being or beyond essence*. Martinus Nijhoff Publishers.

Mead, M., & Métraux, R. (Eds.). (1953). *The study of culture at a distance*. Vol. 1. Berghan Books.

Merleau-Ponty, M. (2004). *The world of perception*. Routledge Press.

Merleau-Ponty, M. (2012). *The phenomenology of perception*. Routledge Press.

Murphy, A.V. (2018). "The will to live and the meaning of life": Hunger as vulnerability in French existential phenomenology. *Journal of the British Society for Phenomenology*, 49(3), 193–204. https://doi.org/10.1080/00071773.2018.1434960.

*New York Times*. (2020). New York City COVID map and case count. Retrieved 19 November 2000, from. https://www.nytimes.com/interactive/2020/nyregion/new-york-city-coronavirus-cases.html.

Newman, A. (2020, April 10). What New York City sounds like every night at 7. *New York Times*. https://www.nytimes.com/interactive/2020/04/10/nyregion/nyc-7pm-cheer-thank-you-coronavirus.html.

Paybarah, A. (2020, April 3). Coronavirus in New York: How to help. https://www.nytimes.com/2020/03/12/nyregion/nyc-coronavirus-help-volunteer.html.

Portillo, S., Rudes, D.S., Viglione, J., & Nelson, M. (2013). Front-stage stars and backstage producers: The role of judges in problem-solving courts. *Victims and Offenders*, 8(1), 1–22. https://doi.org/10.1080/15564886.2012.685220.

Reissman, C.K. (1990). Strategic uses of narrative in the presentation of self and illness: A research note. *Social Science and Medicine*, 30(11), 1195–1200. https://doi.org/10.1016/0277-9536(90)90259-U.

Rose, J. (2020, April 12). As an ER doctor, I fear our era's defining symbol will be the refrigerator truck. https://www.washingtonpost.com/outlook/2020/04/11/refrigerated-truck-morgue-coronavirus/.

Rothfield, M., Sengupta, S., Goldstein, J., & Rosenthal, B. (2020, March 25). 13 deaths in a day: An "apocalyptic" coronavirus surge at an N.Y.C. hospital. https://www.nytimes.com/2020/03/25/nyregion/nyc-coronavirus-hospitals.html.

Sanger-Katz, M. (2020, May 8). Why 1.4 million health jobs have been lost during a huge health crisis. https://www.nytimes.com/2020/05/08/upshot/health-jobs-plummeting-virus.html.

Saxena, A.K., & Johnson, J.L. (2020, May 31). Cues for ethnography in pandamning times: Thinking with digital sociality in the COVID-19 pandemic. *Somatosphere: Science, Medicine, and Anthropology*. http://somatosphere.net/2020/ethnography-in-pandamning-times.html/.

von Scheve, C. (2012). Emotion regulation and emotion work: Two sides of the same coin? *Frontiers in Psychology*, 3(2012), 496. https://doi.org/10.3389/fpsyg.2012.00496.

Wells, P. (2000, March 30). Restaurants find hope in delivering donated meals to hospitals. https://www.nytimes.com/2020/03/30/dining/restaurants-hospitals-coronavirus.html.

Wolf, M., & Hall, K. (2018). Cyborg preparedness: Incorporating knowing and caring bodies into emergency infrastructures. *Medical Anthropology*, 37(6), 486–98. https://doi.org/10.1080/01459740.2018.1485022.

# 9 A Spatial Snapshot of the Relationship between the COVID-19 Pandemic and Selected Crimes in California

BY GABRIELE PLICKERT AND EMILY COOPER

## Introduction

The occurrence of crime is not a matter of uniform or random organization in space and time (Ratcliffe, 2010). To understand changes in the timing and space (e.g., location) for certain types of crime or variation in crime rates, criminologists and social scientists alike have examined, for example, economic insecurities (Cantor & Land, 1985; Kleck & Chiricos, 2002; Thornberry & Christenson, 1984; Young, 1993) and natural disasters (Elmes at al., 2014; Frailing & Harper, 2017) to untangle reasons for changes. During the process of writing this chapter in late 2020, the COVID-19 pandemic has caused another series of lockdowns across the United States, and infections and deaths are increasing rapidly, the likes of which we have not seen before. The initial goal of this chapter was to explore the occurrence of property crime and robbery in California at the start of the pandemic. However, in its current iteration, our chapter compares the initial period, or the first 10 months of the pandemic, in selected Californian cities during the pre-vaccination period.

In March of 2020, in response to the COVID-19 pandemic, enforced health measures of social distancing and the shelter-in-place order gradually transformed everyone's daily routines in the United States and around the world. While the lockdowns in March of 2020 had a sweeping impact on everyone's personal and professional lives, questions emerged among researchers whether and to what extent these imposed measures might have altered or curbed the motivations for crime and criminal deviance. Diverse social science literature provides evidence of a relationship between periods of mass emergencies (i.e., natural disasters, economic emergencies) and elevated crime rates (LeBeau, 2002; Frailing & Harper, 2017). However, the COVID-19 pandemic with its enforced spatial constraints, paired with the experience of emotional and economic distress, appears unlike any previous natural disaster or economic disruption. The nature of the current pandemic is also overwhelmingly different from what regions, cities, and countries around the world have experienced before. Emerging research investigating the effects of the COVID-19 pandemic on crimes suggests that the enforced lockdowns and opening of only essential businesses have, in fact, transformed the map of people's daily routines and therefore changed the occurrence of committed crimes (Boman & Gallupe, 2020; Felson et al., 2020; Mohler et al., 2020). Although the pandemic remains ongoing at the time of this writing, recent studies seeking to investigate the pandemic-crime relationship do offer an initial understanding of the dynamics between the pandemic, social environments or locations, and the reporting of crimes.

Adding to this early work, we are motivated to explore the following assumptions: On the one hand, stay-home orders might reduce property crimes in residential areas because of increased vigilance over personal belongings and space, while, on the other hand, low-risk areas (e.g., non-essential businesses) in the absence of efficient guardians might become places of motivation for vandalism and other types of property crimes and robbery. Considering these crime scenarios during the uncertainties of a pandemic, this chapter seeks to investigate the link between social distancing and stay-home orders owing to COVID-19 and the occurrence of property crimes and robberies across selected Californian cities. We evaluate the selected crimes in the context of time, examining the months before and during the pandemic.

There is no doubt that the worldwide pandemic, with its impact on every organizational, structural, and individual aspect of society, provides a novel ground for research in a wide range of areas. Initial research on the subject shows that early lockdowns emptied streets and stopped non-essential business operations and leisure activities (Stickle & Felson, 2020). For instance, country-wide police reports reveal a decline in property crimes and robberies. However, it remains unclear whether the decline for selected crimes is similar across all city sizes and whether the frequency and distribution of selected crimes are altered but not different overall from the previous year.

In the next section, we briefly reflect on the occurrence of crime in times of crisis. Following this section, the chapter outlines the theoretical framework supporting a relationship between social distancing and crime. The third section discusses the crime rates in California before the COVID-19 pandemic. The chapter then examines selected property crimes and robbery during the coronavirus's initial months and 10 months into the pandemic across specific Californian locations. We also compare the 2020 results to those for 2019. The final section considers the future of crime analysis with data collected during the pandemic. This chapter excludes crimes of assaults and domestic violence, for which research studies have provided evidence of dramatic increases during the time of the pandemic (Ashby, 2020; Leslie & Wilson, 2020; see also discussion in Adur and Narayan's chapter).

## Crime Trends in Times of Crisis

Interrelated with the perspective of social and ecological environments are approaches that seek to explain changes in crime by examining associations between the impact of natural disasters (Elmes et al., 2014; Fritz, 1961; Leitner et al., 2011; Prelog, 2016; Zahran et al., 2009) or economic insecurities (Cantor & Land, 1985; Cohen et al., 1980; Land & Felson, 1976; Thornberry & Christenson, 1984) and various types of crime. Although natural disasters, compared to a worldwide pandemic, have a significant short-lived impact on specific areas, communities, or places, the aftermath of natural disasters often brings longstanding economic insecurities and a wide range of social and emotional interruptions to everyday life, which are potentially comparable to experiences of the current pandemic. While the 1918 Influenza Pandemic, erroneously referred to as the "Spanish Flu," had a worldwide impact in bringing shocks to society, mainly affecting the working-age population, the majority of existing research has focused primarily on the medical causes of the flu (Tumpey et al., 2005; Bos et al., 2011) or examined its societal and economic effects (Gupta et al., 2005; Karlsson et al., 2014). However, research on the 1918 Influenza Pandemic or other later major outbreaks (e.g., 2006 bird flu, or SARS in 2002–4) has not disentangled the relationship between a pandemic and changes in crime rates (Ashby, 2020). In this section, we direct our attention to existing research on natural disasters and economic instability to unveil crime trends in times of crises.

Natural disasters represent a major threat to safety, health, and property (Mileti, 1999). While the thoroughness of analytical data on natural disasters is challenging, natural hazards not only cause human and property loss but also interrupt social order and community life altogether. In fact, natural disasters have been described as offering a social laboratory for analysing basic questions on how individuals and communities respond and adjust to conditions of loss and dispossession (Fritz, 1961, p. 654). Aside from severe environmental damage, natural disasters trigger deaths and extensive evacuations, followed by major interruptions of work and leisure activities, which ultimately lead to changes in one's daily routine activities. While the media often cover natural disasters as the cause for social disorder, looting, and criminal deviance (Fischer, 1998), the empirical evidence linking natural disasters and crime remains mixed within crime and disaster research.

Studying the aftermath of hurricanes in the United States, researchers found increases in burglaries (LeBeau, 2002; Frailing & Harper, 2017), explained by the absence of guardians over unattended properties and the hospitalization or evacuation of people to shelters away from their homes (Leitner & Helbich, 2011). Investigations of patterns of property and violent crime following natural disasters show that both violent and property crime either remain unaffected or decrease after disasters (Lemieux, 1999; Quarantelli, 1994; Zahran et al., 2009).

Similarly, Varano et al. (2010) examined violent and non-violent crime rates in cities that accommodated large numbers of Hurricane Katrina evacuees and found no significant changes in crime rates due to the hurricane. In contrast, Prelog (2016) examined the effects of natural disasters at the county level and noted that disasters were associated with higher incidents of property crime.

Leitner and Helbich (2011) employed spatial and temporal analysis to examine the impact of Hurricane Katrina on reported crimes in communities of Louisiana. Using Louisiana Uniform Crime Report (UCR) data from January 2000 to June 2006, the authors observed an increase in reported crimes for two Louisiana communities, while the majority of neighbourhoods did not experience a change in violent and non-violent crime rates relative to pre-Katrina trends (Leitner et al., 2011). Overall, the findings on Hurricane Katrina did not show any apparent impact on changes in crime rates.

Moving the research focus from Louisiana to Florida, Zahran et al. (2009) examined changes in crime patterns in Florida from 1990 through 2005. Using data from the Spatial Hazards Events and Losses Database for the United States (SHELDUS) and crime data from the Florida Department of Law Enforcement's UCR, the authors found that natural disasters had a modest effect on the decline of violent and property crime (Zahran et al., 2009). In contrast to violent and property crime, Zahran et al. found evidence of an increase in domestic violence associated with disasters.

Leitner and Helbich (2011) studied Hurricane Rita in Texas and found a significant short-term increase in burglaries and auto thefts. Their findings identified highly significant spatio-temporal burglary clusters in the northeastern area of Houston, though these burglary clusters were only short-lived, occurring a few days before and after the landfall of Hurricane Rita. The mandatory evacuations explained the short-term spike in burglaries before the hurricane's arrival. The occurrence of burglaries during the short time frame before and soon after the arrival of the hurricane was attributed to individuals who did not follow the evacuation order but instead took the opportunity to burglarize the residence of neighbours who did evacuate.

In light of these mixed findings, one potential limitation suggests that decreases in reported violent and property crime following a disaster may be due to the disaster's impact on citizens reporting crimes and changes in law enforcement's reporting practices during the period immediately following a disaster. As Barsky et al. (2006) observed when examining looting during the immediate post-hurricane period, the reporting of criminal activities declined as law enforcement personnel prioritized search, rescue, and crime control activities over record keeping.

Another question central to the understanding of changes in crime has been the relationship between economic insecurities and crime, or more specifically the effect of unemployment or poverty on crime. Cantor and Land (1985) were among the first to empirically assess the link between unemployment and crime rates by stating the possibility that unemployment could both increase motivation and decrease opportunity for crime. For example, they found a negative association between unemployment and property crime in the United States. They argued that unemployment decreases the opportunity for property crime, as it generates a slowdown in production and consumption activities and at the same time increases the ability to guard property because of more time spent at home.

From a sociological approach, Merton's strain theory asserts that crime results when people are unable to achieve their goals through legitimate means, thus predicting a strong positive relationship between unemployment and crime (Merton, 1938). In view of Merton's assumptions, and contrary to Cantor and Land's findings, contemporary research found evidence of higher unemployment rates (Doyle et al., 1999; Raphael & Winter-Ebmer, 2001; Gould et al., 2002) and larger income inequality (Kelly, 2000; Choe, 2008) being linked to increases in crime. However, the magnitude of this relationship was found to differ depending on the type of crime. For instance, findings consistently indicate that unemployment is an important factor for property crime to occur but less so for violent crime (Raphael & Winter-Ebmer, 2001).

Although there is extensive literature on the unemployment-crime relationship, findings show a mixture of conflicting claims between income inequality, unemployment, and crime, ranging from strong positive associations (Land & Felson, 1976) to negative correlations (Cohen et al., 1980), and even non-significant associations among the measures (Kleck & Chiricos, 2002; Thornberry & Christenson, 1984; Young, 1993).

In light of the existing research, the current COVID-19 pandemic is somewhat different from natural disasters, mainly because the pandemic has had an immediate impact on human activities and, more generally, on people's personal lives at the regional, state, national, and international levels. However, as the current pandemic continues into 2021, we already observe an impact of various degrees on the economic environment, with mandated closures of businesses and increases in unemployment across sectors. Given the continuing nature of the current pandemic, the attempt to explore crime trends across various environments, times, and stages of the pandemic will contribute to our knowledge of disaster and crime research.

**Theoretical Explanations**

Seeking to investigate to what extent the consequences of COVID-19 have impacted spatial and temporal patterns of crime, research thus far primarily explored the theoretical premises of routine activity theory (Cohen & Felson, 1979), crime pattern theory (Brantingham & Brantingham, 1984), and general strain theory (Agnew, 1992).

To understand the impact of a disastrous event or crisis on crime, researchers have examined the assumptions of *Routine Activity Theory*, which states that disaster or crisis events potentially generate changes in the structure of a community's routine (Cohen & Felson, 1979; Felson & Clarke, 1998). These structural changes increase the likelihood of suitable targets, motivated offenders, and lack of suitable guardians (Cromwell et al., 1995), and thus will more likely increase crime in affected communities. However, the opposite can be possible when communities act as suitable guardians over their space (e.g., protecting their properties, neighbourhoods), and thus the theory would predict a decline in crime following a crisis event. Although the coronavirus pandemic is different in magnitude from natural disasters, it nevertheless can be seen as a crisis event with significant impact on communities and the professional and personal lives of people locally and worldwide. This chapter explores the impact of mandated lockdowns and social distancing – restrictions implemented as a result of the magnitude of the pandemic – on property crimes and robbery.

Property crimes are crimes in which property is the offender's target; these are constituted by *auto theft, residential burglary* and *commercial burglary*, and *vandalism*. Overt *vandalism* (e.g., graffiti, mailbox destruction) is associated with property crimes. However, compared to auto theft and burglaries, it is largely committed by more than one offender (Boman & Gallupe, 2020). Typically, property crimes occur in the absence of witnesses, and thus, with stay-home orders in place, early studies have reported a shift from solely residential to more non-residential areas for property crimes (Campedelli et al., 2020).

In comparison, *Crime Pattern Theory* seeks to understand how people come together in space and time within crime settings. Criminal events (e.g., robbery) most likely occur in areas where potential offenders' activity space overlaps with the activity space of potential victims. An individual's activity space becomes familiar through everyday activities, since this is where the individual works, lives, and shops. A crime event involving an offender and victim is only realized by the intersection of their activity spaces. Thus, crime occurrence is the result of people frequenting an area where victims and potential offenders interact. However, in times of lockdowns and social distancing, victims and potential offenders are less likely to intersect in activity spaces; thus, Crime Pattern Theory would point to a general reduction in robberies and burglaries, for example. Taking into account people's daily activities of life, individuals place themselves into situations that can likely increase or decrease their risk of criminal victimization (Degarmo, 2011). The interactions or the absence of interactions with other people, places, and activities potentially impact the likelihood of criminal behaviour.

Overall, considering the impact of a crisis event on crime, Routine Activity Theory and Crime Pattern Theory most likely point in the direction of declines in the occurrence of crime. In contrast to these theories stands Agnew's *General Strain Theory*, which most likely supports the assumptions of an increase in crime in light of a crisis event. For example, in situations of limited freedom of movement, increased social isolation, financial distress, and uncertainties related to shelter-in-place orders, people may be exposed to higher levels of stress (e.g., negative stimuli), which may lead them to experience a range of negative emotions (Agnew, 1992). If these negative emotions remain unchecked, this may in fact lead to the engagement in criminal behaviour. While limited interactions and reductions in movements can affect crime in the short term, lengthy shelter-in-place orders and regulations may generate spikes in certain types of crimes in the future (Campedelli et al., 2020).

Whether and how crime changes because of the coronavirus pandemic has been explored so far at different societal and environmental levels. Similar to single-region or city studies, Ashby's (2020) comprehensive analysis of cities across the United States points to no consistent pattern (i.e., increases, declines, or no changes) of crime during the pandemic. Overall, the current initial findings on the pandemic are in flux, yet they may expand existing explanations and inform future theoretical implications.

**Crime Rate Trends in California before the COVID-19 Pandemic**

Before we explore property crimes in California during the coronavirus pandemic, we briefly reflect on California's crime rates over the past 10 years: According to the Center of Juvenile and Criminal Justice, during the first six months of 2019, the rates for property crimes in 69 Californian cities with populations of 100,000 or more fell by 10.4% when compared to data for the same time period in 2010 (Federal Bureau of Investigations (FBI), 2020; California Department of Finance (DOF), 2020; California Department of Justice (DOJ), 2020). Comparisons in crime rates between the first six months of 2018 and the first six month of 2019 show declines of 0.8% for property, 9.1% for motor vehicle theft, and 5.5% for burglary. Overall, crime rate trends over the past 10 years for property crimes have decreased by 9.9% between the first six months of 2010 and the first six months of 2019.

The decline in crimes is not only a phenomenon for California. Across the United States, crime rates dropped significantly between 1993 and 2019. For example, using FBI data, violent crime rates fell 49%, robbery rates fell 68%, and property crimes fell 55% with significant declines in burglary (69%) and a decline of 64% for motor vehicle theft. Declines are even greater when using the Bureau of Justice Statistics, with decreases of 74% for violent crimes and 71% for property crimes between 1993 and 2019 (Morgan & Oudekerk, 2019; Gramlich, 2020). No matter the data, overall, nationally and regionally, we see declines for the majority of crimes since the mid-1990s.

In California, crime rates have stayed low and stable for the past 10 years. The crime rate for property crime fell during the 1980s and 1990s and was down approximately 63% by 2011. Although California's crime rate showed a decline, nationwide the state ranked 25th among all states for property crimes. Of the reported property crimes, 17% were auto thefts, 18% were burglaries, and 65% were larceny thefts (Lofstrom & Martin, 2016).

During these 10 years, California also employed some significant criminal justice reforms – the introduction in particular of Public Safety Realignment, Proposition 47, and Proposition 57. The 2011 Public Safety Realignment focused on the overcrowding in California's prisons and shifting the responsibilities for individuals with non-violent, non-sexual, and non-serious convictions from state to counties (*Brown v. Plata,* 2011). Proposition 47 was passed in 2014 and addressed six minor drug and property offences, from possible felonies to misdemeanours, which led to the resentencing and release of thousands of people from jails and prisons (Proposition 47, 2014). In 2016, the enactment of Proposition 57 increased opportunities for individuals in state prison to earn credit towards early parole consideration by taking part in rehabilitative programs (Proposition 57, 2016).

While most of the largest cities in California registered a decline in crimes between early 2018 and 2019, it is important to note that crime rates vary by cities and counties. For instance, among the major Californian cities, Oakland and San Jose reported an increase in violent crimes, while Fresno, Long Beach, Sacramento, San Diego, and San Francisco reported declines. The biggest variation in violent crime across regions has been observed for robbery. For example, the robbery rate in 2017 for Los Angeles County was more than five times higher than the rate in the Sierras, which combines seven counties altogether (Lofstrom & Martin, 2016). Furthermore, Los Angeles reported declines in violent and property crimes between the early months of 2019 and the same time in 2018.

Despite the scepticism of some groups regarding implemented justice reforms, crime rate trends have remained historically low during large-scale reforms in California. Moreover, no relationship has been established between justice reforms and the continuous low crime rates across the state. Instead, the Center on Juvenile and Criminal Justice (CJCJ) reports indicate that fluctuation in certain crime rates appears to be a rather localized matter, suggesting that decreases or increases in crimes may result from differences in public safety approaches at the city or county level (CJCJ, 2018, 2019).

Considering the relationship between crime and natural disasters, economic insecurities, and the historic low in property crimes across California, in the next section we attempt to explore how property crime and robbery varied during the early progression of COVID-19 and 10 months into the pandemic, comparing both periods to 2019.

## Progression of COVID-19 and Crime in California

At the start of 2020, the COVID-19 pandemic spread with deleterious effects across the globe in a matter of months. The United States experienced a dramatic diffusion of the virus, and many states started to respond to the outbreak with regulations and health measures around March 2020 (Mervosh et al., 2020). In California, infections and deaths due to the pandemic surfaced during the first two weeks in March (Figure 9.1), and, within these first weeks, cumulative confirmed infectious cases rose from 281 to 4,679.

The spread of COVID-19 in early March forced the state's Governor Newsom to implement restrictive statewide interventions to contain it. The first response of interventions included limiting group gatherings to less than 250 people on 11 March, followed by school closures, promulgated on 13 March, and further closures of gyms, health clubs, and movie theatres as well as closures of restaurants, shifting service to take-out only, on 16 March. With the drastic increase of cumulative infectious cases, the state government began to implement a statewide stay-at-home order, enforcing the closure of all non-essential businesses on 19 March, followed by closures of parks and beaches by 29 March (see Figure 9.1). Except for essential workers, the stay-at-home order transformed businesses and workplaces, and where transitions to online were not possible, work was placed on hold entirely. Practically, the mandated order brought a halt to day-to-day travels, eliminated commutes from home to work, and limited many other outdoor activities. The statewide stay-at-home order originally mandated in March started to be relaxed during May 2020 and gradually continued to adjust to the reopening of non-essential businesses.

While the pandemic has triggered substantial changes to individual routine activities as well as to political agendas and economies, whether and how the frequency and distribution of crime activities have been affected by the pandemic and its implemented statewide health measures quickly became the focus across various fields of research (Ashby, 2020; Campedelli et al., 2020; Leslie & Wilson, 2020; Piquero et al., 2020; Stickle & Felson, 2020).

Thus far, researchers have investigated the COVID-19 pandemic-crime relationship in the context of cities or neighbourhoods (Ashby, 2020; Mohler et al., 2020). Overall, findings of these studies reveal declines in crime of certain types, yet declines are not consistent across all contexts and types of crime. Exploring the initial crime trends during the early months of the COVID-19 pandemic, findings show mixed results regarding the direction of changes in crime activities (Stickle & Felson, 2020). For California, Shayegh and Malpede (2020) found a 43% drop in crime in San Francisco and a 50% drop in Oakland at the beginning of the mandated shelter-in-place order during March 2020. In comparison, research focusing on specific crimes provides evidence of increases in

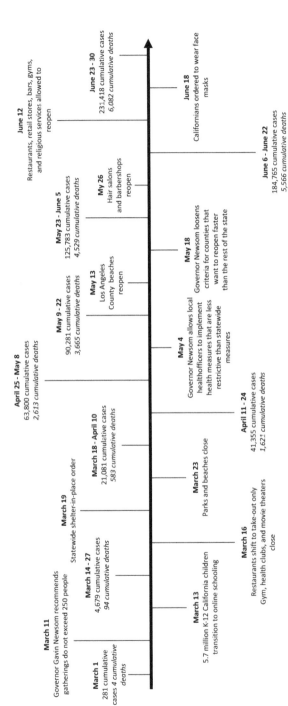

Figure 9.1. California COVID-19 progression and implemented health measures at the time of crime analysis, March–June 2020.
Sources: Cases/Deaths: Cumulative total on the last day of the two-week period; Source: Johns Hopkins – data from JHU CSSE Health measures: https://calmatters.org/health/coronavirus/2020/04/gavin-newsom-coronavirus-updates-timeline/; Cal Matters: https://www.gov.ca.gov/2020/03/11/california-public-health-experts-mass-gatherings-should-be-postponed-or-canceled-statewide-to-slow-the-spread-of-covid-19/; Cal Matters: https://calmatters.org/projects/schools-shut-down-in-massive-numbers-across-california-amid-coronavirus-fears/; Cal Matters: https://www.latimes.com/california/story/2020-03-16/gavin-newsom-california-restaurants-theaters-closures-coronavirus

commercial crimes (e.g., retail burglary) in Los Angeles by 64% for the same time period (Pietrawska et al., 2020).

Research focusing on crime trends across US cities points to mixed results, with declines for certain crimes while other crimes show no variation in trends (Ashby, 2020; Piquero et al., 2020) or even increases in some communities over others (Campedelli et al., 2020). While the pandemic continues to impact economies, politics, and personal lives, efforts to understand the impact of the pandemic on crime across time (e.g., years and months) and location (e.g., urban vs. rural) remain critical.

Research in crime and delinquency provides evidence that crime does not occur randomly across time and space (Johnson, 2010; Weisburd, 2015). Instead, crime events are more likely clustered and occur based on different types of social, economic, demographic, and ecological environments (Sampson & Lauritsen, 1994; Shaw & McKay, 1942; Weisburd, 2015). Considering these differences, emerging research has started to explore various forms of the environment to understand the dynamics of the pandemic-crime relationship.

The current research explores publicly available crime data focusing on selected cities in California before and during the pandemic, seeking to examine whether property crimes have changed as a result of pandemic-related restrictions or whether property crime rates continue to remain low in spite of health emergencies caused by the pandemic.

*Early Progression of COVID-19 and Crime (March–June 2020)*

In the first part of this section, property crimes and robbery are explored during the early progression of the COVID-19 pandemic (March through June 2020). We selected data from three of the largest five cities in California: Los Angeles, San Francisco, and San Diego. These three cities share similarities for reporting higher crime rates than the national average across all communities in the US, from the largest to the smallest. Based on 2018–19 FBI crime data, San Diego had a crime rate that was higher than 63% of the state's cities and towns of all sizes. Similarly, Los Angeles' crime rate was higher than 86% of the state's cities and towns of all sizes. In contrast, the majority of communities in California, more than 99%, had a lower crime rate than San Francisco. With a population of 883,305, San Francisco had a combined rate of violent and property crime that was very high compared to other places of similar population size.

We also extend the analysis to three cities of the greater Los Angeles area (more informally referred to as the Southland): Oxnard in Ventura County, Long Beach in neighbouring Los Angeles County, and Irvine as part of Orange County. Based on 2018 and 2019 FBI crime data, Oxnard and Long Beach shared higher crime rates compared to Irvine. Relative to other cities in California, Oxnard had a crime rate that was higher than 77% of the state's cities and towns of all sizes. Similarly, Long Beach had a crime rate that was higher than 86% of the state's cities and towns of all sizes. In contrast, Irvine's crime rate was lower than approximately 67% of California communities.

Before the outbreak of the pandemic, property crime rates and robberies in the USA showed a decrease in victimizations proportional to the population count (see also Argun & Daglar, 2016). This study focuses on robbery, burglary upon residential and non-residential places, vandalism, and motor vehicle theft. Considering the nature of these crimes, prior to the pandemic, these crimes were likely to be reported to police (see also Ashby, 2020). As Routine Activity Theory tries to explain why crime occurs in certain situations, it could be expected that the pandemic has also impacted the occurrence and frequency of selected property offences and robberies.

As we investigate the initial period of lockdowns and social distancing health measures between March and June 2020, we find declines in robberies (including residential and commercial robberies) for San Francisco (26%), San Diego (26%), Oxnard (23%), Long Beach (14%), and Irvine, with the lowest reports of robberies over the four-month period. Only Los Angeles reveals an increase in robberies of 17% between March and June 2020.

For burglaries, we note stark differences across the cities under examination. Mainly we find increases in commercial burglaries, and to a lesser extent, in residential burglaries for San Francisco (53%), Los Angeles (21%), and Irvine (44%) and a rather dramatic increase for Long Beach (94%) within the first four months of the coronavirus outbreak. Exploring the addresses for burglaries, Los Angeles and Long Beach show a higher frequency of commercial burglaries while San Francisco's increase of burglaries applies to both commercial and residential areas. One crime that notably increased across all cities during the first four months of COVID-19 was motor vehicle theft, with the highest increase for Oxnard (92%), Irvine (85%), and Long Beach (61%). In comparison, among the three larger cities, Los Angeles had an increase in motor vehicle theft of 55%, followed by San Francisco with an increase of 31% and San Diego 26%. These findings correspond with Ashby's (2020) research, who also noted significant increases in vehicle theft after the implemented shelter-in-place orders across cities in the United States. While data do not provide consistency in the reporting of locations for crime occurrence, researchers have suggested that vehicle theft likely happens in places of unattended cars, parking lots in residential areas (Ashby, 2020), or nearby unsupervised parking lots of businesses.

For vandalism, described more as an activity likely involving groups of peers (Boman & Gallupe, 2020), we found increases in mostly unattended areas – mall areas, parking lots, or unattended buildings – at the start of the first mandatory lockdown. For cities with available information on reported vandalism, Los Angeles reported a 12% increase and San Diego a 33% increase in vandalism within the first four months of lockdowns. All other cities of the

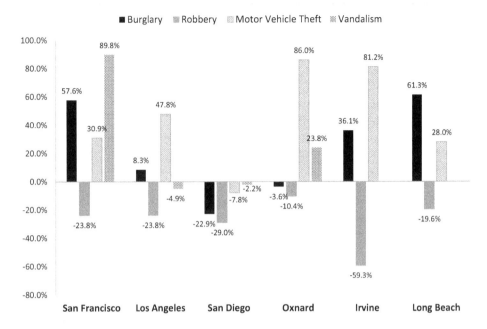

Figure 9.2. Percent change of crimes by selected California cities, March–June 2019 and 2020.
Note: Graph shows California's three largest cities and the greater Los Angeles area (Southland) with Ventura County (Oxnard), Los Angeles County (Long Beach), and Orange County (Irvine). Violent crime includes commercial and residential robbery. Property crime includes burglary (residential and commercial), motor vehicle theft (listed as auto theft in some databases), and vandalism (not available for Irvine and Long Beach at time of data analysis).

study showed no differences across the four months in vandalism. Overall, the findings for the first four months of initiated lockdowns and social distancing measures show some changes in crimes, such as declines in robberies and residential burglaries. In contrast, commercial burglaries and motor vehicle theft show increases and differences by location. Noting these findings, we could assume that the lockdowns are, to some extent, the reason for differences in robbery and selected property crimes. However, can the mandated lockdowns from the pandemic impact crime trends in a state that has seen historic lows in crime rates for the past decade?

In Figure 9.2, we compare selected crimes during the initial four months of the pandemic (between March and June 2020) to the same first four months of 2019. The only city to experience declines across all types of property crimes and robbery is San Diego. For all other cities, we find declines in robberies during the first four months in 2020 compared to the same time in 2019. As for burglaries, Long Beach, Irvine, and San Francisco show increases in commercial and residential burglaries compared to 2019. As already noted for the period between March and June 2020 comparison, motor vehicle theft has increased across all selected cities.

Compared to 2019, Oxnard and Irvine show an increase of over 80% in motor vehicle theft in 2020 compared to 2019. Los Angeles, San Francisco, and Long Beach follow; however, they show increases below 50% from 2019 to 2020 for motor vehicle theft. Data for vandalism were not equally available for each of the cities of study. However, we find the largest increase in vandalism in San Diego, with an increase of almost 90% compared to 2019.

While the descriptive data do not test for a direct interaction between mandated lockdowns and crime occurrences, we nevertheless insinuate that the early restrictions owing to the coronavirus pandemic have affected activities of commercial and residential robberies more than burglaries and motor vehicle thefts. The comparison of time periods between years reveals a slightly different picture, suggesting that location and type of crime activity might be interconnected. Initial areas of frequent property crime may have shifted because of tougher restrictions in one area over another, thus potentially moving crimes from one place to another; not necessarily stopping the crime activity, but instead, allowing for increases in crime in areas where restrictions were less enforced. For example, Shayegh and Malpede (2020) provide evidence of shifts in crime hot spots for Oakland and San Francisco before and after COVID-19 struck the area with restrictions. For some cities, common hot spots disappeared with implemented restrictions; however, other areas emerged, exhibiting a more even spread of crime across cities (Felson et al., 2020). Campedelli et al. investigated community-level trends in burglaries and robberies to understand how implemented health restrictions affected crime across neighbourhoods of one city. The researchers found that higher numbers of people at home in fact presented higher levels of capable guardianship and thus lowered opportunities for crime. Communities that strictly followed COVID-19 health orders saw fewer robberies or burglaries. Moreover, researchers found evidence of a link between the occurrence of robberies and burglaries, showing that declines in burglaries were strongly associated with declines in robberies and vice versa (Campedelli et al., 2020).

*A Snapshot over Time: Crime Occurrence 10 Months before and 10 Months into the Coronavirus Pandemic*

While we have seen declines and spikes, especially in burglary and motor vehicle thefts during the initial period of COVID-19, we extend our analysis and explore crimes over the first 10 months of the pandemic compared to the first 10

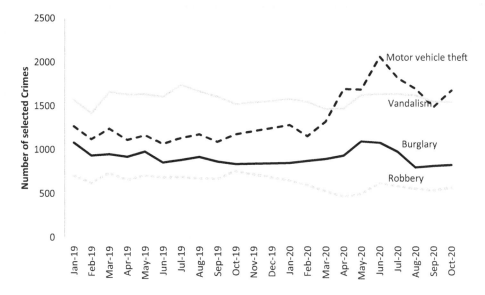

Figure 9.3. Selected crimes, Los Angeles, 2019–20.

months of 2019. Recalling the argument of the historical low of crime rates for property crime and even robbery, we investigated available crime data for three of the largest cities in California, San Francisco, Los Angeles, and San Diego.

Allowing for a snapshot over time, for Los Angeles, Figure 9.3 illustrates that all types of property crimes and robbery, except motor vehicle theft, are somewhat steady between 2019 and 2020. We might see some small spikes of an increasing trend in 2020, when, compared to 2019, the frequencies of these selected crimes are not significantly different from the previous year.

For San Francisco (Figure 9.4), we observe increases in burglary, especially commercial burglaries, between April and June of 2020 compared to 2019. Followed by burglaries is the frequency of motor vehicle theft, which spiked particularly after June through August 2020. For robberies, compared to 2019, we see declines starting in March 2020, likely corresponding with the stay-home restrictions and shifts in daily routine activities, suggesting fewer crime opportunities owing to possibly greater guardianship over homes and within communities. For vandalism we do not find significant changes between 2019 and 2020.

Among the three cities, San Diego shows no significant change for any of the selected crimes between 2019 and 2020 (see Figure 9.5). Similarly, in the initial months of the pandemic, we continue to see frequencies of lower property crimes and robbery. In comparison to Los Angeles and San Francisco, the data for San

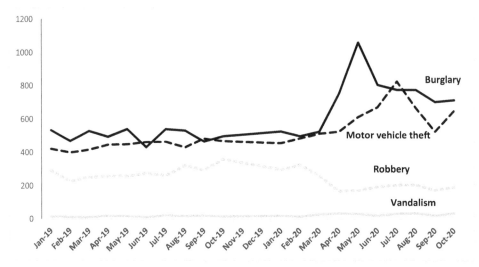

Figure 9.4. Selected crimes, San Francisco, 2019–20.

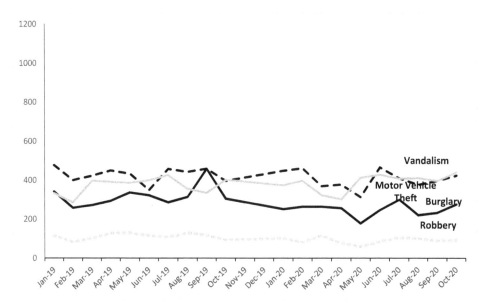

Figure 9.5. Selected crimes, San Diego, 2019–20.

Diego suggest no striking variation in property crimes when compared to the previous year. While stay-in-shelter orders have been implemented throughout the state, the question is whether citizens in the three cities responded differently to these implemented orders. We find no changes or increases in these non-violent crimes in San Diego compared to Los Angeles or San Francisco.

Figures 9.3, 9.4, and 9.5 offer a snapshot of the differences in burglary, robbery, motor vehicle theft, and vandalism before and during the pandemic. The view across months provides an overview of the selected crime categories. The descriptive results indicate some differences for certain offences across the three cities. Future research should consider further attending to the changing distribution in property crimes and robbery accounting for the critical periods of newly implemented lockdowns, including non-compliance with these mandates by businesses and citizens. Do these shifts in stay-at-home orders change the effectiveness of guardianship over property from residential to non-residential areas or will crimes move from larger cities to adjacent cities/areas? The short- and long-term responses by businesses and citizens to social distancing and stay-at-home orders could add to the analysis of crime occurrence across location and time. To regain control over social environments (i.e., the spread of the coronavirus), behaviours of compliance or non-compliance may offer insights into the shift of non-violent crimes within cities and across cities.

Our analyses depended on publicly available crime data. Although residential and commercial offences for burglary and robbery allowed for the untangling of locations and smaller areas within cities, for the selected cities of study, existing gaps of locations and geographical records across the major cities and the cities of the greater Los Angeles area have somewhat limited the current comparison among the selected crimes of study. Moreover, inconsistencies in recording of crime events by date reported versus date occurred add artificial variation in crimes, as the dates of occurrence and reporting often do not coincide. Crimes may have occurred months before they were officially reported. We tried to detect these gaps in our analyses. In light of our findings and the observed limitations, we conclude the chapter with future considerations for using the pandemic's impact to expand analyses of intervening factors to further understand the short- and long-term effects of the relationship between the pandemic and frequency and patterns of crime.

**Crime Analysis of the Pandemic: Future Considerations**

While the findings of this study are only descriptive in nature, they nevertheless do not provide support for a substantial decline in crime across selected Californian cities. Similar to the initial studies examining the relationship between COVID-19 and crime, we find that for some cities, the implemented shelter-in-place order seems to be connected to declines in certain crimes. Yet, for other areas and categories of crime, the view across time does not show a significant overall change in crimes.

Early research findings on the topic report that, nationwide, cities have seen changes in offence patterns (Campedelli et al., 2020) and declines in overall crime (Ashby, 2020; Mohler et al., 2020; Shayegh & Malpede, 2020). However, nationwide declines in crime might be clouded by important differences in crime data collection across cities, states, and regions. As we see differences in coronavirus infections across states and regions because of differences in health measures, crime trends may also be different, not because of stricter policies, but because of differences in policing and reporting by region, city, and neighbourhood. The Routine Activity approach taps into understanding the influences of social context on opportunities of crime. Thus, the implications of stricter health measures in one state or county may provide different crime results compared to states or counties with more relaxed health orders. Alternatively, in states or areas where individuals – as a personal choice – oppose statewide mandated orders, perhaps guided by their poor risk assessment of disease transmission and influenced by mixed messages from various formal and informal information channels, the opportunities for property crimes may persist.

The current findings on the selected cities in California invite further discussion about the relationship between effects of the pandemic and the occurrence of crime. For example, when we compare property crime and robbery across cities and between years, the results do not tell a consistent story of declines in crime. Describing crime trends from one year to the next or across months, we employ percentage changes to reflect on the actual increase or decrease in crime before and during the pandemic. As presented in Routine Activity Theory, crimes change based on changes in behaviour at the level of society at large (Felson & Boba, 2010). What does this mean for the collection of crime data during the current pandemic?

While opportunities for crime have shifted for certain crimes during the pandemic (e.g., from residential to commercial or vice versa), crime reporting may have also experienced changes. At the time of this study, in-person encounters with the police were required in order to report a stolen vehicle or other stolen property. Thus, fearful of contracting the virus or wanting to avoid in-person contact, people (i.e., victims of crime) may be less willing to report property crime victimization to the police, which will consequently impact the reporting of property crimes and crime rates. As the 2018 data reveal, reporting of property crimes has already been less, even under non-pandemic circumstances (Morgan & Oudekerk, 2019). Currently, official crime data come from law enforcement agencies. Records of crimes are reported to or by the police. Thus, the reporting of crime and its documentation are dependent on police practices. The implemented restrictions of the pandemic may have already affected police departments and crime reporting. On the one hand, how citizens take the risk (or not) and report property crimes in person potentially impacts the reporting of crimes. On the other hand, police officers or those who have themselves fallen ill might minimize unnecessary contact with the public and may consequently respond to certain offences differently than before the pandemic. All these factors have continued and will continue to impact the reporting of non-violent crimes such as property crimes and the reporting of violent crimes like armed robberies, for example.

Since the pandemic, the reporting of crimes may not entirely correspond with the reality of existing crimes: some of the declines found in COVID-19-related crime research might already be the result of changes in policing rather than changes in criminal behaviour. Crime reporting will not be diminished; however, it may take time to prioritize and implement more efficient reporting practices that meet the new normal for catching crime and providing valuable data. These anticipated changes in reporting and data collection during the pandemic may ultimately affect data output and analysis. Moreover, as the pandemic's current direction continues to interrupt our everyday lives, crimes may increase with time, not because of the pandemic but because of rising economic insecurity and growing economic inequalities. Thus, it will be essential to assess multiple data sets that account for a range of criteria (i.e., including event criteria) to allow for a comprehensive analysis of current and future crime trends.

## Conclusion

Looking ahead, many states in the US have relaxed their constraints on daily life. It will remain of interest, and not only to criminologists, to see if crime rates are returning to previous levels. Or might we encounter a new normal in activities of crime? The pandemic offers a unique opportunity to examine in more detail the crisis-crime relationship over time. It also allows for investigating the interplay between social forces and the varying levels of crime patterns across countries, states, cities, communities, and areas within the same neighbourhood. This pandemic has undoubtedly brought a shock to society, and to understand its role in the impact of crime, each of the current studies potentially contributes to the puzzle we have yet to attend – explaining the pandemic-crime relationship. The currently existing studies already inform us about limitations within the available data, but they also invite researchers to delve into new methodologies to describe the pandemic-crime relationship. Furthermore, the pandemic also calls for innovation in crime prevention that potentially prepares us for an era of a new normal in crime analysis and beyond.

*Acknowledgments*: I am grateful for my collaboration with Emily Cooper, who co-wrote and led the undergraduate research team for this project and is currently an MA student at George Mason University in the Department of Criminology, Law and Society. We also acknowledge former Cal Poly Pomona students Jaymee Montoya (MA UC-Irvine), Nicole Goforth, and Julian Perales, who worked diligently on compiling and preparing the final data for this project.

REFERENCES

Agnew, R. (1992). Foundation for a general strain theory of crime and delinquency. *Criminology*, *30*(1), 47–88. https://doi.org/10.1111/j.1745-9125.1992.tb01093.x.

Argun, U., & Dağlar, M. (2016). Examination of routine activities theory by the property crime. *Journal of Human Sciences*, *13*(1), 1188–98. https://doi.org/10.14687/ijhs.v13i1.3665.

Ashby, M.P. (2020). Initial evidence on the relationship between the coronavirus pandemic and crime in the United States. *Crime Science* *9*(6), 1–16. https://doi.org/10.1186/s40163-020-00117-6.

Barsky, L., Trainor, J., & Torres, M. (2006). Disaster realities following Katrina: Revisiting the looting myth. National Hazards Center, Institute of Behavioral Science, Boulder, CO, 53, 215–34. http://udspace.udel.edu/handle/19716/2367.

Boman, J.H., & Gallupe, O. (2020). Has COVID-19 changed crime? Crime rates in the United States during the pandemic. *American Journal of Criminal Justice*, *45*, 537–45. https://doi.org/10.1007/s12103-020-09551-3.

Bos, Kirsten I., Schuenemann, V.J., Golding, G.B., Burbano, H.A., Waglechner, N., Coombes, B.K., McPhee, J.B., DeWitte, S.N., Meyer, M., Schedes, S., Wood, J., Earn, D.J.D., Herring, D.A., Bauer, P., Poinar, H.N., & Krause, J. (2011). A draft genome of Yersinia pestis from victims of the Black Death. *Nature*, *478*(7370), 506–10. https://doi.org/10.1038/nature10549.

Brantingham, P.J., & Brantingham, P.L. (1984). *Patterns in crime*. Macmillan.

Brown v. Plata. (2011, May 23). Supreme Court of the United States. N o. 0 9 - 1233. https://www.supremecourt.gov/opinions/10pdf/09-1233.pdf.

California Department of Finance (DOF). (2019, May 2019). E-4 population estimates for cities, counties, and the state, 2011–2019 with 2010 census benchmark. Sacramento, California. http://www.dof.ca.gov/Forecasting/Demographics/Estimates/E-4/2010-19/.

California Department of Justice (DOJ). (2020). Crimes & clearances. *Open Justice*. Retrieved 1 November 2020, from https://openjustice.doj.ca.gov./data.

Campedelli, G.M., Favarin, S., Aziani, A., & Piquero, A.R. (2020). Disentangling community-level changes in crime trends during the COVID-19 pandemic in Chicago. *Crime Science*, *9*(21), 1–18. https://doi.org/10.1186/s40163-020-00131-8.

Cantor, D., & Land, K.C. (1985). Unemployment and crime rates in the post–World War II United States: A theoretical and empirical analysis. *American Sociological Review*, *50*(3), 317–32. https://doi.org/10.2307/2095542.

Center on Juvenile and Criminal Justice (CJCJ). (2018). Crime in California cities remains stable through justice reform era (2010–2017). http://www.cjcj.org/news/11934.

Center on Juvenile and Criminal Justice (CJCJ). (2019). Urban crime declines in 2018: A positive trend in California's justice reform era. http://www.cjcj.org/news/12529.

Choe, J. (2008). Income inequality and crime in the United States. *Economics Letters*, *101*(1), 31–3. https://doi.org/10.1016/j.econlet.2008.03.025.

Cohen, L.E., & Felson, M. (1979). Social change and crime rate trends: A routine activity approach. *American Sociological Review*, *44*(4), 588–608. https://doi.org/10.2307/2094589.

Cohen, L., Felson, M., & Land, K.C. (1980). Property crime rates in the United States: A macro dynamic analysis, 1947–1977; with ex ante forecasts for the mid-1980s. *American Journal of Sociology*, *86*(1), 90–118. https://doi.org/10.1086/227204.

Cromwell, P., Dunham, R., Akers, R., & Lanza-Kaduce, L. (1995). Routine activities and social control in the aftermath of a natural catastrophe. *European Journal on Criminal Policy and Research*, *3*, 56–69. https://doi.org/10.1007/BF02242928.

Degarmo, M. (2011). Understanding the comparisons of routine activities and contagious distributions of victimization: Forming a mixed model of confluence and transmission. *International Journal of Criminology and Sociological Theory*, *4*(1), 584–603.

Doyle, J.M., Ahmed, E., & Horn, R.N. (1999). The effects of labor markets and income inequality on crime: Evidence from panel data. *Southern Economic Journal*, *65*(4), 717–38. https://doi.org/10.1002/j.2325-8012.1999.tb00196.x.

Elmes, G.A., Roedl, G., & Conley, J. (Eds.). (2014). *Forensic GIS: The role of geospatial technologies for investigating crime and providing evidence*. Geotechnologies and the Environment, 11. Springer.

Federal Bureau of Investigation (FBI). (2020). Table 4: Offenses reported to law enforcement, by state by cities 100,000 and over in population. *Preliminary Semiannual Uniform Crime Report, January–June 2019*. https://ucr.fbi.gov/crime-in-the-u.s/2019/preliminary-report.

Felson, M., & Boba, R.L. (2010). *Crime and everyday life*. 4th ed. Sage.

Felson, M., & Clarke, R.G. (1998). *Opportunity makes the thief: Practical theory for crime prevention*. Police Research Series Paper No. 98. Research, Development and Statistics Directorate, London. https://popcenter.asu.edu/sites/default/files/opportunity_makes_the_thief.pdf_.

Felson, M., Jiang, S., & Xu, Y. (2020). Routine activity effects of the COVID-19 pandemic on burglary in Detroit, March (2020). *Crime Science* 9(10). https://doi.org/10.1186/s40163-020-00120-x.

Fischer, H.W., III. (1998). *Response to disaster: Fact verses fiction and its perpetuation: The sociology of disaster*. University Press of America.

Frailing, K., & Harper, D.W. (2017). *Toward a criminology of disaster: What we know and what we need to find out*. Springer.

Fritz, C.E. (1961). Disaster. In Merton, R.K., & Nisbet, R.A. (Eds.), *Contemporary social problems: An introduction to the sociology of deviant behavior and social disorganization* (pp. 651–94). Harcourt, Brace & World.

Gould, E.D., Weinberg, B.A., & Mustard, D.B. (2002). Crime rates and local labor market opportunities in the United States: 1979–1997. *Review of Economics and Statistics*, *84*(1), 45–61. https://doi.org/10.1162/003465302317331919.

Gramlich, J. (2020, November 20). What the data says (and doesn't say) about crime in the United States. Pew Research Center. https://www.pewresearch.org/fact-tank/2020/11/20/facts-about-crime-in-the-u-s/.

Gupta, A.G., Moyer, C.A., & Stern, D.T. (2005). The economic impact of quarantine: SARS in Toronto as a case study. *Journal of Infection*, *50*(5), 386–93. https://doi.org/10.1016/j.jinf.2004.08.006.

Johnson, S.D. (2010). A brief history of the analysis of crime concentration. *European Journal of Applied Mathematics*, *21*(4–5), 349–70. https://doi.org/10.1017/S0956792510000082.

Karlsson, M., Nilsson, T., & Pichler, S. (2014). The impact of the 1918 Spanish flu epidemic on economic performance in Sweden: An investigation into the consequences of an extraordinary mortality shock. *Journal of Health Economics 36*, 1–19. https://doi.org/10.1016/j.jhealeco.2014.03.005.

Kelly, M. (2000). Inequality and crime. *Review of Economics and Statistics*, *82*(4), 530–9. https://doi.org/10.1162/003465300559028.

Kleck, G., & Chiricos, T. (2002). Unemployment and property crime: A target-specific assessment of opportunity and motivation as mediating factors. *Criminology*, *40*(3), 649–80. https://doi.org/10.1111/j.1745-9125.2002.tb00969.x.

Land, K.C., & Felson, M. (1976). A general framework for building dynamic macro social indicator models: Including an analysis of changes in crime rates and police expenditures. *American Journal of Sociology*, *82*(3), 565–604. https://doi.org/10.1086/226354.

LeBeau, James L. (2002). The impact of a hurricane on routine activities and on calls for police service: Charlotte, North Carolina, and Hurricane Hugo. *Crime prevention and community safety*, *4*(1), 53–64. https://doi.org/10.1057/palgrave.cpcs.8140114.

Leitner, M., Barnett, M., Kent, J., & Barnett, T. (2011). The impact of Hurricane Katrina on reported crimes in Louisiana: A spatial and temporal analysis. *Professional Geographer*, *63*(2), 244–61. https://doi.org/10.1080/00330124.2010.547156.

Leitner, M., & Helbich, M. (2011). The impact of hurricanes on crime: A spatio-temporal analysis in the city of Houston, Texas. *Cartography and Geographic Information Science*, *38*(2), 213–21. https://doi.org/10.1559/15230406382213.

Lemieux, F. (1999). The impact of major social disruption on criminal opportunities and relative frustration: A case study. *Canadian Journal of Criminology and Criminal Justice*, *46*(1), 45–72. https://doi.org/10.3138/cjccj.46.1.45.

Leslie, E., & Wilson, R. (2020, May 14). Sheltering in place and domestic violence: Evidence from calls for service during COVID-19. *Journal of Public Economics*, forthcoming. Available at SSRN: https://doi.org/10.2139/ssrn.3600646.

Lofstrom, M., & Martin, B. (2016, April 9). Crime trends in California. Public Policy Institute of California, San Francisco, CA. Retrieved 9 April 2020, from https://sbud.senate.ca.gov/sites/sbud.senate.ca.gov/files/FullC/PPIC_Crime_Trends_JTFFinal.pdf.

Merton, R. (1938). Social structure and anomie. *American Sociological Review*, 3(5), 672–82. https://doi.org/10.2307/2084686.

Mervosh, S., Lu, D., & Swales, V. (2020, April 20). See which states and cities have told residents to stay at home. *New York Times*. https://www.nytimes.com/interactive/2020/us/coronavirus-stay-at-home-order.html.

Mileti, D.S. (1999). *Disasters by design: A reassessment of natural hazards in the United States*. Joseph Henry Press.

Mohler, G., Bertozzi, A.L., Carter, J., Short, M.B., Sledge, D., Tita, G.E., Uchida, C.D., & Brantingham, P.J. (2020). Impact of social distancing during COVID-19 pandemic on crime in Los Angeles and Indianapolis. *Journal of Criminal Justice*, 68(May–June), 101692. https://doi.org/10.1016/j.jcrimjus.2020.101692.

Morgan, R.E., & Oudekerk, B.A. (2019, September). Criminal victimization, 2018. Bureau of Justice Statistics. NCJ 253043. https://bjs.ojp.gov/content/pub/pdf/cv18.pdf.

Pietrawska, B., Aurand, S.K., & Palmer, W. (2020). COVID-19 and crime: CAP's perspective on crime and loss in the age of COVID-19. *CAP Index*, Issue 2. https://capindex.com/wp-content/uploads/2020/05/CAP-Index-and-COVID-19-Issue-2-Los-Angeles-Crime-and-the-Routine-Activities-Theory-040720.pdf.

Piquero, A.R., Riddel, J.R., Bishop, S.A., Narvey, C., Reid, J.A., & Piquero, N.L. (2020). Staying home, staying safe? A short-term analysis of COVID-19 on Dallas domestic violence. *American Journal of Criminal Justice*, 45(4), 601–35. https://doi.org/10.1007/s12103-020-09531-7.

Prelog, A.J. (2016). Modeling the relationship between natural disasters and crime in the United States. *Natural Hazards Review*, 17(1), February. https://doi.org/10.1061/(ASCE)NH.1527-6996.0000190.

Proposition 47. (2014). Criminal Sentences. Misdemeanor penalties. https://repository.uchastings.edu/ca_ballot_props/1323.

Proposition 57. (2016). Criminal sentences. Parole. Juvenile criminal proceedings and sentencing. Initiative constitutional amendment and statute. https://repository.uchastings.edu/ca_ballot_props/1350.

Quarantelli, E.L. (1994). Disaster studies: The consequences of the historical use of a sociological approach in the development of research. *International Journal of Mass Emergencies and Disasters*, 12(1), 25–49. https://doi.org/10.1177/028072709401200102.

Raphael, S., & Winter-Ebmer, R.. (2001). Identifying the effect of unemployment on crime. *Journal of Law and Economics*, 44(1), 259–83. https://doi.org/10.1086/320275.

Ratcliffe, J. (2010). *Crime mapping: Spatial and temporal challenges*. In Piquero A., & Weisburd, D. (Eds.), *Handbook of quantitative criminology* (pp. 5–24). Springer. https://doi.org/10.1007/978-0-387-77650-7_2.

Sampson, R.J., & Lauritsen, J.L. (1994). Violent victimization and offending: Individual-, situational-, and community-level risk factors. In Roth, J.A., & Reiss, A.J., Jr. (Eds.), *Understanding and preventing violence, volume 3: Social influences* (pp. 1–114). National Academy Press.

Shaw, C.R., & McKay, H.D. (1942). *Juvenile delinquency and urban areas*. University of Chicago Press.

Shayegh, S., & Malpede, M. (2020, April 2). Staying home saves lives, really! *SSRN*. https://dx.doi.org/10.2139/ssrn.3567394.

Stickle, B., & Felson, M. (2020). Crime rates in a pandemic: The largest criminological experiment in history. *American Journal of Criminal Justice*, 45, 525–36. https://doi.org/10.1007/s12103-020-09546-0.

Thornberry, T.P., & Christenson, R.L. (1984). Unemployment and criminal involvement: An investigation of reciprocal causal structures. *American Sociological Review*, *49*(3), 398–411. https://doi.org/10.2307/2095283.

Tumpey, T.M., Basler, C.F., Aguilar, P.V., Zeng, H., Solórzano, A., Swayne, D.E., Cox, N.J., Katz, J.M., Taubenberger, J.K., Palese, P., Garcia-Sasre, A. (2005). Characterization of the reconstructed 1918 Spanish influenza pandemic virus. *Science*, *310*(5745), 77–80. https://doi.org/10.1126/science.1119392.

Varano, S.P., Schafer, J.A., Cancino, J.M., Decker, S.H., & Greene, J.R. (2010). A tale of three cities: Crime and displacement after Hurricane Katrina. *Journal of Criminal Justice*, *38*(1), 42–50. https://doi.org/10.1016/j.jcrimjus.2009.11.006.

Weisburd, D. (2015). The law of crime concentration and the criminology of place. *Criminology*, *53*(2), 133–57. https://doi.org/10.1111/1745-9125.12070.

Young, T.J. (1993). Unemployment and property crime: Not a simple relationship. *American Journal of Economics and Sociology*, *52*(4), 413–16. https://doi.org/10.1111/j.1536-7150.1993.tb02564.x.

Zahran, S., Shelley, T.O., Peek, L., & Brody, S.D. (2009). Natural disasters and social order: Modeling crime outcomes in Florida. *International Journal of Mass Emergencies and Disasters*, *27*(1), 26–52. https://doi.org/10.1177/028072700902700102.

# 10 Employing Lyn Lofland's and Ray Oldenburg's Urban Sociology to "Read" the Emptying of Los Angeles' Publics

BY JACK FONG

## Introduction

At the time of this writing, during April 2020, the COVID-19 infection has spread across the United States, rendering it the epicentre of infections after Italy and Spain as well as major countries in Europe such as Germany and the United Kingdom. Much of Asia has already responded to the onslaught of COVID-19, whether through hardcore measures, seen in China, or through comparatively less stringent measures, seen in South Korea, India, Vietnam, and Thailand, and also elsewhere, such as in Sweden, Germany, and New Zealand. For a dysfunctional response to the coronavirus, the United States claims top prize, however, since its narrative regarding the virus's effects on biology and physiology has blatantly neglected to address the sociological implications of how COVID-19 has reconfigured urban spatiality for the worse. Not only has the corporeal dimension of human existence been affected by the pandemic, but also communities, economies, and the social fabric of life – interpersonal communication *and* recreation – have all been exponentially stunted, if not arrested during our regressive period of social distancing. I thus employ the greater metropolitan area of Los Angeles, California, where I am living during the pandemic, to draw some extrapolations on urban dynamics when systemic crisis stunts its operations and reconfigures its spatiality.

My chapter forwards the view that, during this period of social distancing, revitalized ideas of urban sociologists Lyn Lofland and Ray Oldenburg can offer a reading on how to understand the components and dynamics of the public realm and its ability to influence urban residents' interactions with one another (Lofland, 1973, 1998). The process will also make operative the concept of "third places," Oldenburg's term for those areas of society designed to enable a controlled release of the angst and tensions of communities through recreative and communicative interaction (Oldenburg, 1999). Although many scholars have since expanded on Lofland's and Oldenburg's timeless analyses, the fact that they were able to tease out the contours of urban forces allows us to systematize urban dynamics into variables that show varying degrees of human interaction in urban contexts. Moreover, both Lofland's and Oldenburg's insights are prescient, allowing for cautious extrapolations that will inform how this chapter illuminates a variety of distancing dynamics that have taken place in the context of the Greater Los Angeles area.

By illuminating timely narratives about how urban and community dynamics potentially deform during a systemic crisis that mandates exceptional social regulation and physical distancing, a sociological imagination that can explain the process by which our public is emptied will have great implications for understanding how societies cope during systemic crises of a non-conflict nature. Staring into the hollowing publics of societies and cities at the time of this writing, a condition where the social actor primarily emerges only to attend to material culture that mitigates any absolute deprivations, offers us cues that we are on the verge of understanding about how social life is emaciated when social systems are, if only temporarily, rendered catatonic. The sociological imagination for my chapter is thus shaped by these reflections, and I hope readers will consider it useful, if only because there is no "public" at the time of this writing to serve as an antipode of alternative contexts. That is, such an absence of the social is itself epistemological in that it reveals how crisis reconfigures urban social relations under duress. Such a reconfiguration has already ushered in for many actors the existential threat to the social self in relation to demographics (e.g., particular age groups most vulnerable to the pandemic), urban dynamics (the desocializing of the interactive urban experience, a process that includes being unemployed), and/or racialization (as in the "coughing as Asian" stigma, for example, that has befallen those phenotypically Orientalized as Far East Asian in many urban areas of the United States).

My chapter examines the implications of the aforementioned factors by analysing social distancing in the United States during the pandemic. It will harness Lofland's and Oldenburg's analyses of urban life to make visible specific social forces and urban spaces that, under pandemic conditions, have been emptied from our various publics. This chapter thus attempts to draw some rudimentary conclusions about how processes of deurbanization appear during systemic crisis. By illuminating the content and contours of deurbanization during health-related systemic crisis, we can consider new insights about the destructive and restorative consequences of pandemic distancing. My chapter hopes to convey such considerations in a manner that informs public policy orientations that can anticipate future outcomes and discontents related to similar crises.

To undertake such a task, however, requires us to analyse the urban experience through its material consequences. In a normative prescription, I write this chapter to respond to social conditions that have an immediate impact on one's corporeal, urban existence. Given acute liminalities and urgencies framing our pandemic, the flow of my argument aims to exhibit a somewhat frenetic and urgent momentum, slowed only by our discussion and amalgamation of Lofland's and Oldenburg's ideas, which we amplify to read the antipode of a thriving urbanism: deurbanization due to systemic crisis.

## The Relevance of Lofland's Ideas for Understanding the Pandemic

By harnessing Lofland's ideas in a manner that can be made to amalgamate with observations made by Oldenburg, my chapter attempts to illuminate how local, state, and federal mandates for social distancing during the pandemic have resulted in removing from civil society the apolitical process of familiarizing the actor with the "unknown" – the "other" – that is, the stranger. I begin with Lofland's classic 1973 work *A World of Strangers*. In later treatment, I consider her other timeless work, *The Public Realm* (1998).

For Lofland, living in the modern or industrial city means to be exposed, if not immersed, in conditions where anonymity exists between urban residents, where urban residents' stress and coping mechanisms are shaped by their interactions with strangers, and where the detached and dispersed spaces in the city reinforce such "strangeness." Lofland did not criticize this condition as much as illuminate it as a taken-for-granted facet of the urban social experience. For Lofland, being a stranger is not a hermetically sealed experience, since some communication continues unabated: the city dweller is always "picking up information about the other" and "simultaneously … giving off information about themselves which the other codes and acts upon" (1973, p. 97). Such a skill, if one can use the term, is born from one's immersion in city life. In contrast, rural communities enhance the private and parochial; they are spaces where community prevails, a world of "the neighborhood, workplace or acquaintance networks" (Lofland, 1998, p. 10), but not of strangers.

Lofland defines the stranger as someone whom an actor does not know, someone with "whom we are not acquainted, whom we have never met, even though we may possess a great deal of biographic information about the person" (Lofland, 1973, p. 17). That is, a stranger is "any person who is personally unknown to the actor of reference" (p. 18). "Stranger," then, is not a pejorative term. In contrast, a non-stranger is one whose social bonds occupy a temporal dimension that is of longer duration than any one moment, requiring the individuals involved to "share … personal worlds" (p. 17). Lofland divides non-strangers into three types: those we know casually, those with whom we are familiar, and those with whom we are intimate (p. 17). Furthermore, she provides further discernment of the stranger; that is, whether an individual "knowing" a celebrity would mean that the notion of stranger is now nullified, a proposition which Lofland denies: "while a celebrity is personally known to many persons, the majority of these are not personally known" to the celebrity (p. 18).

For Lofland, many of our accumulated stressors living in the modern city stem from interactions we have with strangers who populate the majority of our lifeworld spaces. Even the automobile is complicit in maintaining urban anonymity, since it "allows its passengers to move through the public sectors of the city encased in a cocoon of private space ... without ever encountering for more than a few brief seconds the world of strangers" (1973, p. 136). Lofland emphasizes outright in the preface of her work that to live in the city is to live "surrounded by large numbers of persons whom one does not know," and that "to experience the city is ... to experience anonymity" (1973, p. vii). Yet it should be emphasized that experiencing anonymity with strangers should by no means be fully equated with a lack of civility. We have as exemplars the good Samaritan who performs CPR on a citizen experiencing a cardiac episode or even a shopper's concession of a place in the supermarket queue when another party arrives at the checkout line. The experience of anonymity cannot always be seen with sanguine eyes, however, as strangers can also hurl verbal insults and engage in physical assaults against other members of the urban environment. Moreover, Lofland reminds us that even in recreation, the majority of instances of human interaction are constituted by our interactions with strangers, a process that roots our daily experiences in large collectives shaped by social structure *and* infrastructure, social environs operated and populated by many people we do not know.

Lofland's later work, the *Public Realm* (1998), extrapolates from *A World of Strangers* by reassuringly presenting outcomes that point to a variety of cooperative social dynamics enabling urban residents – even if they are strangers to one another – to maintain synchronicity in how they spatialize mobility without exacting a currency: territorializing spaces at the expense of other strangers (Lofland, 1998). However, it is in *A World of Strangers* that Lofland asserts that there are consistent behavioural tendencies in urban life, even if urban actors are unaware of them, manifesting between those navigating the public realm. Indeed, many situational awareness tendencies are frequently exhibited by residents not fully mindful of how they manoeuvre publicly in ways that actually add to urban order. By the release of the *Public Realm*, Lofland systematically expands upon our aforementioned traits. Lofland begins her analyses with *cooperative motility*, a term derived from biology and other natural sciences to refer to how organisms move in unison, as seen, for example, when a large flock of geese fly together without crashing into one another in ways that would cause physical harm. In the urban environment, cooperative motility almost always ensures safe physical outcomes for urban residents walking on a crowded sidewalk or pier, or in an airport or stadium. Furthermore, it allows residents to navigate around "inanimate objects, animate objects, and inanimate objects propelled or inhabited by animate ones (e.g., doors, buses, automobiles, elevators)" (1998, p. 29). Lofland colourfully notes:

> Most of the time our movement through the public realm is simply uneventful, and it is so because humans are cooperating with one another to make it so ... most of us get through doors without incident, most pedestrians don't collide with other pedestrians, most buses and cars do not flatten human beings, and most people do not get body parts crushed by closing elevator doors. (1998, p. 29)

The second tendency exhibited by many urban residents is *civil inattention*, a concept wonderfully elaborated by the sociologist Erving Goffman in *Behavior in Public Spaces* (1963) and *Relations in Public* (2010). Civil inattention refers to how urban residents are conditioned to not be fazed by other individuals, so as to avoid "conversational entanglements" (Goffman, 2010, p. 219) and properly go "about one's own separate business" (Goffman, 2010, p. 267). It involves "so delicate an adjustment" (Goffman, 1963, p. 85) of social behaviour that urban "passers-by" nonetheless "owe each other" (Goffman, 2010, p. 204).

*Audience role prominence* is another state of being for urban residents. Somewhat related to civil inattention, this role enables the individual to be engaged primarily in an audience role, a role that reinforces comfortable inconspicuousness. By being inconspicuous – by "blending in" – the urban resident can fully attend to the art of seeing the progression of unstaged urban practices, occurrences, and performances. Whereas civil inattention essentially refers to the art of ignoring people, engaging in a prominent audience role positions the urban resident to be a people watcher. Such a role allows the spectator to witness street musicians, courtship rituals, public arguments, or tragedies resulting from accidents and other health emergencies (Love, 1973; Whyte, 1980; Palmer, 1983; Harrison, 1984; McPhail, 1987).

*Restrained helpfulness* refers to how urban residents deliberately limit themselves from being too helpful. Although such a response appears rather insensitive, many of us living in cities have frequently been practitioners of this gesture. One can argue that in the "malfunction" of restrained helpfulness, we have the notorious bystander effect that enabled the 1964 murder of Kitty Genovese in New York City. In less graphic scenarios, however, restrained helpfulness is a common coping mechanism of urban residents. In practice, restrained helpfulness can still build community in that some urban residents will usually provide directions for someone who is lost. However, long-term friendships will rarely form, and all parties eventually part ways, returning to their role as strangers. What ultimately constrains this temporary assistance is, of course, the fear of the "streets," an American view that Lofland believes spans "two hundred years" and has "generated negative emotions among some proportion of the population" (1998, p. 151). What Lofland explicitly makes visible is that, in the context of the United States, urban residents tend to be afraid of their public spaces even in the best of times. This anxiety shapes their views on urban communities and delimits the degree to which they will assist strangers.

Finally, urban residents navigating the public realm are more tolerant of diversity. Thus, *civility towards diversity* offers the urban fabric its multicultural motifs. Urban residents might actually enjoy such diversity when walking into a restaurant where the proprietor does not speak the language of the patron. In a best-case scenario, Todd Pittinsky (2011) describes those who covet community diversity as exhibiting allophilia (see also Pittinsky & Montoya, 2009, and Pittinsky, Rosenthal, & Montoya, 2011). In such scenarios, patrons visiting different ethnic enclave markets will relish the cultural experience. All these aforementioned dynamics work together to constrain and enhance one another:

> Civil inattention, for example, ensures that people will not become so thoroughly engrossed ... that they are unable to maintain the level of alertness necessary for cooperative motility and will also be available to respond to requests for restrained helpfulness. Audience role prominence is fueled by the environmental scanning and alertness involved in cooperative motility and itself fuels the mind-set required for both restrained helpfulness and civility toward diversity. (Lofland, 1998, p. 33)

In spite of the many challenges that appear to work against communal bonds in the aforementioned scenarios, attempts do exist to minimize the "strangeness" in the other. In *A World of Strangers*, Lofland envisions how urban residents are quite versed in "figuring out" something about an unknown person. Such minimization of strangeness manifests through deduction from locations and public spaces of interaction. This process of *spatial ordering* allows the industrial urban resident to contextualize unknown persons by where they are from as a means of minimizing their strangeness. In the modern city, "a man is where he stands," creating a culture where city dwellers can link "who" to "where" (Lofland, 1973, pp. 82–3).

Spatial ordering is a more informative process for minimizing strangeness in the stranger because in *industrial* society a variety of previously public acts have retreated into the private sphere. In the wake of such a retreat, idealized articulations from communities known for their industrial outputs, ethnic/religious expressions, or linguistic uniqueness surface to spatially enrich the cultural content of urban areas. Experiencing the city in such a manner, one must learn the meanings inherent to locations, "what is expected to go on where and who is to be expected to be doing it" (1973, p. 101). Lofland observed in her research that how subjects understood their city depended on their "being told about this or that location or type of location by this or that family member, friend, or acquaintance" (p. 101). For Lofland, it is possible somewhat to know the other person by simply knowing the urban space occupied by that individual.

Spatial ordering is muted under *preindustrial* conditions, since infrastructural configurations are inexact, thus requiring strangers' appearances (the costuming of the street juggler, the leper, the dishevelled beggar, the knight on horseback, or the priest) to provide for the observer more clues about the observed and their position in life, a process she designates as *appearational ordering*. Through appearational ordering, preindustrial urban residents contextualize the stranger by their appearances through visual markers, a process that can still be seen to take place intermittently in modern urban environments, as we shall later examine. For Lofland, appearances do provide some social cues that minimize strangeness. Although clothing and attire no longer carry the gravity of revealing to strangers one's social status, a variety of items born from the conspicuous consumption of the 20th and 21st centuries do: the type of car one drives, the way one adorns one's body, how hygienic one appears on a daily basis, one's type of hair, or the jersey of one's favourite sport team, to name but a few demotic diacritica. For Lofland appearational and spatial ordering processes are different because "in the modern city, appearances are chaotic, space ordered," while in the preindustrial city, "space was chaotic, appearances ... ordered"; that is "in the preindustrial city, a man was what he wore" (1973, p. 82). Thus, although Lofland situates the appearational process in preindustrial time, I will attempt to highlight how appearational ordering did exhibit its residual dynamics under pandemic conditions, since urban spaces were severely damaged by pandemic mitigation measures, robbing from spatiality its ability to inform urban residents about one another.

For Lofland, because public spaces in the preindustrial city and its activities were disorganized, urban residents had to be informed of others' social status through attire. Because unphotogenic and unhygienic human activities saturated public life in ways that often offended the sensibilities of residents, being easily identifiable as someone complicit or not in such activities was crucial in order for those with important social roles to function effectively. In such scenarios, activities that were seen in the public sphere during the preindustrial period included, for one, the elimination of body wastes. Certainly, some civilizations of antiquity, the Romans and Harappans, for example, had their public bath houses and even rudimentary sewage systems. However, most preindustrial urban residents employed a variety of public environments to relieve themselves, frequently "in full view of anyone who happened to be about" (1973, p. 35). Moreover, city officials frequently allowed pigs to scavenge on urban trash and refuse as a means of waste disposal.

Public executions and torture were also common, only to retreat indoors, as Foucault would later observe in the classic *Discipline and Punish* (1977). Lofland offers accounts of visitors to London during the 1700s who reported that "public executions ... take place every six weeks [when] criminals are hanged on these occasions," while in medieval France, "the 'gibbet' was a permanent fixture in every city and town, and after the hanging, the corpse was allowed to remain until it crumbled" (1973, p. 35). Other activities such as buying and selling, schooling, dissemination of news through a town crier, and entertainment and public pageantry were also chaotically thrown into the spatially disorganized public space, requiring all participants to explicitly declare their orientations and roles through aesthetic markers. Lofland notes how in "cities as diverse in time and place as eighth century Damascus and sixteenth century London, actors, storytellers, musicians, and street singers were ever-present, giving their performances whenever and wherever they could collect a crowd" (1973, p. 39).

By the industrial period, a variety of these dynamics were institutionalized and brought indoors, as both Lofland and Foucault observe. Prisons became more sophisticated, performing a variety of punitive roles out of the public eye. Executions, torture, and incarceration were removed from sight. The public was becoming "cleaner." With the onset of industrialization, the engineering of plumbing infrastructure also allowed for the elimination of body waste to be confined indoors. Communal restrooms and baths were slowly becoming spaces that were rendered intimate and private. Similarly, performances by entertainers, musicians, and thespians often could take place in structures specifically designed for the task of collecting audience members to watch performances. Education, once a not so important requirement for non-elites, took place rather informally in public spaces. With the onset of institutionalized education, schooling was made available for many, and learning activities became entrenched inside schools and universities. Trash collection became institutionalized as well, with the mechanization of waste removal an asset that helped to consistently remove refuse from public areas. All of these activities served to reconfigure the public space, allowing for more organization and predictability about others.

It is thus important to emphasize to readers that Lofland's treatment of strangers in the urban context – be it preindustrial or industrial – has never been one that renders them *impossible* to "know." Here lies the ironic elegance in her work: she clearly notes that people can and do derive rich cues about others they do not know, ultimately minimizing the strangeness of the stranger. Yet, Lofland remains within her analytical parameters, not extrapolating towards urban environments in the context of social and systemic crisis. I am of the view that, in the process of making visible the sequence by which an urban environment like Los Angeles under pandemic conditions slowly denuded its publics of urban forces such as cooperative motility, civil inattention, restrained helpfulness, audience role prominence, and civility towards diversity, urban sociologists can make operative appearational ordering for explaining the unanticipated consequences of pandemic distancing.

In the case of the United States, the notion of social distancing during the pandemic was pitched in the narratives of those serving different apparatuses of the state. In Los Angeles, I saw the governor, mayor, and, on a national scope, the president and his oligarchy use the term, all in hopes of reconfiguring large segments of society under an emergency panopticon meant to ensure that the corporeality of existence – our health, our lungs – took pride of place in shaping the dynamics of social relations for the sake of maintaining social order. Society had to be shut down for the time being in a manner that needed the cooperation of strangers en masse, a situation that ruptured our ability to employ spatiality to root individuals to their physical communities. My chapter attempts to make visible social forces that remained, forces that surfaced in social spaces of duress in the Los Angeles area during the intermittent periods when urban residents were compelled to venture into emptying publics for resources. I also ask us to consider to what degree we can minimize strangeness in another if the public is emptied of residents, its civil society, the link of public to private enterprises, and the energy of economic labour – thriving publics in different spatialities that articulate community and economic energies – when there is systemic crisis, a condition that engenders macro-level angst if not fear among its actors. To offer further considerations, we need to harness the ideas of Ray Oldenburg in regard to the "third place" as elaborated in his important work, *The Great Good Place* (1999).

## The Relevance of Oldenburg's Ideas for Understanding the Pandemic

Oldenburg's observations underscore the significant relationship between third places and community life. I discuss Oldenburg's notion of the third place to remind readers that when such places are rendered inoperative, dynamics upon which the social is nourished – primarily environments where members of community can communicate and recreate – concomitantly evaporate as well. For Oldenburg, an urban sociologist, the third place is "largely a world of its own making, fashioned by talk and quite independent of the institutional order of the larger society" (1999, p. 48), a world that performs the vital function of "uniting the neighborhood" (1999, p. xvii). Bars, hair salons, and coffeehouses allow citizens to engage in talk and "let loose" simply because these places are able to consistently host "the regular, voluntary, informal, and happily anticipated gatherings of individuals beyond the realms of home and work" (1999, p. 16). Indeed, the home is not a third but the "first place," where the conjugal family lives privately and informally, while the work environment constitutes the "second place," where public and public life are formalized around procedural details related to occupation. Both for Oldenburg are unable to enhance community interaction in informal and convivial contexts.

The consequences of the Industrial Revolution, for Oldenburg, ossified the boundaries between first and second places, a distinct shift from a preindustrial context where such a boundary was blurred; that is, work was often done from home, be it through the skilled artisan or the farmer working the land. However, with the advent of industrial output, mass production, and mass consumption, the boundary between place of work and place of residence was demarcated in ways that for many removed "productive work from the home and ... morality, and spirit from family life" (1999, p. 16). Work and private environments thus became clearly distinct, but only in the third place can the perfunctory acts of both spheres be discarded, if only momentarily. For Oldenburg, third places still managed to have tremendous staying power in society in spite of major transformations to the family format and work environments because of the forces of industrialization.

In cities, third places are as diverse as they are plentiful. Cultural, economic, and political dynamics across the present and across time configure the size, the dispersion, and the atmosphere of third places. Oldenburg's global travels and analyses have resulted in some interesting observations: third places in the United States have a very weak presence in urban life, pointing to my earlier observation that the American population tends to fear its public spaces. Even Lofland concedes to such a view, noting that the American view of public spaces tends to render strangers as those who constitute the mob, outcasts, and criminals who "rob, rape, maim and kill; and males whose prey is women" (Lofland, 1998, p. 152). Oldenburg notes, rather unflatteringly, "In newer American communities ... third places are neither prominent nor prolific ... one may encounter people rather pathetically trying to find some spot in which to relax and enjoy each other's company" (1999, p. 17). Oldenburg was impressed by how Ireland, France, and Greece – indeed much of Europe – enabled third places to have an important function in the daily life of their urban residents. In the United States, Oldenburg saw less promise, noting that "many public establishments reverberate with music played so loudly that enjoyable conversation is impossible" (1999, p. 30).

Oldenburg argued that third places have a therapeutic effect upon residents of urban society and laments how little research has been undertaken on this area of social life. Although urban resources that reduce stress through rituals such as yoga and gym activities abound, the third place, in Oldenburg's formulation, involves dynamics that are the "people's own remedy for stress, loneliness, and alienation" (1999, p. 20). Having travelled through Europe to explore Vienna's coffeehouses, pubs in London, cafés in France, the Italian *taberna*, and a variety of venues here in the United States, Oldenburg notes how third places function as neutral grounds in the community where people can be "most alive" and "most themselves" (1999, p. 20). In a prescient and haunting observation relevant to our pandemic period, he argues how urban environments that discourage third place association will prompt many to "withdraw to privacy as turtles into their shells," and people will become lonely "in the midst of many" (1999, p. 203). Oldenburg also incisively notes how third places "remain among the very few places where the generations still enjoy one another's company" (1999, p. xx) and that "nothing contributes as much to one's sense of belonging to a community as much as 'membership' in a third place" (1999, p. xxiii).

Third places are sites of public life that counter notions of the city as being alienating places where, as urban sociologists like Lofland (1973) note, we are forced to cope with strangers on a daily basis. Third places are where community is most "upbeat" and "cheerful," and where the art of conversation constitutes the main activity and is a "lively game" (1999, pp. 20–9). For Oldenburg, the third place is the place where that "lone stranger ... is most apt to become a regular" (1999, p. 35). He discusses how cafés, book stores, and hair salons function as social lubricants to remove many interpersonal inhibitions. Successful third place cafés, bars, or pubs are deliberately cozy and have jovial environments that provide for patrons what Oldenburg calls "spiritual tonic," where talk is "scintillating, colorful, and engaging," where "the joys of association" are found and "the art of conversation is preserved against its decline in the larger spheres, and evidence of this claim is abundant" (1999, pp. 26–8).

Third places are also highly effective environments for "leveling" all patrons' social status so that "honest expression triumphs over sophistication" (1999, p. 125). Those who are experiencing disadvantages and duress in life, for example, can find in third places an environment where the struggle of daily life is made to lose "much of its sting," simply because the "disadvantaged can be accepted as equals" (1999, p. 25). Social status – muted within the third place – does not dictate who is spoken to or when one speaks. Third places are places that are welcoming and informal, and emphasize happiness, humour, and wit, exclusively relied-upon communication styles that engage patrons in "dramatic" conversations. Through such communicative dynamics, patrons escape the daily grind of life. In other ideal situations, levelling is also good for business, as it does not formally engage in a politics of exclusion against certain types of patrons. By functioning to be inclusive of all patrons through levelling, and with social status in third places relatively muted, the establishment of intersubjectivity may lead to mutual consensus. If such a state is not attainable, membership in third places entails "coming to terms with people who, on certain subjects ... one doesn't agree with" (Oldenburg, 1999, p. xxv). Yet Oldenburg maintains that this form of community is healthier than communities that form associations "based on ideology or 'political correctness'" (Oldenburg, 1999, p. xxv).

Drawing from his historical and cross-cultural examination of third places in Western Europe and the United States, Oldenburg concludes that *urban* development has negatively affected *community* development. Third places, then, have an important function in community simply by being rooted, embedded places not of worship or occupation but of joy and relaxation. For Oldenburg, many thus fail to appreciate how "relieving stress can just as easily be built into an urban environment as those features which produce stress"

(1999, p. 10). Oldenburg is critical of urban residents' seeming ignorance about urban stressors that bombard their lives. That Oldenburg considers the dearth of third places in urban environments a deficit needs to be appreciated for its prescience in illuminating the social discontents that emerged in many areas of Los Angeles affected by the pandemic. In this context what has manifested are the emptying publics that take environments of community building with them. Oldenburg presciently notes how "Without such places, the urban area fails to nourish the kinds of relationships and the diversity of human contact that are the essence of the city. Deprived of these settings, people ... will grow ever more apart from one another ... a retreat from society [rather] than a connection to it" (1999, p. xxix).

Extending the critique of non-places into the televisual realm – by implication a critique that applies to online activities as well – Oldenburg exhibits concerns parallel to those of Jürgen Habermas (1991, 1984, 1987) regarding the effects of a colonized lifeworld upon a community's communicative action dynamics, especially in what Habermas saw as the public sphere. Such a sphere is being abandoned, according to Oldenburg: "Currently, Americans spend about 90% of their leisure time in their homes. Is the figure so high because home life is so attractive or is it because we have created a world beyond the home that no longer offers relaxed and inexpensive companionship with others, a commodity once as easily obtained as a stroll down the street?" (Oldenburg 1999, p. 214). Oldenburg envisions the solution as beginning with rejecting components of mediation such as television, radio, and news, and immersing oneself in a community's third places, for there,

> people get to know one another and to like one another and then ... care for one another. When people care for one another, they take an interest in their welfare; and this is a vastly superior form of welfare than that obtained by governmental programs. It is based on mutual consent, genuine empathy, and real understanding of people's situations. Nobody is a "case." (Oldenburg, 1999, p. xxi)

Such concerns are not new, however, for even during the Cold War, Frankfurt School theorists such as Herbert Marcuse and his coterie of Critical Theory thinkers, of which Habermas is an intellectual scion, critically engaged with issues of systemic regulation by corporatocracies like the United States. They warned about ubiquitous advertisement and how people were being transformed into "one-dimensional" beings who had lost all ability for independent thinking. Indeed, one can even see parallels between Habermas's notion of the public sphere (1991) and Oldenburg's third places, but there is a key difference: Oldenburg infuses joy and recreation into third place environments. Habermas did not observe affective expressions of community, since he examined the public sphere during early industrial Europe, when it was oriented towards the economic and political. Where Habermas converges with Oldenburg is in the former's view that the public sphere's health is contingent upon communicative action that inspires citizens to contest oppressive macro-level institutions. Only in such a manner could democracy complete itself as a project. Oldenburg's formulation gave pride of place to being "present" within community establishments, even if those gathered did nothing else but exhibit nostalgic memories from halcyon days. In this regard, Oldenburg was as critical of failing and/or eroding third places as Habermas was of public spheres increasingly being colonized by social systems.

Also pertinent for our discussion of public spheres/third places is how both Habermas and Oldenburg consider them as environments that are indicative of a healthy democracy. Under authoritarian systems, should patrons at public spheres/third places such as cafés or bars contest the state too aggressively, there will indeed be reprisals by the state. Habermas provides a seventeenth-century account found in Emden's (1956) work *The People and the Constitution*:

> Already in the 1670s the government had found itself compelled to issue proclamations that confronted the dangers bred by the coffee-house discussions. The coffee houses were considered seedbeds of political unrest: "Men have assumed to themselves a liberty, not only in coffee-houses, but in other places and meetings, both public and private, to censure and defame the proceedings of the State, by speaking evil of things they understand not, and endeavoring to create and nourish universal jealousies and dissatisfaction in the minds of all His Majesties good subjects." (cited in Habermas, 1991, p. 59)

And herein lies the utility of Oldenburg's elaboration: not only do third places at the very least have political inclinations, but they are also often informed by "philosophy, geography, urban development, psychology, [and] history" in ways where urgent debate and communicative action behoove participants to "air their notions in front of critics" (1999, p. xxv). Oldenburg offers a penetrating insight: "If Americans generally find it difficult to appreciate the political value of third places, it is partly because of the great freedom of association that Americans enjoy. In totalitarian societies, the leadership is keenly aware of the political potential of informal gathering places and actively discourages them" (Oldenburg, 1999, p. 66). In this regard, Oldenburg's ideas, amalgamated with observations forwarded by Lofland, resonate with possibilities for highlighting how one can conceptualize urban forces that can no longer make operative their third places, a condition that frames how the pandemic has been informing community relations and reconfiguring urban spatialities in Los Angeles, California.

**The Emptying of Los Angeles' Publics**

The imminence of self-isolation observed in the Los Angeles area along with the rest of California began to be felt by 4 March 2020. One day prior, the White House released guidelines that mandated for all Americans to "avoid social gatherings in groups of more than 10 people," and to "avoid eating or drinking at bars, restaurants, and food courts," as well as using "drive-thru, pickup, or delivery options" (President's Coronavirus Guidelines for America, 3 March 2020). With this pronouncement, I began to examine the city's COVID-19 dynamics drawn primarily from the governor's and the mayor's nightly updates during the period under examination, as well as popular news and cultural websites that focus on Los Angeles, with data primarily derived from journalists of the beloved LAist local webpage that reports on the communities and neighbourhoods of the city. From such sources, I hope to render how Los Angeles' publics have been emptied.

On the same date, California's governor, Gavin Newsom, declared a state of emergency. Newsom urged all residents over 65 years and others with ongoing health complications to self-isolate, an edict that would affect millions in the state (see also LAist, 2020a). By Sunday, 15 March, the mayor of Los Angeles, Eric Garcetti, enacted bans on recreational and community environments like dine-in restaurants (takeout and drive-through transactions were exempt from the ban), bars, and other entertainment/recreational facilities such as nightclubs, movie theatres, arcades, gyms, and fitness centres. Museums and libraries in Los Angeles County were similarly ordered to cease operations. Institutions vital to the functioning of the city, such as hospitals and facilities for first responders, were allowed to "operate with modified services to minimize in-person contacts," with groceries, pharmacies, and food banks exempt (Denkmann, 2020).

Garcetti also urged churches and places of worship to close. The first efforts to assist renters resulted in a moratorium on collecting rent. Within the week, Garcetti issued the "Safer at Home Order for Control of COVID-19" for Los Angeles County, while Newsom followed with similar orders for the entire state of California, except as needed to "maintain continuity of operations of the federal critical infrastructure sectors" (LAist, 2020c). Non-critical businesses that could not operate remotely were ordered to cease operations. Garcetti emphasized: "I want to be clear about this … that the only time you should leave your home is for essential activities and needs – to get food, care for a relative or a friend or child, [or to] get necessary health care" (LAist, 2020c). During the same period, public officials were sloganeering the need to "practice social distancing" – more accurately, *physical* distancing. Not all urban residents complied: during a brief stint of sunny weather in mid-March, many thousands of Southern Californian residents visited beaches and outdoor venues such as hiking trails and community parks. The over five million seniors – many of whom were asthmatic, diabetic, or had issues with the heart, kidney, and/or the respiratory system – complied, however.

By this stage in the pandemic's development, finer discernments of personnel needed to offer community and public health services emerged. These included emergency personnel, first responders, government employees, medical personnel, vital infrastructure workers, healthcare providers, transportation services workers, grocery store workers, and restaurant workers (who offered delivery or take-out services only). Additional personnel exempt from restrictions were those employed at news outlets, hardware stores, gas stations, banks and financial institutions, plumbers, electricians, and those that operated dry cleaners and laundromats. The pattern of institutional preservation and dispatch, from an Oldenburg perspective, is explicitly clear: the emptying out of urban publics will still preserve certain sectors of second places that attend to existential threats to society.

By mid-April, the state of New York, one most afflicted by COVID-19, had already recorded over 181,000 cases with close to 9,000 deaths; Los Angeles, the US epicentre on the West Coast, had registered fewer, with close to 300 deaths from approximately 9,000 cases in Los Angeles County, while the state of California recorded over 22,000 cases during the same period (Los Angeles Times, 2020). With *segmented* second places in operation, the visual media not only in Los Angeles but around the world were able to glorify the efforts of their first responders and other essential workers engaged in the provision of health care and the distribution of resources, respectively. That their roles as heroes of society were extolled, however, overlooks how the glorification process also had the unanticipated consequence of embedding more existential stressors back into the community, since these second place personnel attend to life and death issues that, during the early phase of the pandemic in Los Angeles, appeared manageable. Whether first responders or grocery store workers, such persons now shared overlapping urban spaces in Los Angeles as the city lost its recreational/community establishments: the restaurants that did not offer delivery, pickup, or drive-through transactions, the bars, the theatres, the coffee shops, the hair salons, the manicurists, to name but a few (see also LAist, 2020b). The textures of third place closure also sliced across all ethnic enclave markets, with Los Angeles' Chinatown and other Asian American outlets bearing the undesirable consequences being shunned by those who were germophobic and xenophobic, even though many were compliant in social distancing practices mandated by California and Los Angeles County.

Even as early as March, Los Angeles' normally thriving Chinatown began to experience an acute emptying of its public. A 2 March *Los Angeles Times* article documented the denudation: a famous restaurant, Yang Chow, saw a 30% drop in business compared to the previous year, while Foo Chow, rendered famous in the 1998 film *Rush Hour* starring martial arts phenom Jackie Chan and comedian Chris Tucker, saw only nine people dine in its large two-story environment on the date journalists visited. A day earlier, in Koreatown, an online hoax implicated a restaurant (which at the time was closed for repairs) as being a spatial vector spreading COVID-19. Han Bat Shul Lung Tang, along with five other restaurants in the ethnic enclave market, allegedly were patronized by an infected Korean Air flight attendant who visited the area during a layover. The owner of the said restaurant, John Kim, denied the allegation and had the Los Angeles Department of Public Health confirm his defence. The Republic of Korea consulate in Los Angeles also got involved, noting that the flight attendant was not contagious while in the city; that is, she "did not develop symptoms of the illness … until after leaving LA" (Melley & Associated Press, 2020). It was too late, however, as Koreatown restaurants registered a 50% drop in business because of the rumour. Nearby at Honey Pig, a restaurant with 25 tables, only six customers visited on the day reporters arrived to document the decline in business traffic. Another restaurant, Hanshin Pocha, one that had never closed since its opening in 1998, shut down operations even before March 2020 (Melley & Associated Press, 2020).

During mid-April, the month that public health officials predicted would see California's infection rate peak, over two million residents were unemployed, as seen by unemployment claims released by the Department of Labor, with half of Los Angeles' workers at risk of losing their means of employment (Wagner, 2020b; see also Wagner, 2020c). Media imagery began revealing long queues of cars awaiting food distribution at local food banks. The situation was less hopeful for restaurateurs who had been relying on the emergency model of providing takeout/pickup/delivery to patrons: Kristine Lefebvre, owner of two popular Los Angeles restaurants, notes that such a model only constitutes 1/15th of her type of business (Tseng, 2020).

The National Restaurant Association (NRA) even published a series of creative articles on how to manoeuvre through an industry unable to harness its in-house dining experience while trying to stave off the need to terminate the employment of its workers (NRA, 19 March 2020). However, the state's 800,000 hospitality workers – 12.7% of the state's total workforce – were reeling as numbers predicting that between five and seven million restaurant labourers would be out of a job should social distancing be mandated for three more months (Tseng, 2020). Accompanying such a scenario were the numerous undocumented workers who also faced unemployment, but tragically without the benefits of receiving the federal stimulus cheques referred to as Economic Impact Payments. Furthermore, third place employees of non-essential services working at nail and hair salons, barbershops, golf courses, indoor malls, spas, and Los Angeles' many retail shops located in their "strip malls" were also made to comply with the closure. Other important third places like bars, playgrounds for children, and gyms were also ordered closed, as were quasi-Third Place venues like movie theatres, flea markets, and swap meets.

Gig workers, those engaged in flexible, sometime temporary employment patterns made possible by hyper-connected online networks, suffered greatly from the cessation of operations – drivers of Postmates, Uber, or Lyft, freelancers, project-based workers, musicians, and thespians, to name but a few, encountered exponential declines in demand (Wagner, 2020a). Such an ignoble

Image 10.1. The normally congested 60 and 710 freeway interchange emptied during the coronavirus outbreak in Los Angeles, 5 April 2020 (Alamy/Timothy Swope).

decline of the labour force contrasts with the time when gig workers were seen as a bellwether, changing the business terrain, a terrain seen to be so promising that the National Association of Counties (NACo) praised the gig economy, one that had grown from 10.1% in 2005 to 15.8% in 2015. Indeed, by 2016, a quarter of Americans "reported earning some money from the 'digital platform economy'" (Istrate & Harris, 2017, p. 4). Yet by mid-April 2020, more than two million Californians had already filed for unemployment benefits, with the risk of job loss asymmetrically seen in low income minority neighbourhoods and groups earning low wages (Wagner, 2020b).

    A noticeably less peopled environment became apparent when one observed the lightly travelled highways and arterial roads of the normally congested city and the relatively empty parking lots of its strip malls that not long ago welcomed patrons to Los Angeles' third places like bars, coffee shops, and nail salons (see Image 10.1). Also noticeable were the quieter soundscapes of the city, with less noise pollution stemming from vehicles with loud aftermarket stereos and/or exhausts, buses, semi-trucks, and passenger jets flying overhead – all these diminished during the Los Angeles pandemic period. That video footage from the time showed Los Angeles International Airport with few cars during weekend afternoons added to the surrealism of the city's emptying public spaces as it crawled towards 1 May 2020, when my analyses ended with California Governor Gavin Newsom announcing a slow and phased reopening of the state.

## A City Becomes a Warehouse

The pandemic period of examination as it affected the greater Los Angeles area revealed two distinct patterns by which a large, thriving, urban environment transitions towards systemic crisis. The first pattern suggests that the emptying of the public, even in crisis, will preserve the roles of an urban environment's essential workers. A segment of the second place is thus retained even in systemic crisis. Second place personnel, already situated in a context of a formal public, are thus legitimated, even glorified, when their cultural capital becomes amplified for formally engaging in social change for the betterment of a population's health. Such legitimation in the form of preserving second place urban actors' roles if they are essential workers entailed that first responders, workers in the grocery sector, warehouse workers, food packagers, and workers at food-testing labs, businesses that provide shelter for the disadvantaged, news outlets, gas stations and automotive repair and supply centres, banks, and insurance companies become agents of social repair during the pandemic.

However, the normally frenetic activity of urban life, the site of Lofland's cooperative motility, civil inattention, audience role prominence, restrained helpfulness, and civility towards diversity – a process that also grants urban actors their autonomy in figuring out the "stranger" – became stunted. While Lofland argued that spatial ordering would one day replace appearational ordering as social spaces became organized, the fact that urban actors can no longer access such organized urban spatiality during the crisis – that is, access the spatiality where the dynamics for navigating the public realm take place – suggests that cities actually become *non*-cities during systemic crisis, transitioning themselves into large-scale distribution and receiving centres that now are linked directly to what urban sociologists and geographers term as "break-of-bulk" points – environments where arriving cargo and goods are then redistributed to local facilities.

First rendered popular in urban sociology by Harris and Ullman as far back as 1945, the term *break-of-bulk* refers to the process by which large cargo shipped *in bulk* is broken down into smaller units to be delivered to more specific off-loading points. Break-of-bulk dynamics thus coexist with other urban activities that are not directly related to trade. Break-of-bulk centres are usually large airports, railway hubs, and large seaports where containers from freighters are reloaded onto trucks that take them to particular locales. In these designated areas, bulk is again broken down into smaller units for localized delivery. Such a city usually has a strong railway and/or highway infrastructure and frequently river transport infrastructure to distribute goods to their localized destinations. In the pandemic context, the *entire* city welcomed, to the best of its ability, medicine-related break-of-bulk through resource exchanges that no longer served the recreational needs of community and city, but instead, attended to materials and resources related to health practices. In this regard, the greater Los Angeles area became a warehouse, one quickly accessed without third place logistical activities crowding the mobility and dynamics of the warehouse the "Southland" temporarily became. Such uncluttered access between break-of-bulk places like ports and downtown Los Angeles revealed an important outcome of systemic crisis: during the pandemic period spring this chapter examined, distribution of resources was lacklustre not because of a lack of access to the city (since no infrastructural damage was incurred), but because of a lack of coordination in crisis management set into motion by an incompetent US president.

One major consequence of the changed role of Los Angeles during the pandemic is how a particular segment of second place workers, such as first responders and other essential workers, was able to shape the dynamics of break-of-bulking without interference from the logistics demanded by third places in need of ensuring that their business operations remain intact. Areas once key to third place operations were converted to engage in distribution of food and health resources that hedge against absolute deprivation: parking lots of churches, retail, supermarkets, and health clinics became temporary break-of-bulk distribution centres for the food insecure. Televised images of dozens of vehicles waiting to receive food from places of worship or other facilities flooded the media. Even hospitals became distribution centres as they coordinated closely with public officials to quickly pack and ship ventilators to other jurisdictions within and without the city. The emptying of Los Angeles' public had concomitantly enabled the city to function as a massive open-air warehouse of life-saving bulk. Authoritarian energies propelled food distribution, cleaning materials, medicine, professional protective equipment (PPE), and medical supplies – bulk intended to mitigate any absolute deprivation and sustain health – and very little else.

At this juncture, one can engage with some cautious extrapolations about the process by which a public is emptied: the pandemic that affected Los Angeles saw third places become social liabilities. In spite of such environments being emblematic of urban harmony, replete with their material culture born from the city's multicultural existence, they are, in the final instance, culturally expendable. Third place infrastructure, however, remained intact for revitalization. Such a reconfiguration of community is instructive – a city driven by a segment of the second place, as in Los Angeles during the early months of the pandemic, suggests that we can evaluate and enhance urban responses to systemic crises by focusing on the efficacy of distribution systems as a function of operating statuses of third places.

Thus far, my rendering of emptying publics risks ossifying boundaries between the private and public. Urban actors in the United States, and even in Southern California, *were* allowed out – but, as noted at the outset of the chapter – for essential supplies for living. Yet there were less sanguine examples that evinced reactionary backlash, as can be seen by some urban residents who ventured into community under great personal stress, and in a manner indicating they were no longer beholden to community civility. Subjected to stay-at-home orders that rendered urban residents a captive audience to be exposed to the shock value of coronavirus motifs meted out by the news media, citizens became anxious and upset. Informed further by media sensationalism and political spin by xenophobic pundits of the Trump state apparatus, some began to conceptualize urban life in survival mode. In some cases, paranoia was stoked as leaders of the United States vacillated between demonizing China and enabling the fog of phenotype that subjected many Asian Americans to potential backlash. In others, nurses and doctors experienced backlash in the United States and around the world because many people feared that healthcare workers were exposing others to the virus (Caraccio, 2020).

## The Multicultural Consequences of Social Distancing

Understanding the sociological reconfiguration of community and society during the pandemic requires a revitalization of Lofland's appearational ordering concept that, when inflected by a crisis context, reveals a less sanguine picture of the American experience. In particular, being Thai-Chinese American, and Asian American in general, had become a social deficit during the pandemic period framing my writing, not unlike how Muslim Americans were subjected to hostility and belligerence in the period after September 11. As such, looking like an "Asian" or "Chinese" in Los Angeles during the pandemic fed into belligerents' appearational ordering of those who phenotypically exhibit the attributes of someone imagined to be from the Far East, and thus a potential carrier of COVID-19. A lifeworld under a pandemic without nourishing third places enabled warped sociological imaginations and Orientalisms about people to surface, further inflected by existential angst, fears, and anger.

In the week of 19–25 March 2020, 673 reports of coronavirus discrimination were submitted to the "Stop AAPI Hate" section of the Asian Pacific Policy & Planning Council (A3PCON) website. By early April, close to 1,100 reports were filed by Asian Americans who were exposed to verbal harassment, shunning, and physical assault (Jeung, 2020, p. 1). At the time of this writing, A3PCON and the Chinese for Affirmative Action (CAA), two authoritative local watchdog groups whose data were compiled by Dr. Russell Jeung of San Francisco State University, reported an average of 100 daily incidents. More egregiously, at least 5.5% of incidents stemmed from other minorities hostile to Asians who phenotypically appear Far Eastern. The report also documents how 61% of anti-Asian discrimination reports came from Asians who were not ethnically Chinese yet experienced the same backlash and enmity. Subjected to verbal harassment and name calling, actions that constituted 2/3 of reports, Asian Americans are thus more likely to "face coronavirus discrimination in businesses, especially stores, rather than at schools and public transit as previously observed" (Jeung, 2020, p. 1).

Appearational ordering during systemic crisis, then, can be a process of racialization if it feeds into one's desire to minimize strangeness due to acute fears of the public. Indeed, minimizing strangeness by "knowing" that a particular Asian individual may be a vector of COVID-19 transmission became a marker as effective as attire in communicating aesthetic attributes of a culture. Appearational ordering under systemic crisis thus accommodated racialization born from careless deductions about cultural dynamics. In the pandemic, visually looking like someone imagined to be from Wuhan is Hester Prynne's scarlet letter. And, like Prynne, Asian Americans can be shunned by community: an analogue to Prynne's community of condemning Puritans can be found in the pandemic context by those who envision themselves as cleaner, and thus employ hostile physical distancing in the lifeworld, acrimony included, as a symbolic political hammer against the foreigner. The "perpetual foreigner" motif is reawakened to frame Asian Americans in yet another atavism that subjects them to having their legitimacy as Americans retracted from their lifeworld (see also Fong, 2008).

In the A3PCON and CAA report, 89.5% of those who engaged in the act of discrimination did so purely on the basis of judging the other person's race. More surprisingly, close to 50% of discriminatory acts took place at businesses, followed by public parks and public transit. More telling that racialization as appearational ordering is primarily a visual event that does not discern among those who are attacked, 94.5% of the respondents who experienced discrimination spoke English as their first language, with women bearing the brunt of prejudice at 73.6%. Sadly, for all its sloganeering of being a state enlightened with a consciousness of diversity, California also reported the highest rates of discriminatory acts at 31.8%, with the state of New York registering only 12.9%. Jeung documents a variety of accounts in the A3PCON and CAA (2020, pp. 5–6): One filed report notes how "2 separate Uber drivers would not pick me up due to my race. Each time the driver would arrive, look at me, and speed off quickly … The second driver had a face mask on, and he slowed down enough to look at me and shake his head and wave a hand at me rejecting the ride." Some were coughed on: "I took a walk with a friend of mine in Visalia, California. While we were passing a group of 4 men, one of them coughed into me, not once, but TWICE, without covering his mouth. As I turned my head back, they all burst out laughing. Then they biked away." Another individual reported: "I was walking home and someone in a pickup truck threw a bottle at me really hard. He missed" (p. 6). Accounts of shunning include experiences such as one that befell an Asian American public transit passenger: "As I was

Image 10.2. Hundreds rallied in Manhattan, New York City on 27 February 2021, to condemn COVID-19-related anti-Asian hate crimes that have sprung up around the country. Many accuse President Trump of fanning the flames of hate by referring to COVID-19 as "kung flu" or the China flu (Alamy/David Grossman).

walking to my bus, a white, middle-aged man screamed at me to 'wear a respirator' because I'm Asian. And when I was on the bus, a middle-aged woman sitting across from me kept staring at me while holding a rosary in front of her. After a few minutes of this, she moved a few seats away from me while maintaining eye contact with me." Even as early as late February 2020, a Chinese individual in San Francisco collecting cans for recycling was attacked by a golf-club-wielding individual, who, after being filmed in the attack, was arrested with his accomplice. Too many physical incidents such as these can be found in the formaldehyde that is YouTube, and none of these have been included in A3PCON's and CAA's diligent reporting of anti-Asian American racism, hinting at the ubiquity of how such practices have gone unreported. More violent assaults have taken place since 2020 and ghastly murders have also ensued, as when an Asian American woman was pushed onto a subway track in New York City, to name but one horrifying incident. As a response, numerous public protests condemning hatred towards the Asian American community surfaced in urban centres throughout the United States and abroad, continuing into 2021 (see Image 2). In the case of Los Angeles, innovative defence practices such as the use of an online tracking tool that traces racist incidents towards Asian Americans was launched (Huang, 2020).

Although first responders' ethos of assisting the public has rendered them heroes in Los Angeles as well as around the world – one need only search for news footage of cities designating certain evening hours to clap for their first responders – many while in uniform have faced discrimination and attacks. Looking like a nurse became a marker that in many instances minimized their strangeness, prompting the fear-ridden to view them as emanating from an environment where they are seen as vectors able to spread COVID-19. A close friend of mine, pulmonologist Dr. Elbert Chang, based in the Greater Los Angeles city of Montclair, had obscenities, recriminations, and vitriol hurled at him by Trump fanatics while they were, ironically, being treated and returned to health for their COVID-19 infections. Elsewhere, being seen in scrubs or being known for having a profession that involves providing physical health care can be an appearational liability. For example, on 10 April 2020, Daniel R. Hall slashed the tires of 22 vehicles at New York's Presbyterian Hudson Valley Hospital (Koulouris, 2020). The *Miami Herald*'s David Caraccio documents how other healthcare facilities have seen doctors and nurses encounter belligerent patients, with some destroying the lobbies of their facilities while others have been "coughing up and spitting on health care workers." This prompted many hospitals in the area to increase security. In Oklahoma during the same period of reporting, a nurse experienced a violent attack on the way to work (Caraccio, 2020).

Such appearational liabilities are global: In India on 1 April 2020, the BBC's Vikas Pandey reported backlash against doctors and nurses in Delhi, Hyderabad, and Indore. Some male patients pranced nude in front of female doctors while other healthcare workers were spat on. In Indore, doctors visiting areas of the city to attend to persons suspected of infection were violently chased away – a scene filmed and shared with the world. One of the doctors who escaped, Dr. Zakiya Sayed, noted with aplomb that such acts "won't deter me from doing my duty," while she proceeded to note how she had "somehow fled from the mob" with her colleague, and that she was "injured but not scared at all" (Pandey, 2020). During the following week, the *Washington Post* also reported on hostilities towards healthcare workers in Mexico, Colombia, the Philippines, and Australia: "people terrified by the highly infectious virus are lashing out at medical professionals – kicking them off buses, evicting them from apartments, even dousing them with water mixed with chlorine" (Sheridan et al., 2020). The same report notes how in Australia and Mexico, hospitals have urged nurses not to wear uniforms in public settings for fear their uniforms would invite hostilities. A uniformed nurse working in Mexico's Guadalajara Civil Hospital recalled how her bus decided to stop at the next block; the nurse, Maria Luisa Castillo, noted, "It was clear they didn't want to pick me up" (Caraccio, 2020).

One case in the Philippines was particularly brutal. As note the administrators at St. Louis Hospital in Tacurong City:

> On 27 March 2020, at around 5:00 p.m., a St. Louis Hospital personnel, who was on his way to report for duty, was violently attacked ... when a mob of five individuals ganged up on him. Outnumbered and alone, he was helpless as these vile individuals splattered ZONROX all over his face, which could have caused irreparable and permanent damage to his sight. Fortunately, he was able to rush himself to the hospital where he was given prompt treatment. (St. Louis Hospital Public Advisory, 28 March 2020)

The discussion in this chapter attempted to make visible how macro-level health emergencies, such as that ushered in by the pandemic, influence urban transformations. The process as it unfolded in the Los Angeles example included the cessation of third place operations and, indirectly, the soul of community itself. With the spatiality for community building and socializing slowly denuded of life and monetary flows, the economic and cultural consequences that followed reflected the angst of a population trapped in state-mandated liminalities. Moreover, the cessation of numerous third place operations resulted in mass unemployment that continues at the time this chapter was concluded during May 2020, when Los Angeles removed its mandate for social distancing. Yet in such sociological deficits, the role of the second place roared into operation, if initially on a non-coordinated basis, propelled by a humanitarian drive to make our bodies stronger than circumstances. In the process, a repository of information based on health updates infused with sociocultural narratives for tolerance in the midst of xenophobia was collected, and a shared humanity was built by workers and the people in spite of the frenetic conditions that drove the emptying of Los Angeles. In the context of the city's liminality during the pandemic, and one day in its reanimation, I hope that its residents will realize that they have had privileged insight into *and* memories of an urban machine that could have, nonetheless, performed more efficaciously in its social contract, lest we again regress as a society when the next systemic crisis surfaces.

REFERENCES

Caraccio, D. (2020, April 13). Slashed tires and violence: Health care workers face new dangers amid COVID-19 battle. *Miami Herald*. https://www.miamiherald.com/news/coronavirus/article241967281.html.

Denkmann, L. (2020, March 15). Garcetti orders eviction moratorium and more closures as LA county shuts down offices. LAist. https://laist.com/latest/post/20200315/los-angeles-government-closures.

Emden, C.S. (1956). *The people and the constitution*. Clarendon.
Fong, J. (2008). American social "reminders" of citizenship after September 11, 2001: Nativisms and the ethnocratic retractability of American identity. *Qualitative Sociology Review, 4*(1), 69–91. https://doi.org/10.18778/1733-8077.4.1.04.
Foucault, M. (1977). *Discipline and punish: The birth of the prison*. Pantheon Books.
Goffman, E. (1963). *Behavior in public places*. Free Press of Glencoe.
Goffman, E. (2010). *Relations in public*. Transaction Press.
Habermas, J. (1984). *The theory of communicative action, volume 1: Reason and the rationalization of society*. Beacon Press.
Habermas, J. (1987). *The theory of communicative action, volume 2: Lifeworld and system – A critique of functional reason*. Beacon Press.
Habermas, J. (1991). *The structural transformation of the public sphere: An inquiry into a category of bourgeois society*. MIT Press.
Harris, C.O., & Ullman, E.L. (1945). The nature of cities. *Annals of the Academy of Political and Social Sciences, 232*(November), 7–17.
Harrison, S. (1984). Drawing a circle in Washington Square Park. *Visual Communication, 10*(Spring), 68–83.
Huang, J. (2020, March 26). Online tool tracks racist incidents toward Asian Americans during COVID-19 crisis. LAist. https://laist.com/latest/post/20200326/coronavirus-xenophobia-racism-asian-american-stop-hate-aapi.
Istrate, E., & Harris, J. (2017, November). The future of work: The rise of the gig economy. NACo Counties Futures Lab. https://www.naco.org/featured-resources/future-work-rise-gig-economy.
Jeung, R. (2020, March 25). Incidents of coronavirus discrimination March 19–25, 2020: A report for A3PCON and CAA. Asian Pacific Policy and Planning Council (A3PCON) and Chinese for Affirmative Action (CAA). http://www.asianpacificpolicyandplanningcouncil.org/wp-content/uploads/A3PCON_Public_Weekly_Report_3.pdf.
Koulouris, C. (2020, April 11). Why? Peekskill man arrested for slashing tires of 22 cars parked outside hospital. *Scallywag and Vagabond*. https://scallywagandvagabond.com/2020/04/daniel-r-hall-peekskill-man-arrested-slashing-tires-22-cars-parked-outside-cortland-hospital/.
LAist. (2020a, March 19). Governor orders all Californians to stay home to help contain spread of coronavirus. https://laist.com/latest/post/20200319/coronavirus-california-stay-home-order-march-19.
LAist. (2020b, March 21). Why many Californians are still confused about what they can and can't do. https://laist.com/news/why-many-californians-are-still-confused-about-what-they-can-and-cant-do.
LAist. (2020c, April 9). Coronavirus in LA: Your no-panic guide to daily life and the new (and changing) rules. https://laist.com/2020/04/09/coronavirus-los-angeles-covid-19-rules-explained.php.
Lofland, L.H. (1973). *A world of strangers: Order and action in urban public space*. Waveland Press.

Lofland, L.H. (1998). *The public realm: Exploring the city's quintessential social territory.* Aldine de Gruyter.
*Los Angeles Times.* (2020, April 12). Tracking coronavirus in California. https://www.latimes.com/projects/california-coronavirus-cases-tracking-outbreak/.
Love, R.L. (1973). The fountains of urban life. *Urban Life and Culture, 2,* 161–209. https://doi.org/10.1177/089124167300200202.
Marcuse, H. (1964). *One dimensional man.* Beacon Press.
McPhail, C. (1987). Social behavior in public places: From cluster to arcs and rings. Paper presented at the Annual Meetings of the American Sociological Association, Chicago.
Melley, B., & the Associated Press. (2020, March 1). Coronavirus rumors spread on an app cripple LA's Koreatown restaurants. *Fortune.* https://fortune.com/2020/03/01/coronavirus-rumors-koreatown-restaurants-los-angeles/.
National Restaurant Association (NRA). (2020, March 19). Operators navigate a business decline during the coronavirus outbreak. https://www.restaurantnewsresource.com/article109873.html.
Oldenburg, R. (1999). *The great good place: Cafés, coffee shops, bookstores, bars, hair salons, and other hangouts at the heart of a community.* Marlowe & Company.
Palmer, C.E. (1983). Trauma junkies and street work: Occupational behavior of paramedics and emergency medical technicians. *Urban Life, 12*(2), 162–83. https://doi.org/10.1177/0098303983012002003.
Pandey, V. (2020, April 3). Coronavirus: India's doctors "spat at and attacked." BBC. https://www.bbc.com/news/world-asia-india-52151141.
Pittinsky, T.L. (2011). Allophilia: A cornerstone for citizenship education in pluralistic countries. *Intellect, 2*(6), 175–87. https://doi.org/10.1386/ctl.6.2.175_1.
Pittinsky, T.L., & Montoya, R.M. (2009). Is valuing equality enough? Equality values, allophilia, and social policy support for multiracial individuals. *Journal of Social Issues, 65*(1), 151–63. https://doi.org/10.1111/j.1540-4560.2008.01592.x.
Pittinsky, T.L., Rosenthal, S.A., & Montoya, R.M. (2011). Measuring positive attitudes toward outgroups: Development and validation of the allophilia scale. In Tropp, L.R., & Mallett, R.K. (Eds.), *Moving beyond prejudice reduction: Pathways to positive intergroup relations* (pp. 41–60). American Psychological Association.
President's Coronavirus Guidelines for America. (2020, April 1). 30 days to slow the spread. Coronavirus.Gov. Retrieved on 12 May 2020 from https://doc.louisiana.gov/04-01-20_30days_to_slow_the_spread/.
Sheridan, M.B., Masih, N., & Cabato, R. (2020, April 8). As coronavirus fears grow, doctors and nurses face abuse, attacks. *Washington Post.* https://www.washingtonpost.com/world/the_americas/coronavirus-doctors-nurses-attack-mexico-ivory-coast/2020/04/08/545896a0-7835-11ea-a311-adb1344719a9_story.html.
St. Louis Hospital Public Advisory. (2020, March 28). Facebook. Retrieved 15 April 2020, from https://www.facebook.com/SLHTacurongCity/photos/a.1295321333934547/1877781249021883/?type=3&theater.

Tseng, E. (2020, March 23). How LA's restaurant industry is trying to save itself – and what you can do to help. LAist. https://laist.com/2020/03/23/how_the_restaurant_industry_is_trying_to_save_itself_and_what_you_can_do_to_help.php.

Wagner, D. (2020a, March 26). What the coronavirus stimulus package could mean for LA gig workers. LAist. https://laist.com/latest/post/20200326/unemployment-insurance-coronavirus-self-employed.

Wagner, D. (2020b, April 9). More than 2 million Californians are out of work, but some can't yet file for unemployment. LAist. https://laist.com/2020/04/09/self_employed_gig_freelance_workers_unemployment_pua_california_edd.php.

Wagner, D. (2020c, April 10). Coronavirus puts nearly half of LA workers at risk of losing their jobs. LAist. https://laist.com/latest/post/20200410/coronavirus_la_study_economic_roundtable_jobs_unemployment_layoffs.

Whyte, William F. (1980). *City: Rediscovering the center*. Doubleday.

**European Union**

# 11 Trust between Citizens and State as a Strategy to Battle the Pandemic: Were Senior Citizens Collateral Damage in the Swedish Government's Plan to Flatten the Curve?

BY ANN-CHRISTINE PETERSSON HJELM

### Introduction: Elderly Care and Social Measures in the Initial Phase of the Pandemic

Early on, Sweden had a "do no harm strategy" (Hippocrates) that stood out in the wake of the COVID-19 pandemic. The Swedish welfare state builds on strong confidence in the authorities, made apparent, for example in the motto "Freedom with Responsibility" conveyed by the *Folkhälsomyndigheten* (Swedish Public Health Agency), and is a state that ensures its population is protected against communicable diseases, practises social distancing, and takes responsibility for complying with guidelines (Swedish Public Health Agency, 2020b). This approach intended to reduce infections and make it possible for health services to care for those infected by COVID-19, while also ensuring that the population was able to live according to necessary restrictions during the time required to flatten the curve so that Sweden's healthcare system did not exceed its capacity (Government and Government Offices of Sweden, 2020b).

In March 2020, the Public Health Agency clarified that Sweden was entering a new phase, and that measures were being intensified to prevent the spread of infection, especially in areas affected, such as providing health care for the elderly (Swedish Public Health Agency, 2020a). Senior citizens in Sweden were seriously impacted during the early phase of the pandemic, clearly apparent from the statistics at the time. Indeed, after three weeks under general recommendations from authorities not to visit homes for the elderly owing to a very high risk of infection, these recommendations were replaced by a ban on 20 March 2020. The ban on visits was one of the few direct bans issued in Sweden while the population was still compliant with recommendations on social distancing issued at the beginning of the pandemic.

During April 2020, Sweden's Public Health Agency confirmed that the country had failed to protect its elderly according to the Health and Social Care Inspectorate (IVO) (2020b). Particularly vulnerable, according to pandemic statistics for the deceased, were Sweden's elderly who received home care services and those who resided in retirement homes. More than 5,600 people had died from COVID-19 in Sweden by the end of July 2020, and nearly 5,000 of the deceased were over 70 years in age, according to the Swedish National Board of Health and Welfare. Half of the deceased lived in retirement homes and a quarter of them received home care services (National Board of Health and Welfare, 2020b).

Sweden's plight attracted international attention during the early phase of the pandemic when most countries elsewhere, as shall be discussed, implemented extensive lockdowns. The handling of the pandemic became a watershed moment where critical voices in Sweden, in particular, set the tone of the narrative. The main line of criticism was directed against the state's lack of strong actions: that is, why the actions and decisions of Sweden's crisis preparedness authorities and politicians were delayed, why Sweden took such weak action, and why the strategies of general testing and tracing were abandoned during the early stages of the pandemic.

Why then was there no powerful pandemic distancing in Sweden during the onset of the pandemic? This chapter answers the question by first discussing the problems of Swedish civil crisis preparedness, since insight is required to understand the approach of a country with a strong and stable welfare sector, especially in the area of elder care – a country whose last major disruption was war in 1814. Sweden's civil crisis preparedness is based on a peacetime administrative structure with no special laws designed for enforcement in a crisis – that is, laws that apply during socially stable conditions are understood accordingly to apply in situations of crisis. There is no general law on states of emergency or any "crisis law" that regulates the responsibility for decision making in civilian emergencies emanating from the national level as stipulated in Chapter 15, "War and Danger of War," in the Instrument of Government (1974: 152).[1] The government must make decisions "collectively" in a crisis, and ministerial government is prohibited from fully operating in a crisis as it normally does under non-crisis situations. Such practices applied, among other things, to recommendations to avoid unnecessary visits to facilities offering elderly care, recommendations that were replaced by a ban on visiting retirement homes by Sweden's national government on 1 April 2020.

Given such challenges, the elderly care narrative is important for understanding Swedish provision of health care to its seniors. Criticism, particularly of the high mortality rate in Swedish elderly care, but also of the approval of the relatively open Swedish strategy, has been expressed in various fora. There have been discussions on many differences between Sweden's strategy and that of neighbouring Nordic countries such as Finland, for example, which has legislation and a prohibition on ministerial government similar to the Swedish model, but which applied significantly more restrictive social distancing practices when it dealt with COVID-19.

*Description of Problem*

Given such complexities, this chapter questions how, in the case of Sweden, its government handles the country's coronavirus strategies in ways that impact the elderly in residential care and those who have received home help services. In approaching the topic, my chapter provides insights into the role of the country's Public Health Agency and its evolving set of recommendations, which include prescriptions on social distancing, further framed within Sweden's reliance on its strong welfare state ideology, its tradition of personal independence, and its people's trust in the Swedish state, municipalities, and authorities. It is also relevant to draw attention to how the Swedish Constitution was restrained at the outset of the pandemic, and that historically, in peacetime or in crisis situations, Sweden has had no harsh regulations analogous to a "state of emergency" even though the government decided to prohibit visits to the *entirety* of the country's retirement homes to prevent the spread of COVID-19. Moreover, the pandemic compelled the government to authorize the Communicable Diseases Act (2004: 168), which remained effective until June 2020, with the understanding that such a period would allow the Swedish government to ascertain the pandemic situation as well as engage in immediate action, if need be, to cease operations of key areas in the country.

This chapter underscores the trade-off between individual freedom and authoritarian rule in regard to how Swedish senior citizens are affected. According to statistics from the country's National Board of Health and Welfare, the elderly have also been more vulnerable in certain regions such as Stockholm. Such considerations point to questions concerning the autonomy of regions and municipalities experiencing crisis in relation to the nation in general (Chapter 14, "Local Authorities," Instrument of Government (1974: 152)). If municipal autonomy is restricted, this restriction may not be greater than necessary, considering the purpose at hand (Chapter 14, "Local Authorities," Section 3, Instrument of Government (1974: 152)). Similarly, ministerial government was not permitted to fully operate during the pandemic in Sweden. This chapter thus engages in further discussion about how these circumstances have had consequences for welfare law, for the image of Sweden, and for its people's confidence in society during a civilian crisis.

The legal security principle and individual rights are accordingly examined in relation to the Swedish Constitution and the restrictions that limited individual rights during the pandemic. Here, it is important to emphasize that Chapter 2, "The Individual's Obligation to Prevent the Spread of Infection," sections 1–2 of the Communicable Diseases Act (2004: 168), obliges the individual to prevent the spread of infection. Even though no sanctions are set forth, there are penalties in the Swedish Penal Code for anyone infecting another person or putting another person in danger, and there are further regulations concerning personal injury in Chapter 2, "Liability for Damages due to Personal Fault," Section 1, of the Tort Liability Act (1972: 207), that may entail an obligation to pay damages. The legal frameworks and the administrative law structures are thus of great importance for discussing Sweden's COVID-19 strategy. It is important initially to clarify the legal decisions made and then to problematize and analyse primarily the prerequisites given to handle the spread of COVID-19 that was occurring at retirement homes and through home care services (Government Bill 2019/20: 155). Such decisions concern, in particular, the possibility that the government could issue special regulations on infectious disease prevention, but also the problematizing of the provisions – primarily general guidelines – which were applied during the period.

The actions of public sector actors led to the formation of the Corona Commission, one appointed to evaluate the actions of the government and the affected administrators of the municipalities to limit the spread of COVID-19. The Corona Commission was appointed on 30 June 2020, with the aim of examining actions in Sweden based on several assignments, especially in light of the country's subscription to the *principle of responsibility*. The overall goal that the Corona Commission emphasized for Sweden was "the handling of the virus that causes the disease COVID-19 so as to limit the spread of infection, to protect people's health, and to secure healthcare capacity" (Ministry of Health and Social Affairs, 2020b, p. 2). The Swedish strategy in the handling of the virus is such that the actions to be undertaken should be based on knowledge and well-tried experiences, as well as undertaken at the appropriate time. As such, the Swedish welfare state (rooted in the Social Democratic model) builds on the people's confidence in authorities.

> Sweden is one of the few countries in the world where people do not regard government as an "evil," but see it as an agent of development, and that this is the case demonstrates that the welfare state can successfully serve as a "political stability and security" project in a country that has suffered from chronic political instability and insecurity. (Svensson et al., 2012, p. 81)

*Chapter Objective*

The overall objective of this chapter is to examine the legal consequences, in a welfare context, of Sweden's COVID-19 strategy for the elderly between 1 March 2020 and 30 June 2020. The process will also be based on discussing the following three perspectives: the responsibility principle in crisis situations, legal security, and elderly rights. The objective includes clarifying the legal preconditions that existed and how they were handled in elderly care and home care services during the period examined. The Corona Commission's pandemic evaluation served as a starting point for my discussion of the difficulties the government experienced in their response to COVID-19. Selected issues are highlighted accordingly to identify topics relevant to senior citizens that have been subject to the Commission's examination. In this chapter, the following questions will be addressed more specifically:

1 How did Sweden handle the general limitations made to its senior citizens' freedom of movement because of COVID-19 in relation to fundamental rights, that is, the legal conditions for lockdown?
2 How did Sweden during its crisis handle the consequences of its constitutional law, which is based on ordinary administrative structures rather than laws stipulated under a state of emergency?
3 How did Swedish authorities realize the responsibility principle in a manner that attended to individuals' legal security and rights, at both national and regional levels, as different municipalities attended to the pandemic and their senior citizens?
4 What consequences of welfare law will this have for the Swedish narrative on responding to the pandemic, not least in the form of social and humanitarian effects on elderly care?

*Delimitations*

My discussion is limited to what here is considered the early phase of the pandemic during the pre-vaccination period of 2020. The chapter is therefore based on examining the various recommendations and rules introduced on 1 March 2020, such as visiting restrictions in elderly care, until 30 June 2020, when the legal changes became effective – and when COVID-19 was deemed a disease dangerous to the public and society as conveyed in the Communicable Diseases Act (2004: 168) on 1 July 2020 and in the Corona Commission's evaluation of public sector actions at the beginning of the COVID-19 crisis (Government Bill 2019/20: 144).

This chapter primarily analyses senior citizens at retirement homes and those who receive home care services, that is, some form of social service provided by authorities. It should be noted at the outset that Sweden's Public Health Agency and the National Board of Health and Welfare are not responsible for stopping the spread of infection at retirement homes per se. The responsibility rests with the homes' operators, that is, the municipalities or private companies. Consequently, extensive findings on the organizational sustainability of elderly care, such as personnel continuity of its hourly employees and acute shortages of nurses (compared with Norway), are not addressed, nor are issues of how a higher number of tests or more access to protective equipment could have affected developments addressed in this chapter. It should be noted that the Corona Commission's mandate also comprises an examination of, for example, basic hand hygiene and personnel turnover, which is not a focus of our discussion but is of great importance for the overall evaluation of the conditions of care affecting Sweden's senior citizens. Lastly, the examples provided will illustrate differing strategic actions that, among other things, were applied in other neighbouring Nordic countries, but no comparative studies were undertaken.

*Method and Materials*

To clarify the legal framework in Sweden, a traditional legal research approach is applied through studies of relevant legal sources. Moreover, the pandemic has compelled the government to authorize the Communicable Diseases Act (2004: 168), which was to remain effective until June 2020, with the understanding that such a period would allow the Swedish government to ascertain the pandemic situation as well as engage in immediate action, if need be, to cease operations of key areas in the economy. Focus will also be on investigating the temporary, national visiting ban that was implemented in elderly care during the period studied. The aim is to further identify what the Swedish government perceived as more central to evaluate, a process undertaken through an overview of the Corona Commission's directives and its concerns for implementation strategies that uphold health practices for enhancing the quality of life for Sweden's elderly during the pandemic period. By shedding light on areas the Commission examines in the report – the outcome of the authorities' actions – the country's strategies can be contextualized with a focus on the regulation of elderly care. This underscores the authorities' central role in the evaluation of the pandemic and how they have been able to keep the country operating in crisis situations without a ministerial government.

By employing a socio-legal approach in the later part of the chapter, which analyses how municipalities handled the pandemic in their care of Sweden's senior citizens at the national and municipal levels, I approach the matter from a constitutional and more individual basis, an approach that reveals the architecture of Sweden's legal system, legal culture, and how these relate to justice for its senior citizens. Additionally, the chapter's theoretical framework highlights humanitarian issues, primarily their legal consequences in a Swedish welfare context of COVID-19 strategies pertaining to the elderly. The highlighting process includes discussion of the potential consequences of welfare law in shaping the Swedish narrative about authorities' handling of the pandemic, not least in how it affects social and humanitarian efforts to care for the elderly. The significance of the responsibility principle, notions of legal security, and the rights of the elderly, during the initial phase of the current crisis, are discussed throughout, based on sources that address how the perspectives manifested during the pre-vaccination period of the pandemic in 2020.

*Theory*

The legal preconditions during the pandemic are analysed based on a welfare state perspective. Also addressed is the significance of theoretical approaches such as the responsibility principle, the legal security principle, and elder rights. From these themes, we ask what significance these legal cornerstones have had for the Swedish corona strategic response. The theories are briefly clarified below to illuminate this chapter's use of perspectives as well as to clarify the welfare state perspective. Finally, the chapter also analyses how these frames of reference are interpreted. Indeed, the outcomes of social distancing can further be framed within Sweden's reliance on its strong welfare state ideology, its tradition of personal independence, and its people's trust in the Swedish state, municipalities, and authorities (Greve, 2020). The approach extends Greve's confirmation that the Swedish population has traditionally been confident in its authorities' abilities to operate on behalf of the Swedish welfare state.

Swedish crisis preparedness builds on ordinary administrative structures and on the responsibility principle, which means that the party that is responsible for administrative operations under normal situations has equivalent operational responsibilities in a crisis. According to the Committee Terms of Reference, such responsibilities also include evaluating actions that handle the outbreak of the virus that causes COVID-19 (Ministry of Health and Social Affairs, 2020b, p. 4). Similarly, the legal security principle in Swedish law ensures fundamental guarantees to the individual in relation to society (Vahlne Westerhäll, 2002; Gustafsson, 2002; Rönnberg, 2020). Formal legal security lies in the ability of the individual to appeal a legal decision. Yet in the case of Sweden's elderly, if their actual needs are not made visible, it will be difficult for them to get their cases tried in court (Petersson Hjelm, 2018, p. 15). Similar cases being handled in a consistent manner is an important legal security issue in Sweden (Vahlne Westerhäll, 2017). It is also crucial that a decision is ethically defensible, that is, that "substantive" legal security exists (Petersson Hjelm, 2018). Finally, social rights as fundamental goals for enhancing individual outcomes in society are articulated in the Instrument of Government (1974: 152), Chapter 1, "Basic Principles of the Form of Government," Article 2, Paragraph 2. It addresses goal-oriented legislation, for example, that attends to the means of distributive social justice stipulated in the Social Services Act (2001: 453), or that upholds national fundamental values towards elderly care as conveyed in the Instrument of Government (1974: 152). Such a legal right to assistance and support, especially in the context of residential care and the provisioning of home care services, can be considered a claim right. It entitles a person to counsel and guidance if care cannot be provided for, in our case, the elderly, either by themselves or someone else (Gunnarsson, 2007). In such circumstances, enforcement often rests with local authorities.

## The Legal Framework during the Pandemic

The governing minority coalition in Sweden, consisting of the Social Democratic Party and the Green Party, deferred to centralized objectives during the pandemic. This provides insight into the architecture of Swedish public administration, upon which an account of how pandemic events impacted the Swedish legal situation will be offered. Moreover, central statutes that regulate the situation and circumstances of the elderly during COVID-19, derived from the Instrument of Government (1974: 152), the Social Services Act (2001: 453), and the Communicable Diseases Act (2004: 168), as well as selected regulations on control by authorities of the elderly area, will be discussed. The Communicable Diseases Act (2004: 168), for example, makes possible coercive measures towards individuals, and there must thereby be legal support, as Sweden does not have laws that allow compulsion of elderly people (except for self-defence and/or through its Compulsory Psychiatric Care Act). The legal conditions that were enacted to regulate visiting bans at retirement homes, such as through social distancing measures, are addressed separately below.

### Crisis Laws: Limited Authority According to Constitutional Law

Sweden has not been in a prolonged war or crisis situation for an extended historical period, and today has no direct civilian crisis legislation equivalent to a "state of emergency." In contrast to war or the danger of war, there is no special regulation for civilian crises in constitutional law, as noted in the Instrument of Government (1974: 152). In the current Instrument of Government (1974: 152), Sweden's preparedness is based on what is known as *fullmaktslagarna*, or emergency powers legislation, as stated in Sweden's Government Official Investigations 2008: 61 (Statens offentliga utredningar (SOU) 2008: 61). The Instrument of Government (1974: 152) is designed to enable fast decision making and the setting of standards, which among other things can be used in civilian crisis situations.

Safeguards against forced physical intervention, and the right to have one's freedom of movement protected, are constitutionally ensured in Sweden and prevent general restrictions, such as a general curfew, from constraining citizens' freedom of movement (Chapter 2, "Fundamental Rights and Freedoms," Articles 6 and 8 of the Instrument of Government (1974: 152). According to the Instrument of Government (1974: 152), special legislation is required for prohibiting citizens' from moving freely in the country or certain parts of the country, restricting citizens' rights to leave their hometown, and sealing off large areas (such as border zones). It is possible to delimit a certain geographic area for quarantining purposes if there is a suspicion that a contagion dangerous to society has begun to spread. The measures must then be proportional, necessary, and effective. The restrictions must also be reasonable in relation to the negative effects that accompany the stipulations.

The political decision that was made in the early phase of the pandemic – through Government Bill 2019/20: 155 – entailed a temporary expansion of the government's authority to issue rules for handling the spread of the coronavirus. The authorizations gave the government greater manoeuvrability in responding to pandemic conditions and were intended to operate within constitutional limits and be a supplement to applicable rules. Such parameters meant that restrictions could never be formulated in such a way as to limit Swedish citizens' constitutionally established freedoms. Moreover, such restrictions can only be decided by the Swedish Parliament through legislation. Thus, regulations the government issued with support of the authorizations were accordingly and immediately subjected to the Swedish Parliament's review, resulting in a legal amendment to the Communicable Diseases Act (2004: 168) that paved the way for coercive measures to be undertaken because of COVID-19.

In the preparatory work, it was emphasized that the authorization did not provide support for deciding on the need for curfews, which had already been introduced in many other countries responding to COVID-19. The general concern was that the decision to mandate curfews or quarantines for entire communities would probably entail restrictions to fundamental rights that required a new legal amendment (Government Bill 2019/20: 155, p. 17). However, the Instrument of Government (1974: 152) floated the possibility, albeit hypothetical, of introducing curfews similar to those that had been enforced in many parts of Europe and even the United States. Such prerequisites for restrictions to fundamental rights are communicated in Chapter 2, "Fundamental Rights and Freedoms," Article 20, to the extent permitted through legislation according to Articles 21 to 24. According to the principle of proportionality, restrictions can only be established to achieve goals that are acceptable in a democratic society – that is, the restrictions may never exceed what is necessary considering the purpose that gave rise to it.

Even if some restrictions were introduced during the early phase of the pandemic and legislative changes were made, for example, to close upper-secondary schools and universities while preschools and compulsory schools remained open, public recommendations became the dominant strategy in Sweden for informing the people. In general, Swedish authorities provide recommendations on the application of a statute, which state how somebody can or should act in a certain respect; that is, the general guidelines issued by the authorities functioned as a behavioural tool for other authorities and actors in the public sector to abide by. With the onset of the pandemic, the general guidelines that the Public Health Agency prepared were offered to citizens as guiding principles. The guidelines were applied to the elderly over 70 as well as their relatives. Regulations that authorities adopt are binding, in contrast to the general guidelines, which do not have the same status (see also Chapter 11, "Administration of Justice," Article 14, and Chapter 12, "Administration," Article 10, of the Instrument of Government (1974: 152)). General guidelines – which can hardly be seen as written laws – will nevertheless sometimes, during implementation, achieve a position and practice that are reminiscent of laws.

*The Communicable Diseases Act (2004: 168)*

If diseases are classified as dangers to the public and society, the Communicable Diseases Act (2004: 168) provides support for certain extraordinary infectious disease prevention measures that entail restrictions with the help of compulsion. On 1 February 2020, the government decided that regulations conveyed in the Communicable Diseases Act (2004: 168) should be applied to COVID-19, which was classified in the annexe as a disease dangerous to public health and society (Government Bill 2019/20: 155, p. 9). Such a status means that Sweden's Public Health Agency can issue regulations and take actions, and that their regional infectious diseases officers can act based on the mandate provided by the Communicable Diseases Act (2004: 168). According to Chapter 1, "General Provisions," Section 4, of the Communicable Diseases Act (2004: 168), infectious disease prevention shall be based on science and well-tried experience. Furthermore, the actions undertaken may not be more extensive than is defensible considering the danger to human health, and actions are to be taken with equal respect for the integrity of individuals and of the people. The emphasis of infectious disease prevention work in Sweden accordingly lies on voluntary prevention. An example of this orientation is how the Communicable Diseases Act (2004: 168) does not give the government or any other actor authority to make decisions to close private operations that gather several people. This has been motivated by the view that the public experiencing a crisis would voluntarily comply with recommendations from the authorities for infectious disease prevention (Ministry of Health and Social Affairs, 2020b, p. 7).

According to the proposal, the government would be able to issue special regulations on the relationship between individuals and society pertaining to obligations for individuals, or otherwise pertaining to intrusions into the personal circumstances of individuals, but only if they were necessary for halting the spread of COVID-19 (Government Bill 2019/20: 155, p. 1). In Sweden's most acute peacetime crisis in recent memory, the regulations became relevant and were temporarily applied between 18 April 2020, and the end of June 2020. The change was formalized in the Communicable Diseases Act (2004: 168) through three new paragraphs in Chapter 9, "Other Provisions," sections 6a to 6c. The legal proposal provided room for rules on certain coercive measures, which among other things could entail obligations for individuals or intrusions into the personal circumstances of individuals (Government Bill 2019/20: 155, p. 21). According to Chapter 9, "Other Provisions," Section 6, of the Communicable Diseases Act (2004: 168), the government was authorized to

> issue special regulations on infectious disease prevention … if in a peacetime crisis that has a significant impact on the possibilities of maintaining effective infectious

disease prevention, there is a need for coordinated national actions or from a national perspective for other special efforts in infectious disease prevention. (Ministry of Health and Social Affairs, 2020b, p. 7)

The *lagrådet*, or Council on Legislation, emphasized that, were the government to enforce the proposed regulation, it must also observe certain fundamental prerequisites communicated in the Instrument of Government (1974: 152) and the provisions of the European Convention on Human Rights regarding the protection of fundamental rights (see also the Council of the European Union and its "Council Declaration on the European Year for Active Ageing and Solidarity between Generations" report (2012)). The Council on Legislation emphasized that the authorizations did not grant the power to decide on lockdowns, since such delimitations on the right to the freedom of movement can only be decided through legislation by the Parliament (Government Bill 2019/20: 155, p. 38). Yet, the Council on Legislation was concerned that such authorizations allowed government to be invasive towards individuals even if constitutional legislation sets the ultimate limits for the actions that can be taken, and even if the regulations have been subjected to review by Sweden's Parliament.

The Council on Legislation considered that a renewed review was necessary to delimit the actions that the government could take in the relationship between individuals and society. Actions that the government submitted to the Council on Legislation for deliberation were seen to be a suitable delimitation even if it was not possible to fully predict what actions might need to be undertaken. The list of actions was entered as legal text and stipulated what actions the government could undertake, such as enforcing temporary limitations on the number of people in gatherings. These regulations were then to be immediately subjected to the Parliament's review according to Chapter 9, Section 6c. The government proposed that the Parliament should approve its ordinance on regulations in the Communicable Diseases Act (2004: 168), that is, that COVID-19 be entered as a disease dangerous to the public and society in the annexe to the act (Government Bill 2019/20: 144). On 1 July 2020, the act went into effect.

*Framework for the Organization of Elderly Care: The Social Services Act and the National Fundamental Values for the Elderly*

On 30 March 2020, the government decided through its Ordinance 2020: 163 to impose a national visiting ban on retirement homes and a temporary prohibition of visits to special forms of housing for the elderly (Government and Government Offices of Sweden 2020a, p. 30), introduced on a general basis as conveyed in Chapter 16, Section 10 of the Social Services Act (2001: 453). This regulation placed more extensive limitations upon the individual's freedom of movement in order to prevent the spread of COVID-19. Initially meant to be applied until the end of June 2020, the ordinance outlined mitigating conditions for the elderly in need of special forms of housing, services, and care (Chapter 5, "Special Provisions for Different Groups," Section 5, paragraphs 2 and 3). In the ordinance banning visits to the elderly, it was also clarified that visits to the aforementioned types of housing with special needs and services were banned (Chapter 7, "Regulations on Individual Activities and Notifications Obligation," Section 1). The stipulation did not apply to those who, in their professional practice, were required to be in such homes. However, the operationally responsible party for a home had the authority to grant exceptions to the visiting ban if there were special circumstances that motivated an exception, and if the risk of acquiring COVID-19 was minimal inside the facility. Decisions made under the ordinance could not be appealed.

Elderly care in Sweden is based on a rational decision model realized through the Social Services Act (2001: 453), which is a framework law intended to ensure people's sovereignty (Chapter 1, "The Goals of the Social Services," Section 1). With the 2011 promulgation of the National Fundamental Values for elderly people subsumed under Sweden's Social Services Act (2001: 453) in Chapter 5, "Special Provisions for Different Groups," Section 4, the ability of Sweden's seniors to live in dignity, with a sense of well-being, in self-determination, participation, and security, is addressed (see also Government Bill 2009/10: 116, Government Official Investigations 2008: 51, and Government Official Investigations 2017: 21). Sweden's National Fundamental Values articulate the ethics and norms needed to govern how elderly care is conducted. In the law, it is clarified that the activities are to be focused on ensuring seniors' need for dignity as well as confirming that they shall be able to live according to their identity and personality, age in security and with retained independence, live an active life in society, be treated with respect, and have access to good care and welfare. Thus, the fundamental value of a *dignified life* is equated with a senior's private life, integrity, and self-determination. The care bestowed upon such persons shall be individualized and create participation, and the services shall be of good quality. Seniors and their relatives shall be treated well, a fundamental ethical principle. The other fundamental value of *well-being* ensures that seniors shall be able to feel a sense of security and meaningfulness. This ethos is applied nationally in publicly and privately operated services.

In the case of municipalities, they are to offer health care that the individual in special forms of housing may need. *Home nursing* means that health care is provided in the patient's home or equivalent, where medical measures employed shall be coherent over time. In terms of responsibility, home nursing is a complex area and is handled according to the Social Services Act (2001: 453) and the Health and Medical Services Act (2017: 30). An important difference between these laws is that the Social Services Act addresses *rättighetslagstiftning*, or legislation of rights, while the Health and Medical Services Act addresses *skyldighetslagstiftning*, or legislation of obligations. Accordingly, no rights are specified in the latter law.

Efforts by physicians are exempt from their municipality's responsibility, according to Chapter 12, "Responsibility to Offer Health Care," Section 3, of the Health and Medical Services Act (2017: 30). Through reforms at the beginning of the 1990s, medical care and elderly care embarked on separate trajectories in Sweden (Rönnberg, 2020; Petersson Hjelm, 2021). Consequently, the closeness between their institutions decreased, a contrast when compared to Sweden's neighbour Norway where elderly care involved more nurses in the staff and there is a closer connection to doctors and medical care.

At the same time, in Sweden and internationally, the system of elderly care is facing a multitude of challenges with regard to a vast, aging population (see also Greve, 2020). Compounding matters, research already indicates a dismantling of elderly care because of budget restrictions (Szebehely & Meagher, 2018). Despite the declining percentage of social services, retirement homes and home care services are a municipally established social service and a common form of elderly and healthcare provision in Sweden, one primarily financed by taxes on its citizens. In terms of demographics, one in five residents in Sweden is 65 or older (National Board of Housing, Building, and Planning, 2020). At the beginning of 2020, 108,500 people lived in residential care during the year, and 401,000 people aged 65 or older received home care services. Of the population aged 65 or older, approximately 19% had at least one social service provided, according to the Social Services Act (2001: 453); a relatively large percentage of this is for emergency situations, however (National Board of Health and Welfare, 2020a).

*Case Law and Statements by the Parliamentary Ombudsmen in the First Phase of the Pandemic*

Decisions from legal bodies provide guidance on how old people's possibilities of movement are to be handled. During the onset of the pandemic, two relevant rulings were issued in the area of senior citizens, with one partly a decision from the Administrative Court (lower court, not precedential), and the other a guiding statement from the Parliamentary Ombudsmen. The rulings addressed a legal dilemma when several Swedish municipalities decided not to implement decisions on services from other municipalities according to Chapter 2a, "On the Division of Responsibilities between Municipalities in Terms of Support and Assistance under This Act," Section 6 of the Social Services Act (2001: 453), because of COVID-19 infections during March and April 2020. The legal starting point is that, when an individual spends a brief time in another municipality other than their municipality of residence and needs support and help in the form of home care services, the municipality the person resides in is thus obliged to, upon request, carry out the municipality of residence's decision on aiding the elderly person. A common factor driving the municipalities to undertake these decisions is that they have many seniors who annually, but temporarily, stay in other municipalities. Such seniors, entitled to home care services and home nursing, place an extensive burden on the home and healthcare services of the host municipalities. Compelled by the need to reduce the spread of infection in their respective communities while also protecting their most fragile members, these municipalities reasoned that the seniors could have their care needs better attended to in their home municipalities. Such a dilemma in care and nursing was further burdened by COVID-19, since the resistant municipalities had difficulty staffing their own care services because of high sickness figures.

The Administrative Court in Gothenburg decided that a municipality cannot choose whether it wants to follow applicable legislation determined by the municipal executive board; only Parliament can change applicable law (Administrative Court in Gothenburg, Case number 5025-20). The decision was also viewed as incompatible with the Public Health Agency's recommendations on COVID-19, as it entailed the risk of generating greater infections between municipalities. Despite municipal autonomy, the Administrative Court confirmed that the visited municipality was obligated upon request to implement the home municipality's provision of services if an aged person requested care when they spent time, no matter how brief, in the visited municipality. The matter was then subjected to the Parliamentary Ombudsmen (2020). However, the Parliamentary Ombudsmen also confirmed that the visited municipality had an obligation to carry out the home municipality's decisions on services. That is, a visited municipality cannot legally choose whether or not it will follow applicable legislation. The municipal executive board therefore received criticism for having made a decision that lacked legal support, since there is an obligation for a visited municipality to carry out the home municipality's decision on services according to Chapter 2a, "On the Division of Responsibilities between Municipalities in Terms of Support and Assistance under this Act," Section 6 of the Social Services Act (2001: 453).

On 28 May 2020, the Standing Committee on Social Questions submitted an official report to the Swedish Parliament regarding the visited municipalities' responsibilities for the provisioning of social services. The Parliamentary Committee on Health and Welfare, *Socialutskottet* (*SoU*), proposed in its "2019/20: SoU24" official report, "Visited Municipalities' Responsibility for Social Service Efforts" (2020), that the Parliament should notify the government to promptly prepare a legal proposal that allowed a visited municipality the temporary possibility of denying home care services to people from another home municipality. On 3 June 2020, the Swedish Parliament approved the Committee's proposal, and the report was turned over to the Ministry of Health and Social Affairs (2020b). According to that ministry, however, it was not considered relevant for the government under pandemic conditions to submit a legal proposal to realize the content stated in the notification.

## The Authorities Have Command over the Swedish Strategy

According to the Swedish Constitution, government ministers are prohibited from intervening and micromanaging the work of authorities. As such, Swedish authorities can be said to exhibit a form of expert-based governance. Several authorities have played prominent roles during the pandemic while the government remained in the background. Swedish authorities can thus be said to have influenced actions, such as social distancing, through recommendations and general guidelines. These prescriptions emanate from institutional actors that have set the tone for elderly care during the pandemic, such as the National Board of Health and Welfare, the IVO, the county administrative boards (commissioned to prepare a plan to meet the new outbreak of COVID-19 for the autumn season of 2020), the Swedish Civil Contingencies Agency (responsible for issues concerning crisis preparedness, insofar as no other authority has that responsibility), and Sweden's Public Health Agency. We thus examine some of their activities and accomplishments during the pandemic.

### The Public Health Agency of Sweden

The Public Health Agency of Sweden is responsible for the coordination of infectious disease prevention on a national level and takes initiatives to maintain effective infectious disease prevention. Its mission of offering a broader public health perspective on society, and not just on mitigating the effects of communicable diseases, has set the tone in terms of how it harnesses the media. Participating ministers have enabled agency experts to convey COVID-19 information while they themselves have adopted a supportive but more deferential role, one that is in line with the status of not being able to intervene in individual cases or micromanage the authorities' activities. When the Public Health Agency's lawyers investigated the possibility of actions based on the Communicable Diseases Act (2004: 168), they clarified that, although restrictions such as placing individuals in quarantine are possible, they would only seal off small areas (such as a retirement home) and not implement additional extensive lockdowns. They would, however, require COVID-19 to be classified according to the Communicable Diseases Act (2004: 168). Indeed, by February 2020, COVID-19 was classified as a disease dangerous to the public and society. After a temporary legal amendment to the act in spring 2020, the Public Health Agency and an infectious diseases officer could decide on quarantine for a certain building or certain area if verification demonstrated that a disease dangerous to the public and society had begun to spread. The Communicable Diseases Act (2004: 168) stipulates in Chapter 3, "Investigation of Disease Cases," Section 10, that the Public Health Agency can determine how a particular area shall be cordoned off if a disease dangerous to society has begun or is suspected of beginning its spread within a delimited area.

In March 2020, the pandemic situation for Sweden's elderly became strained. Swedish authorities thus intensified the work of isolating retirement homes and implementing restrictions for home care services. On 10 March, the Public Health Agency published information related to infections and broadcast their information daily through the mass media. They advised Swedish citizens, and especially relatives of senior citizens, to avoid unnecessary visits to healthcare and elderly care facilities such as retirement homes. The Public Health Agency issued the recommendations through the so-called general guidelines on 16 March to limit the spread of infection. On 30 March, the Public Health Agency received a government commission to promptly increase the number of COVID-19 tests. The national strategy for COVID-19 during this period categorized elderly care as a priority. By 1 April 2020, general guidelines were issued for people over 70 (Swedish Public Health Agency, 2020b). The general guidelines included instructions that people over 70 and people who belong to other risk groups should limit their social contacts and stay home as much as possible.

## National Board of Health and Welfare

On 13 July 2020, Sweden's National Board of Health and Welfare formally prepared general guidelines for a temporary prohibition of visits to special forms of accommodations for older persons to prevent the spread of COVID-19. The ordinance on a temporary prohibition on visits to special forms of housing placed no obstacles for seniors leaving the home themselves or with help, for example, to take a walk or to meet relatives. It should also be emphasized, even if it is outside the purview of our discussion, that the National Board of Health and Welfare has continuously prepared information, for example on cleaning procedures, to demonstrate the actual work of preventing infection in the area of elderly care. The National Board of Health and Welfare is also responsible for publishing timely information including mortality rates among the elderly in special housing in the wake of COVID-19.

## The Health and Social Care Inspectorate

In Sweden, the Health and Social Care Inspectorate (IVO) supervises operations in elderly care and health care. In the 1990s, through the Freedom of Choice Act, municipalities were subject to competition, and elderly care was increasingly conducted through private contractors instead of by the public sector (Proposal 2008: 962). Municipalities and regions were granted local control, such as when private providers of elderly care were tasked with carrying out the municipalities' services. Moreover, transparency was expected, and municipalities had to provide public insights into their operations when these were attended to by private providers, according to the Local

Government Act (Government Bill 2008/09: 29). In the event of a needed inspection of the condition of elderly people who received home care services or had a special form of housing, it was required that the individual consent to such a need as stipulated in Chapter 13, "Supervision," Section 6, of the Social Services Act (2001: 453).

The IVO has released some findings, such as how the deaths at Sweden's retirement homes in 40 municipalities accounted for 67% of deaths from COVID-19 in such facilities.[2] In their examination, 91 among Sweden's 1,700 retirement homes had deficiencies *of a serious nature*. The IVO deemed that only 60% of the 40 worst affected municipalities had complete conditions for an individual assessment and treatment of COVID-19; however, at these homes only a general assessment was undertaken (Health and Social Care Inspectorate, 2020a). The facilities were thus given notice that if the shortcomings were not corrected voluntarily by the party responsible for facility operations, orders to resolve the shortcomings could be issued by the IVO.

At the beginning of July 2020, a new assignment was given to the IVO because of the indicated deficiencies: the committee was to conduct an in-depth examination of the aforementioned 91 special housing units with deficiencies of a serious nature, one that was completed in late autumn 2020. The IVO report showed that no region had taken full responsibility for individual care in the home care service system. Massive public criticism followed the findings (Health and Social Care Inspectorate, 2020b).

## Corona Commission's Key Mission to Evaluate during the First Phase of COVID-19

What were the Corona Commission's evaluation directives? At the beginning of the chapter, an account was provided of the Commission's strategies and goals for pandemic evaluation as they are formulated in directives (Ministry of Health and Social Affairs, 2020b, p. 2). According to impact assessments, the Commission shall generally report on the consequences of actions as they pertain to human rights and individual freedoms. In the following discussion, a selection of the various assignments presented to the Commission will be examined. The study of the Commission's effectiveness highlights its focus on the rights of the elderly, one that offers an entry point for analysing pandemic evaluation's legal and humanist aspects in the concluding sections of this chapter. The Corona Commission has been given an extended period of time to complete its report on how the authorities' decisions affected the plight of senior citizens in Sweden. The report, originally set to be released in November 2020, remains incomplete at the time of this writing in later summer 2020 because of the continuing unfolding dynamics of COVID-19 and is thus not included in our discussion. Nonetheless, in the following list, we see how primary assignments given to the Commission exhibited a direct connection to, as well as prioritization of, the provisioning of elderly care.

1. Evaluate how the *responsibility principle* and the geographic areas of responsibility functioned during the crisis.
2. Evaluate the effects of social distancing on the population as a whole and especially on the elderly.
3. Analyse the causes of the spread of infections in special housing for the elderly and in the process of provisioning home care services.
4. Evaluate affected administrative authorities, their regions, and their municipalities' actions for limiting infection among Sweden's elderly, as well as enact actions to handle the virus outbreak and its effects *with a focus on the elderly care area*.
5. Undertake a collective assessment of Sweden's strategic attempts to limit the spread of infection *with a focus on the elderly care area*.
6. Evaluate how suitable the Public Health Agency's recommendations have been during various phases of the pandemic.
7. Analyse if infectious disease prevention legislation provided society with suitable conditions and resources to limit the spread of infection.

## Analysis of the Restrictions on Senior Citizens during COVID-19 in Sweden

The legal circumstances for senior citizens during the early phase of the pandemic are analysed in this section based on a welfare state perspective. Sweden, which has frequently been presented as unique because of its continued openness, is analysed herein as a country not having to contend with legal premises that would compel its government to declare a state of emergency, unlike many countries afflicted by the COVID-19 outbreak. To provide a more complex picture of the early phase of the pandemic, my chapter attends to more specific questions about citizens' constitutionally established rights and freedoms. It also examines the legal prerequisites attended to by Sweden's administration as it responded to issues affecting elderly care in times of crisis. The task is to analyse the significance of the following crucial tenets in Swedish law: the responsibility principle, the legal security principle, and rights of the elderly, based on the discussion undertaken in this chapter. My examination thus addresses how the handling of COVID-19 was implemented in Sweden at structural and institutional levels in the area of elderly care.

In line with the prohibition on ministerial government, Swedish society was thus governed by regulations carefully directed by public authorities such as the Public Health Agency, authorities that issued regulations and updated citizens frequently via the media about ongoing developments with the pandemic.

To prevent the spread of COVID-19, for example, the virus was classified as a contagion in the Communicable Diseases Act (2004: 168), which made it possible to place infected individuals or those exposed to infection in temporary quarantine. The population's freedom of movement could thereby be restricted through legal amendments. The Council on Legislation thus emphasized the significance in constructing what could be covered by these restrictions to make the delimitations more acceptable, that is, whether or not there was proportionality between benefits and damages the restrictions entailed.

A review of the preparatory work for this law, Government Bill 2019/20: 155 regarding temporary authorizations in the Communicable Diseases Act (2004: 168), shows that the situation of the elderly was not brought up specifically, which is an interesting omission, since they constitute a prominent risk group with a high mortality rate (Government Bill 2019/20: 155, p. 15). Yet regulations can be assumed within the social services sectors that are needed to protect the lives and health of older persons (Chapter 16, "Other Provisions," Section 10, Paragraph 1, Social Service Act (2001: 453)). Furthermore, Government Bill 2019/20: 155 highlighted the significance of effective inspection and control as well as problems identifying critical areas during the pandemic. As such, changes to the aforementioned act were introduced temporarily, as stipulated in Chapter 9, "Other Provisions," sections 6a to 6c, of the Communicable Diseases Act (2004: 168). These amendments made it possible for the government to be given authority to close certain businesses and limit public gatherings in response to COVID-19 infections. The changes that were implemented enabled the Swedish civil service to introduce a visiting ban in elderly care at a national level, one of the more significant legal measures that has been implemented in modern times. Other changes were also issued. Prohibitions on public gatherings and public events with more than 50 participants received particular attention, yet direct bans in Sweden were relatively limited. The authorities' prominent role in conjunction with prohibitions of ministerial governance were subject to discussion, but surprisingly few discussions attacked the government for having exceeded its authority.

At this juncture, we can consider whether Swedish regulations during the pandemic have been able to ensure the individual formal legal security – especially considering the situation of the elderly. That is, as Vahlne Westerhäll (2002) queries: can there be equality through legal statutes and was substantive legal security based on guidelines that were *ethically acceptable*? As such, have elders' rights been infringed during the pandemic, based on this chapter's analysis?

Few cases have been subject to legal review during the early phase of the pandemic, but the possibilities available for elderly mobility have been subject to both lower court judgment and criticism by the Parliamentary Ombudsmen, both of which support the view that municipalities must comply with applicable laws regardless of the logistical difficulties set into motion by the pandemic.

Demandable rights such as those granted to people who seek home care services provide insights into the issue of elderly mobility. Despite these few rulings during the spread of the pandemic, the legal bodies of Sweden did not find justification for municipalities to evade offering support to visiting seniors who spend time in them, even if the visited municipalities are motivated by preventing the spread of COVID-19. Furthermore, also placed on hiatus was the Standing Committee on Social Questions' report that advocated for a legal proposal to allow visited municipalities to deny home care services to people residing in another home municipality.

Whether or not general social rights, such as the National Fundamental Values, which are not directly demandable, have been effectively handled during the onset of the pandemic is a question for the Corona Commission to address. This chapter problematizes how the strategy – which to a great extent prescribes recommendations in a manner that is open to interpretation – relates to the direct national ban on visiting retirement homes. The authorities' encouragement of voluntary pandemic social distancing was the crucial strategy for the population in the beginning of the pandemic. Indeed, even though the experience of such stipulations is but a tool for authorities to interpret rules, the pandemic crisis forced Swedish authorities to provide guidelines directly to the public as recommendations to abide by. This was especially applicable to people over 70, who were encouraged to avoid contact with other people while they concurrently were still able to experience mobility in society, as in shopping, going for walks, etc. That said, the effect of this social distancing is clearly problematic to measure and evaluate. The relatively high mortality rates of persons in special housing and among those who received home care services during the early phase of the pandemic have been highlighted in this chapter, since Sweden's early attempts at mitigating the pandemic were deemed "a failure" by evaluating authorities.

## Conclusion of the Swedish Pandemic Strategy during Its Early Phase

The Swedish pandemic strategy during its early phase and its current ongoing pandemic mitigation strategies have not *per se* been subject to extensive critical analyses in this exploratory chapter, one that instead made visible the contours of Sweden's crisis response institutions. That said, it is important to understand that Swedish law prevented its government from acting too strongly when compared, for example, to many countries' crisis management approaches; that is, Sweden's open strategy during the onset of the pandemic was based on *författningsberedskap*, anticipatory statutorification in extraordinary situations. During the early phase of the pandemic in Sweden, the Swedish approach, through a variety of statutes, entailed sending delegations to government and sub-delegations to administrative agencies. In the current pandemic crisis, the laws that have surrounded infectious disease prevention in relation to the elderly and home care services were quickly drafted. Such actions compensated for the government experiencing a civilian crisis in a manner where they were unable to implement extensive shutdowns. This scenario is fundamentally important for understanding why Sweden did not introduce general restrictions to constrain its citizens' freedom of movement because of COVID-19. In short, a state of emergency and its expected concomitant actions are simply not compatible with Swedish law.

Two forms of power dynamics manifested during the initial phase of Sweden's pandemic mitigation efforts:1. Authorities and experts stepped to the foreground while the government remained relatively invisible, a situation made possible by how Swedish law prevented ministerial government from interfering with pandemic mitigation practices (in contrast to Denmark and Norway, for example).2.    Authorizations based on the Communicable Diseases Act (2004: 168) meant that the Swedish Parliament, acting within the limits set by the country's Constitution, considered delegating standard-setting power to the government, a legitimate manoeuvre according to Sweden's Instrument of Government (1974: 152) and its Government Bill 2019/20: 155.

A legal amendment, motivated by COVID-19 not being classified as a contagious virus, ultimately paved the way for coercive measures to be undertaken via the stipulations of the Communicable Diseases Act (2004: 168). The Instrument of Government (1974: 152) is extremely cautious about limiting citizens' fundamental rights. However, after deliberations with the Council on Legislation, a more concrete proposal was adopted with the understanding that such authorization did not grant extraordinary powers that allowed pandemic mitigation practices to deviate from Sweden's constitutional guarantees. Thus, the general recommendations for social distancing in Sweden, insofar as the elderly are concerned, can be traced to legislative processes, and constitute but pieces of a puzzle for understanding Sweden's strategies for pandemic mitigation.

Has it been possible to ensure welfare, particularly for the elderly during the pandemic? General guidelines on self-isolation for people over 70 and a national ban on visits to retirement homes primarily constituted the Swedish strategy for senior citizens, one that existed alongside the Swedish strategy of freedom with responsibility. The Swedish government and affected authorities confirmed that elderly care had not been adequately prioritized in the initial phase. Thus, the extensive spread of the virus in communities meant that some retirement homes and elderly recipients of home care services were severely impacted, further adding to the industry's woes. Even before the pandemic, elderly care was a much-discussed topic where shortcomings, such as high personnel turnover and reduced access to places at residential care units, were criticized. Sweden's high standard of living and high life expectancy also constitute an important parameter for considering why elderly and sick individuals were also particularly vulnerable at care homes or when seeking home care services. A multitude of factors thus contributed to the extensive mortality and sickness rates that befell the elderly, Sweden's most at-risk group and one that was most severely impacted by COVID-19.

Through this analysis of the pandemic's initial phase, it became self-evident that the responsibility principle, the legal security principle, and elderly rights have been challenged, and that challenges remain. Just the appointment and formation of the Corona Commission is *per se* in line with the view that the responsibility principle will need to be examined in future research on Sweden's ability to forward new regulations and/or respond to civilian crises of a non-warfare nature. Such an undertaking is an important one, not least because of how Sweden's vulnerable, the elderly and those seeking healthcare services, suffered during the pandemic.

Finally, some concluding thoughts on the welfare state of Sweden during the pandemic and Swedish elderly care, primarily in the form of social and humanitarian effects upon the welfare of the elderly. Several researchers show that the Swedish welfare state is built on its *confidence* in Swedish authorities and their actions. These actions included, for example, authorities holding daily press conferences during the early phase of the pandemic at Sweden's national public television station, Sveriges Television, which had the task of informing society about pandemic developments on behalf of the public sector. More concrete information on the current legalities of civilian crisis situations, of which the COVID-19 outbreak was one, would probably have increased the understanding of both the strategy of freedom with responsibility and of Sweden's national restrictions as they affected how its society responds to elderly care. This chapter indicates that the flagrant criticism directed at the Swedish strategy, nationally and internationally, is to some extent due to the lack of legal and expert input that could explain the Swedish strategy to observers unfamiliar with its system. Authorities have avoided discussions of why no shutdown of Swedish society occurred, motivated by legal preconditions not being covered, for example, by the Public Health Agency's area of expertise. To what extent such differences in operations will reduce public trust in the Swedish welfare state's confidence in its national institutions will be a topic of great importance for future examination.

*Postscript*

After work on the chapter was completed, the National Board of Health and Welfare presented the Corona Commission's mission to the Ministry of Health and Social Affairs on 31 October 2020, and continued with subsequent investigations that lie within its mission. In addition, on 8 June 2020, the National Board of Health and Welfare was tasked to pay SEK[3] 462 million in government grants to municipalities as part of the Elderly Care Improvement Initiative, one that was slated to continue through 2021 (Ministry of Health and Social Affairs, 2020a). The grants should make it possible for employees in municipally financed care and nursing for the elderly to receive training to become nurse's aides and assistant nurses during paid working hours, thus contributing to improving expertise and ideally attracting more to apply for these professions so critical to the well-being of society. The participants have since – through an agreement with the Swedish Association of Local Authorities and Regions and the Swedish Municipal Workers' Union – along with those who participate in the Elderly Care Improvement Initiative, been offered permanent full-time employment. This action directly responded to criticisms on shortcomings in both municipal and private elderly care services concerning large personnel turnover and less skilled staff. Additionally, Sweden's national ban on visits was rescinded on 1 October 2020. The discussion on whether restrictions to freedom of movement can be permitted in a democratic society, not least in relation to a need for a new crisis law to combat potential problems in the social structure, will continue in Sweden in the coming years.

NOTES

1  The value 152 in "(1974: 152)" of the Instrument of Government refers to the specific proclamation for the 1974 year and not a specific page number. In this chapter, a year followed by a colon and number (enclosed with or without parentheses) refers, in Swedish law, to year *and* proclamation, *not* year and page number. Page numbers will be signified by "p." or "pp."
2  The National Board of Health and Welfare, Cause of Death Certificate, 29 June 2020.
3  The Swedish krona, the country's currency.

REFERENCES

Administrative Court in Gothenburg, Case number 5025-20. (2020, May 4). https://www.domstol.se/globalassets/filer/domstol/forvaltningsratten_goteborg/nyheter/dom-fr-5025-20.pdf.
Communicable Diseases Act (2004: 168*)*. (2004, July 4). https://www.riksdagen.se/sv/dokument-lagar/dokument/svensk-forfattningssamling/smittskyddslag-2004168_sfs-2004-168.
Council of the European Union. (2012). Council declaration on the European year for active ageing and solidarity between generations (2012): The way forward. 17468/12, SOC 992, SAN 322.
Government and Government Offices of Sweden. (2020a, April 1). Introduction of a national ban on visits to care homes for older people. Key acts and ordinances

entering into force around the second half of 2020. https://www.riksdagen.se/sv/dokument-lagar/dokument/svensk-forfattningssamling/forordning-2020163-om-tillfalligt-forbud-mot_sfs-2020-163.

Government and Government Offices of Sweden. (2020b, April 7). Strategi med anledning av det nya coronaviruset [Strategy due to the new coronavirus]. https://www.regeringen.se/regeringens-politik/regeringens-arbete-med-anledning-av-nya-coronaviruset/strategi-med-anledning-av-det-nya-coronaviruset/.

Government Bill 2008/09: 29. (2008, October 1). Lag om valfrihetssystem [Law on freedom of choice system]. https://www.regeringen.se/49bbdc/contentassets/ad60d10b5a6f48c083f7a5768b44d197/lag-om-valfrihetssystem-prop.-20080929.

Government Bill 2009/10: 116. (2010, March 2). Värdigt liv i äldreomsorgen [Dignified life in elderly care]. https://www.regeringen.se/49bbd8/contentassets/375c5289fb3b434b8aba108a38d6e1f4/vardigt-liv-i-aldreomsorgen-prop.-200910116.

Government Bill 2019/20: 144. (2020, April 6). COVID-19 och ändringar i smittskyddslagen [COVID-19 and amendments to the communicable diseases act]. https://www.regeringen.se/4a334c/contentassets/c3e6374019444379b7f33c8ccfc40b32/covid-19-och-andringar-i-smittskyddslagen-prop.-2019-20-144.pdf.

Government Bill 2019/20: 155. (2020, April 7). Tillfälliga bemyndiganden i smittskyddslagen med anledning av det virus som orsakar COVID-19 [Government bill, temporary authorizations in the communicable diseases act due to the virus causing COVID-19]. https://www.riksdagen.se/sv/dokument-lagar/dokument/proposition/tillfalliga-bemyndiganden-i-smittskyddslagen-med_H703155/html.

Government Official Investigations 2008: 51 (Statens offentliga utredningar (SOU) 2008: 51). Värdigt liv i äldreomsorgen [Dignified life in elderly care]. https://www.regeringen.se/49b6a8/contentassets/8be1e3c98afd48b4bcb4c0f5e5444d7a/vardigt-liv-i-aldreomsorg-sou-200851.

Government Official Investigations 2008: 61 (Statens offentliga utredningar (SOU) 2008: 61). 2008. Krisberedskapen i grundlagen – Översyn och internationell utblick [Crisis preparedness in the constitution – Review and international outlook]. https://data.riksdagen.se/fil/0F0C434A-C5EF-4F1A-B9E1-F542F770912C.

Government Official Investigations 2017: 21 (Statens offentliga utredningar (SOU) 2017: 21). Läs mig! Nationell kvalitetsplan för vård och omsorg om äldre personer (Del 1 och 2) [Read me! National quality plan for care and nursing for the elderly (Part 1 and 2)]. https://www.regeringen.se/rattsliga-dokument/statens-offentliga-utredningar/2017/03/sou-201721/.

Greve, B. (2020). *Welfare and the welfare state: Present and future.* Routledge.

Gunnarsson, Å. (2007). Gender equality and the diversity of rights and obligations in Swedish social citizenship. In Gunnarsson, Å., Svensson, E., & Davies, M. (Eds.), *Exploiting the limits of law: Swedish feminism and the challenge to pessimism* (pp. 191–211). Ashgate.

Gustafsson, H. (2002). *Rättens polyvalens: En rättsvetenskaplig studie av sociala rättigheter och rättssäkerhet* [The polyvalence of the court: A jurisprudence study of social rights and the rule of law]. Lund University, Department of Sociology.

Health and Medical Services Act (2017: 30). (2017, April 1) https://www
.global-regulation.com/translation/sweden/9620334/health-and-medical-service
-act.html.

Health and Social Care Inspectorate. (2020a, July 7). IVO fördjupar granskningen av vård och behandling på särskilda boenden för äldre [IVO deepens the review of care and treatment in special housing for the elderly]. https://www.ivo.se/publicerat
-material/nyheter/2020/ivo-fordjupar-granskningen-av-vard-och-behandling
-pa-sarskilda-boenden-for-aldre/.

Health and Social Care Inspectorate. (2020b, November 24). Ingen region har tagit sitt fulla ansvar för individuell vård och behandling [No region has taken full responsibility for individual care and treatment]. https://www.ivo.se/publicerat-material/nyheter
/2020/ingen-region-har-tagit-fullt-ansvar-for-individuell-vard/.

Instrument of Government (1974: 152). Up to and including Swedish Code of Statutes (SFS) 2018: 1903. (2021). https://www.riksdagen.se/globalassets/07.-dokument
--lagar/regeringsformen-eng-2021.pdf.

Ministry of Health and Social Affairs. (2020a, June 9). Uppdrag att betala ut statsbidrag till kommuner för kostnader till följd av satsningen äldreomsorgslyftet [Assignment to pay state subsidies to municipalities for costs as a result of the initiative elderly care boost]. Record Number: S2020/05025/SOF. https://www.regeringen.se
/regeringsuppdrag/2020/06/uppdrag-att-betala-ut-statsbidrag-till-kommuner
-for-kostnader-till-foljd-av-satsningen-aldreomsorgslyftet/.

Ministry of Health and Social Affairs. (2020b, June 30). Utvärdering av åtgärderna för att hantera utbrottet av det virus som orsakar sjukdomen COVID-19 [Evaluation of measures to deal with the outbreak of the virus that causes COVID-19 disease]. Directive, 2020: 74. https://www.regeringen.se/49f46d/contentassets
/593c32df14114d9c81eeba9c96e26e41/dir2020_74.pdf.

National Board of Health and Welfare. (2020a, April 4). Statistik om socialtjänstinsatser till äldre [Statistics on social services for the elderly]. https://www.socialstyrelsen.se
/statistik-och-data/statistik/statistikamnen/socialtjanstinsatser-till-aldre/.

National Board of Health and Welfare. (2020b, July 24). Statistik om COVID-19 bland äldre efter boendeform [Statistic on COVID-19 among senior citizens by type of housing]. https://www.socialstyrelsen.se/statistik-och-data/statistik/statistik
-om-covid-19/statistik-om-covid-19-bland-aldre-efter-boendeform/.

National Board of Housing, Building, and Planning. (2020, February 12). Allt fler 80+ i befolkningen [More and more 80+ in the population]. https://www.boverket.se/sv
/samhallsplanering/bostadsmarknad/olika-grupper/aldre/.

Parliamentary Committee on Health and Welfare. (2020, May 28). Vistelsekommuners ansvar för socialtjänstinsatser [Visited municipalities' responsibility for social service efforts]. 2019/20: SoU24. https://www.riksdagen.se/sv/dokument-lagar/arende
/betankande/vistelsekommuners-ansvar-for-socialtjanstinsatser_H701SoU24.

Parliamentary Ombudsmen. (2020, June 23). Dnr 3063-2020. https://lagen.nu/avg
/jo/3063-2020.

Petersson Hjelm, A.C. (2018). Between power and empowerment: Trust and dependence as a legal strategy in elderly care in Sweden. *Nordic Journal on Law and Society*, *2*(1), 1–31. https://doi.org/10.36368/njolas.v2i01.12.

Petersson Hjelm, A.C. (2021). Socialtjänstens ansvar för äldre [Social services' responsibility for the elderly]. In Montoya, T.F. (Ed.), *Juridik för socialt arbete* [Law for social work] (pp. 197–233). Gleerups Utbildning AB.

Rönnberg, L. (2020). *Hälso- och sjukvårdsrätt* [Health care law]. Studentlitteratur AB.

Social Services Act (2001: 453). (2001, July 6). https://www.riksdagen.se/sv/dokument-lagar/dokument/svensk-forfattningssamling/socialtjanstlag-2001453_sfs-2001-453.

Svensson, M., Urinboyey, R., & Åström, K. (2012). Welfare as a means for political stability: A law and society analysis. *European Journal of Social Security*, *14*(2), 64–85. https://doi.org/10.1177/138826271201400201.

Swedish Public Health Agency. (2020a, March 13). Ny fas kräver nya insatser mot COVID-19 [New phase requires new efforts against COVID-19]. https://www.folkhalsomyndigheten.se/nyheter-och-press/nyhetsarkiv/2020/mars/ny-fas-kraver-nya-insatser-mot-covid-19/.

Swedish Public Health Agency. (2020b, April 16). Folkhälsomyndighetens föreskrifter och allmänna råd om allas ansvar att förhindra smitta av COVID-19 m.m. [The Swedish public health agency's regulations and general advice on everyone's responsibility to prevent COVID-19 m.m.]. HSLF-FS 2020: https://www.folkhalsomyndigheten.se/contentassets/0ac7c7d33c124428baa198728f813151/hslf-fs-2020-12u.pdf.

Szebehely, M., & Meagher, G. (2018). Nordic eldercare – Weak universalism becoming weaker? *Journal of European Social Policy*, *28*(3) 294–308. https://doi.org/10.1177/0958928717735062.

Tort Liability Act (1972: 207). (1972, February 6). https://www.riksdagen.se/sv/dokument-lagar/dokument/svensk-forfattningssamling/skadestandslag-1972207_sfs-1972-207.

Vahlne Westerhäll, L. (2002). *Den starka statens fall? En rättsvetenskaplig studie av svensk social trygghet 1950–2000* [The fall of the strong state? A forensic study of Swedish social security 1950–2000]. Norstedts Juridik.

Vahlne Westerhäll, L. (2017). *Lex Maria-anmälningar och klagomål vid suicider – Patientsäkerhet, integritet och rättssäkerhet* [Lex Maria reports and complaints in the case of suicides – Patient safety, integrity and legal certainty]. *Socialmedicinsk Tidskrift*, *94*(5), 593–602. https://socialmedicinsktidskrift.se/index.php/smt/article/download/1674/1557.

# 12  The German Reaction to Corona: The Interplay of Care, Control, and Personal Responsibility within the Welfare State

BY ALBERT SCHERR

## Introduction

The COVID-19 pandemic not only was and is a health threat and a challenge for medical professions and healthcare systems, but should also be considered, in the words of Marcel Mauss (2010, p. 137; see also Wendling, 2010), as a "total social fact." This means that the spread of the pandemic itself, and not least the social reactions to it, had an impact on all areas and subsystems of society and all forms of social interaction. Of central importance were various forms of social distancing, such as distance rules, bans on meeting as groups, and curfews, as these not only had an impact on the local, regional, and national spread of the virus, but also changed the conditions and possibilities of social coexistence, of cooperation, interaction, and communication (see also Lehrer, Juhl, Blom, et al., 2020). Consequently, the pandemic and the reactions to it led to a multidimensional crisis. This includes the macro-level of world society, that is, the dynamics of the global economy and international political relations, as well as the micro-level of everyday life; for example, the ways of living together in families and as couples.

In such a crisis, it is a task and challenge for the social sciences to provide analyses that contribute to the understanding of societal impacts of crisis in all dimensions, making knowledge and information available that can contribute to a rational foundation of political decisions and public opinion formation. In addition, however, such crises also present a special opportunity for the social sciences to gain insight into the structures and dynamics of different societies. Indeed, by examining the effects of crisis on the subsystems of societies and in their reactions to them, the structures that characterize them become particularly clear. If we take even a superficial look at the course of the pandemic crisis, considerable differences between national societies become obvious, not least between the wealthy national societies of the global North and the economically less developed societies of the global South. Even between national societies of the global North, which show numerous similarities – they are characterized, among other things, by capitalist-dominated market economies, democratically constituted politics, the rule of law, and a high level of prosperity by international standards – considerable differences can be seen with regard to the effects of the pandemic and reactions to them. Sociological analyses of the societal dealing with the pandemic will therefore have to minimize the level of abstraction of concepts such as modern society, capitalist society, or functionally differentiated society, and additionally address the particularities that distinguish national societies within the world society.

This chapter's orientation, however, does acknowledge a methodological problem: in order to understand the effects of the pandemic in different national societies, comprehensive knowledge about the political order and culture, economic conditions, structures of the healthcare system, the education system, the welfare state, the legal system, and the significance of the public space is required. With such a large panorama, however, readers of international publications usually are exposed to limited background knowledge that is linked to their disciplinary preferences. In light of this scope condition, it should be noted that the discussion herein acknowledges such a problem and therefore will give a few exemplary insights into how German society is attending to the pandemic as well as how such practices have affected the country overall. Therefore, I will focus on those aspects of the pandemic in Germany that should provide interesting insights for non-German readers. I base my considerations on the assumption that Germany coped relatively well with the crisis during the pre-vaccination phase, and that its crisis management has been relatively successful when compared to many international efforts at combating COVID-19, especially in regard to economic, democratic, sociopolitical, and medical aspects. However, although the potential and the positive aspects of coping with the pandemic in Germany are more strongly emphasized herein, other social problems such as the effects of social inequalities, which have become visible in Germany as well, are not neglected.

In this chapter, I hope to demonstrate that the pandemic has again highlighted the continuing social importance of national social structures and a benevolent nationalism as a worldview and mindset, and then present the development of the COVID-19 crisis in Germany. Against this background, I will argue that in Germany, a combination of welfare state policies with state restrictions on freedom of movement and individual responsibility has made it possible to limit the number of infected persons, thus ensuring their health care. Furthermore, my chapter will then analyse why measures employed by the German government to mitigate the consequences of the pandemic have found broad social acceptance. Finally, I will discuss how the experience of the pandemic challenges sociology to deal more intensively with the social significance of individual encounters between physically present persons.[1]

## A First Observation: The Global Pandemic and Ordinary Nationalism

A central feature of the processes known as globalization is that national borders and spatial distances have lost their significance. Increasing prosperity, transnational cooperation, digital communication on the internet, and the expansion of air travel have led to far-reaching changes in space-time relations and also to an expansion of the international mobility of peoples (such as workers, students, scientists, and tourists) (Giddens, 1979; Gregory & Urry, 1985; Harvey, 2000). Deriving from such a view, societies can no longer be conceptualized as closed national units or lockable containers (Wimmer & Schiller, 2002) – and such a view applies not only to economic and cultural processes or migration and flight but potentially, among other things, also to the spread of infectious diseases. For proponents of such a view, envisioning states in such a porous manner is helpful and important because it is likely that global citizens will, in the foreseeable future, experience additional global pandemics. Indeed, the COVID-19 pandemic is occurring under the conditions of a globalized world society (Luhmann, 2012, p. 83ff.), but one that is also harbouring many serious regional inequalities. My chapter thus argues how in spite of the promises of globalization, and especially under pandemic conditions, the mitigation of the pandemic – at least in the example of Germany – can still be attributed, to a significant extent, to conditions of nation-state structures that shape its politics and laws.

The structural anchoring of nation-state citizenship in the societal subsystems of politics and law corresponds to the spread of different forms of nationalism as a worldview, a mindset, and an organizing centre of social identities. In the social reactions to the COVID-19 pandemic, the continuing power of nationalist concepts has once again become clear in the case of Germany, particularly in a form that Thomas Pogge has described as "common nationalism" (2008, p. 127). It is important to emphasize that this common, benevolent nationalism renounces an ideological orientation that establishes a superiority of one's own nation in relation to others. Rather, the central feature of common nationalism is the assumption that nation-states and their governments are entitled to attach greater importance to the well-being of their own citizens than to the well-being of citizens of other nations (2008, p. 127); that is, the nation-state is responsible for the welfare of its citizens, first and foremost, and only subordinately – and under certain conditions – for the welfare of citizens of other nations. The merits of such a view, as was made clear at the beginning of this chapter's discussion of the COVID-19 pandemic, can be appreciated regardless of the country in which the citizens are currently residing.

After the shutdown of numerous international air connections in March 2020, the Bundesregierung (the German federal government) decided on 17 March to launch a repatriation operation, which enabled around 240,000 Germans to return to Germany using aircraft chartered by the government at an estimated cost of approximately 94 million euros (Bundesregierung, 2020, pp. 1–2). The government similarly made return trips possible for German citizens located in countries that were not initially affected by the pandemic.[2] This was therefore by no means only a protection against acute threats, but rather a reaction pattern in which it was naturally assumed that, in times of an emerging global crisis, one's own citizens should be returned to the territory of their own state, since only there can it be assumed that they will be given adequate protection. The fact that this was based on a common and unquestionably self-evident benevolent nationalism is shown by the fact that the justification and necessity of this costly retrieval action was at no time questioned, that it did not lead to controversy. This made the power of common nationalism visible, which, as a matter of course, assumes that the nation-state is primarily obliged to its citizens, and which, in the German case, assumed that the protection of its own citizens was best guaranteed on its own territory. In contrast, it became obvious that concern for the nation-states' own citizens also meant that the well-known acute misery of refugees perishing in the Mediterranean Sea or living under unacceptable conditions in camps on the external borders of the European Union during the COVID-19 crisis was relegated to the background of public and political attention. The phrase repeatedly formulated by politicians that every single life counts, in the context of ordinary nationalism, thus applied only to citizens.

The nationalistic framing of the crisis response in Germany and other countries of the European Union was also reflected in the fact that on 17 March 2020, European borders were closed for unnecessary travel from countries outside the European Union. Considerable restrictions were also placed on freedom of travel within the European Union, according to the *Auswärtiges Amt* (Federal Foreign Office) (2020).[3] This was the case even though "the danger zones did not coincide with national borders" (Gosewinkel, 2020, p. 18), as the differences in infection rates between German regions were greater than, for example, those between Germany and France. From an epidemiological point of view, therefore, it would have been more obvious to seal off certain regions within nation-states rather than close national borders. Moreover, border controls between nation-states did not exist in the EU, a status quo leading up to the pandemic. A key historical achievement of the European Union, the abandonment of internal border controls, was then turned back for a few weeks during the pandemic, even though there were no clear national boundaries of the zones where the virus had spread. This had a particularly significant impact in the areas close to the borders, where there are considerable numbers of commuters, for example between Germany and France or Germany and Poland, who were now experiencing considerable disruption to their travel opportunities. In addition, budding nationalistic hostilities were reported at border regions, some directed against French people who used the remaining travel opportunities to enter Germany. Other examples included increase in prejudices against persons with an Asian appearance, a pattern reported by various support centres for the victims of discrimination and groups of people affected (Koreaverband, 2020). It thus became clear that the transitions between ordinary state nationalism and aggressive xenophobic nationalism are highly fluid. However, for Germany, although this initially led to an increase in cases of nationalistic and racial discrimination, which is to be expected in times of social crises, I am convinced that it did not lead to an uncontrollable expansion and consolidation of prejudices and hostilities caused by COVID-19.[4] Furthermore, by examining the course of events in May and June 2020, it can be seen that the successful containment of the pandemic by 15 June also expeditiously led to a withdrawal of border controls and travel restrictions within the European Union.[5] During this same period, a benevolent nationalist framework had already been relativized, also in symbolic terms, by the fact that Germany had admitted some patients from France and Italy when free beds were available in intensive care units in the country, a gesture of European solidarity whose symbolic significance was not lost among fellow EU countries.

## The Course of the Crisis in Germany during Early 2020: Case Numbers and Measures

The development of the pandemic crisis can initially be outlined as follows:[6] The first cases of infection with the COVID-19 virus were diagnosed in Germany in January 2020, initially only a few individual cases. By the beginning of March, only 200 infected persons were recorded. However, the number of newly infected persons rose rapidly to 7 per 100,000 inhabitants by early April. The number of acutely infected persons (i.e., those infected but not recovered) reached its peak in the first week of April with approximately 70,000 cases. Since mid-April the number of new infections and acutely infected persons had been declining and had fallen to approximately 5,000 cases by mid-June 2020.

*Overview of the Course of the Pandemic*

By June 2020, the total number of all registered infected persons in Germany (acute cases and recovered) totalled 194,898, with 8,972 fatalities. At the same time, the authorities registered 2,549,069 infected persons in the USA with 125,803 fatalities. If these figures are put into relation to the total population, a considerable difference becomes apparent in comparison with the USA. If one considers the 4.6 times greater population of the USA, an infection rate per inhabitant that is 2.8 times higher and a case fatality ratio that is 3.1 times higher can already be determined at the time of this writing in December 2020. By the end of June 2020 in Germany, approximately 500 cases were registered daily while in the USA approximately 50,000 cases were registered; moreover, this discrepancy had actually increased over time: by mid-July 2020, slightly *less* than 200,000 cases were registered in Germany, but more importantly approximately 185,000 of these had already recovered, with the number of deaths per 100,000 inhabitants at 10.94. However, during the same period, 3.3 million cases had already been recorded in the USA with the number of fatalities per 100,000 inhabitants at 41.20.

In an international comparison against the background of available data, it can be seen that Germany succeeded in keeping this total number of recorded infections at a relatively low level.[7] At no time did the healthcare system become overburdened. At all times there were more than enough beds available for any intensive medical care that might be required. At no time were the capacities required in severe cases for intensive medical care and artificial respiration in hospitals fully utilized. There was always excess capacity. This was a surprising consequence, since although there were tendencies towards an economization of the healthcare system observed in Germany, a consistent implementation of a neoliberal healthcare economy had not taken place. The comparatively well equipped hospitals in Germany also made it possible to admit a limited number of patients from other European countries (e.g., Italy and France) to fully utilize Germany's medical system. Moreover, media images of corpses that could not be adequately buried had no equivalent in Germany. In this respect, it can be asserted that Germany dealt relatively successfully with the COVID-19 crisis.

Accordingly, a comparative analysis of the course of the pandemic in Europe and the USA was offered by *Süddeutsche Zeitung* (South German Newspaper), one of Germany's largest daily newspapers, with an important section that needs to be quoted in length:

> First, there is a remarkable resemblance. If you look at the course of the corona pandemic in the United States and in the European Union, you see two curves that run parallel for weeks, from the beginning of March to mid-April. The numbers of daily new infections initially rise very steeply east and west of the North Atlantic, and the virus spreads exponentially. In the USA as well as in Europe the danger has long been underestimated, the governments reacted too late. But then the measures take effect and the number of cases reaches a plateau. From then on, the paths diverge: in the EU, case numbers leave the plateau and fall in mid-April. Since the end of May, the figures have been fairly stable at around 4000 new infections per day, with the EU benefiting from the fact that the UK is no longer included. The situation is quite different in the USA. There is no talk of an easing of tension there, the crisis has only shifted. (Endt & Zaschke, 2020)

The development outlined above can be characterized as the result of successful containment of the spread of the virus. In Germany, successful containment can be attributed, to a significant extent, to policies established by the country's federal government, which at the beginning of the crisis declared a slowing of the spread of the virus as its primary goal in order to keep the number of people in need of treatment at the same time as low as possible. In an important public statement on 18 March 2020, German Chancellor Angela Merkel stated the following:

> As long as this is the case, there is only one thing to do, and that is to slow down the spread of the virus, to spread it over the months and thus gain time. Time for research to develop a drug and a vaccine. Above all, time so that those who fall ill can receive the best possible care. Germany has an excellent healthcare system, perhaps one of the best in the world. That can give us confidence. But even our hospitals would also be completely overwhelmed if too many patients were to be admitted in the shortest possible time and suffer a severe course of COVID-19 infection. (Merkel, 2020, p. 2)

In the same speech, the chancellor not only announced government measures but also combined this with an urgent appeal to individual responsibility and solidarity of citizens. This was a consequence of the insight that necessary measures can only be successful if the citizens accept and support them:

> I address you today on this unusual path because I want to tell you what guides me as Chancellor and all my colleagues in the Federal Government in this situation. This is part of an open democracy: that we also make political decisions transparent and explain them. That we justify and communicate our actions as well as possible so that they are comprehensible. I firmly believe that we will succeed in this task if all citizens really do see it as YOUR task. (Merkel, 2020, p. 1)

Merkel's pronouncement is an aspirational and exemplary expression of the attempt to manage the crisis politically in a manner that does not rely solely on legal regulations and sanctions to enforce measures, but at the same time tries to convince citizens of the necessity of these measures. What Merkel conveyed to German citizens was not declared as an expression of a specific political program but as an objective necessity.

*Measures of the Federal Government*

In order to achieve this objective, on 22 March 2020, the national government and the governments of the German Bundesländer (federal states) adopted a series of measures that would have a serious impact on everyday life. In contrast to other European countries like France or Spain, there was no curfew. Nevertheless, henceforth it was forbidden for more than two people to meet outside together who were not living together in the same household. In addition, a physical distance of 1.5 metres (1.53 yards) for people in public space was set. Day-care facilities for children, schools, and universities as well as service providers (such as restaurants) were closed, as were all stores with the exception of grocery stores and pharmacies. The federal government did maintain these measures until 20 April 2020, after which they were gradually relaxed. By the end of May 2020, a situation had already been reached in which the number of new infections per day averaged less than 500 nationwide. In response, day-care facilities and schools were reopened in June 2020, but with a reduced number of hours and in alternation only for a part of the pupils' time so that they could comply with distancing rules. By contrast, universities remained closed. Shops and hotels could reopen, however, but had to comply with strict hygiene rules. In addition, responsibilities were then shifted to the level of the sixteen Bundesländer as well as the cities and districts. This then led to different regional regulations. For the sake of clarity, the regulations for one of the German federal states (Baden-Württemberg) are reproduced here as they were valid from 1 July 2020, that is, at time when the number of new infections was persistently low. It should be noted that the distance requirement (1.5 metres) continued to apply and that it was still obligatory to wear face masks in shops and public transportation.

> As of July 1, 20 people are now allowed to meet in public places just like in private. From July 1, private events with no more than 100 participants no longer require a hygiene concept … This applies for example to birthday or wedding celebrations, baptisms, and family celebrations. From July 1, events with up to 250 people will be possible if the participants are allocated fixed seats for the entire duration of the event and the event follows a program that is determined in advance … Dance events with the exception of dance performances and dance lessons and rehearsals are also prohibited … Clubs and discos are still not allowed to open. Sex work sites, brothels, and similar establishments and any other practice of sex work is also prohibited.[8] (Landesregierung Baden-Württemberg (State Government of Baden-Württemberg), 2020, p. 1)

*Effects on Everyday Life*

In spite of the reduction of stricter measures, Germany's situation, like that of many countries elsewhere, must still contend with how its society has yet to experience a full return to pre-pandemic normality. Nevertheless, at the time of this writing, there does not appear to be a profound existential crisis in the German narrative about the pandemic, but rather a return to limited normality in almost all areas of everyday life.[9] This became evident in the revival of the country's public spaces, which had largely lost their significance at the height of the crisis. In order to understand this, it is necessary to explain briefly that cities in Germany (and many other European countries) have a specific culture of public space configuration shaped by tradition. German cities have historically developed from a core constituted by Christian churches and marketplaces at their centre. Even today, most of the inner cities' numerous smaller shops, cafés, and restaurants typically occupy this centre, with many located in pedestrian zones. In addition, centrally located marketplaces, often in the vicinity of old churches, are part of the characteristic image of German cities or the districts of larger cities like Berlin and Hamburg, as are smaller parks and other public squares. As a result, German towns and inner cities are lively public places that invite people to stroll and shop, and where teenagers and young adults gather in spring and summer.[10]

Despite all the tendencies towards a socio-spatial differentiation of residential areas, places of industrial production, and the expanding commercialized areas of consumption (Michel & Stein, 2015), urban policy in Germany is always aimed at "keeping urban spaces open" (Schäfers, 2006, p. 154) for non-functional and non-commercial use by all citizens. For this reason, German cities are characterized by busy centres, which are used exclusively or predominantly by pedestrians, making it possible for people to wander aimlessly around public spaces and for chance encounters to occur. However, the closure of shops, restaurants, and cafés during the peak of the pandemic, along with the ban on meeting in groups in public spaces, temporarily brought life in public spaces to a standstill. City centres had also been largely emptied at times when strollers, groups of young people, and consumers usually constitute the crowds that patronize their establishments. After the relaxation of the measures, quotidian dynamics of a normalizing urban life could be observed again following June 2020, even if the existing rules on distance had to be observed, which they were, more or less. However, open-air or indoor concerts and theatre performances could still not take place unless they fielded a very limited numbers of participants.

**The Social Impact of the Crisis**

The reactions to the pandemic were centrally determined in Germany by political assumptions about the requirements needed to prevent an uncontrollable increase in the number of people who fell ill, and to ensure that all citizens could receive adequate medical care. It was obvious from the very beginning that this would have a considerable impact on the labour market, including effects of the closure of public facilities such as kindergartens and schools, as well as shops and restaurants, and because of falling consumer demand and slumps in international trade relations. Correspondingly, unemployment figures rose. However, this increase was small in comparison with the USA: while the unemployment rate in Germany was 5.2% in 2018 and then 5.05% in 2019, it had risen to 6.2% in June 2020. In the USA, on the other hand, an increase from 3.8% in February 2020 to 14.4% in June 2020 was recorded (Kochhar, 2020, p. 1). When making this comparison, however, it should be noted that the increase in unemployment figures in Germany was also limited by the fact that companies were able to continue to employ workers with reduced working hours without having to cut wages to the same extent. This was possible because they received state support for continued employment in the form of so-called short-time work compensation, as will be discussed in the following section.

*The Importance of the Welfare State*

In order to be able to assess the social consequences of unemployment and short-term work in Germany, it is important to emphasize that all employees in the formal economy[11] are entitled to compulsory unemployment insurance and compulsory health insurance. These entitlements are not linked to one's employment contract. In Germany, all employees who have become unemployed will initially receive unemployment assistance for one year through its Arbeitslosengeld 1 (ALG 1) unemployment benefits program. The amount of this unemployment benefit depends on the previously earned salary.[12] Unemployment in Germany therefore leads to a considerable reduction in income, but not to a complete loss of income or loss of health insurance and, as a rule, not to a loss of one's home. If unemployment for citizens lasts longer than one year, they are then assigned secondary status that qualifies them for Arbeitslosengeld 2 (ALG 2) benefits, although the monetary benefits are less. Nonetheless, the amount of unemployment benefits should guarantee a sociocultural minimum subsistence level akin to the so-called basic social welfare provision, which is also due to all German nationals who have not previously earned any income from work.[13]

Moreover, that the pandemic did not lead to a significantly higher increase in unemployment in the first half of the year is also due to the German government's provision of short-time work relief for over six million employees since March 2020, according to the Bundesagentur für Arbeit (Federal Employment Agency) (2020a). In order to avoid companies having to lay off employees in times of crisis, under certain conditions 60–80% of wages are reimbursed by the state if employees are only employed on reduced, marginal hours. In response to the COVID-19 pandemic, these regulations have been considerably expanded (Bundesagentur für Arbeit, 2020b). In this context, it should be noted that German legislation obliges employers to continue to pay employees their full salary for a maximum of six weeks in the event of illness, and they may not terminate their employment during this period. In the event of a longer illness, there is a subsequent entitlement to the so-called sickness benefit (Bundesministerium für Arbeit und Sozialordnung (Federal Ministry of Labour and Social Affairs), 2019).

Nonetheless, according to the results of a representative survey conducted by Munich's prestigious Ifo and Forsa institutes, unemployment and shorter work hours led to considerable income losses for 15% of the employees. This was more common among lower income groups, as they were relatively more affected by unemployment (Ifo Institut/Forsa, 2020, pp. 10–12). This is because low incomes in Germany are not a typical feature of workers in the core sectors of industrial production, but rather mainly in the personal service occupations (e.g., hairdressers, sales clerks), which were directly affected by the measures. The crisis also led to serious problems for small self-employed persons such as innkeepers or artists who were unable to earn any income at all or only significantly reduced income during the crisis. This was somewhat mitigated by state aid. Forty-six per cent of the self-employed, however, indicated that they had to resort to their private savings, and about 10% had to take out loans (Ifo Institut/Forsa, 2020, p. 19). A representative survey conducted in July 2020 concluded that 20% of all respondents had seen their disposable household income fall since the beginning of the year and that 10% were very concerned that the crisis would cause them considerable financial difficulties (Ifo Institut/Forsa, 2020, p. 21).

In summary, the importance of the function of the welfare state to reduce the dependence of livelihoods upon the economy, of the welfare state decommodification of the workforce (Offe, 1972; Pintelon, 2012), again became obvious as demonstrated by its actions during the pandemic. States such as Germany, where there has been no comprehensive neoliberal dismantling or maligning of the welfare state, could not fully compensate for negative effects but have thus been able to limit the extent of the social impact.

*Limits of the Welfare State*

Despite the compensatory effects of the welfare state, there remained significant issues born from unresolved social inequalities. This is the case even though, according to World Bank calculations, Germany has a lower level of income inequality, with a value on the Gini Index of 0.29, than, for example, the USA, with a value of 0.38.

Firstly, it became apparent that school closures had specific negative effects on children whose parents were not able to compensate for the lack of education by supporting their own learning. A survey of teachers concluded that children of academics were much better able to learn at home without teacher guidance during the COVID-19 crisis than children of parents with a low educational level (Hurrelmann & Dohmen, 2020); that is, this pandemic has been exacerbated by socially induced differences in achievement. It should be noted that the correlation between the socio-economic status of parents and the educational success of children in Germany is in any case above the OECD average (Reiss, Weis, Klieme, et al., 2019, p. 129f.). Secondly, data revealed that during the period when kindergartens and schools were closed, there was an increase in violence against women and children. This increase was somewhat higher in families with acute financial problems and in families where one partner had experienced only short-term employment or became unemployed because of the pandemic (Steinert & Ebert, 2020). Thirdly, during the pandemic the problematic situation of workers in abattoirs and meat-processing plants became apparent because of increased infection rates traced to their work environments (Weinkopf & Hüttendorf, 2017). These workers are mainly contract workers from Eastern Europe who receive low wages and live in densely cramped multi-bed rooms under poor hygienic conditions. This is a consequence of the fact that so-called *werkverträge* (service contracts) are used to undercut the statutory minimum wage. Such developments have led to a hitherto unresolved political discussion about necessary changes of legal regulations to prevent such forms of exploitation. Finally, refugees were particularly affected by the pandemic: Refugees in Germany are legally obliged to live in so-called reception centres initially, and if legally possible, until the conclusion of their asylum procedure.[14] These are camps for often several hundred people who are entitled to 7.5 square metres (80.7 square feet) of living space per capita and who live in shared rooms; some of the refugees are also housed afterwards in so-called shared accommodations under similar conditions (Wendel, 2014). Under these conditions, they cannot maintain the prescribed minimum distance. As a result, high infection rates were recorded in several initial reception centres and accommodation facilities.

If the COVID-19 pandemic is seen as a test of the efficiency of the German welfare state, it can be seen that its core institutions – unemployment insurance, basic social security, health insurance, and the healthcare system – have proved capable of considerably mitigating severe negative outcomes. At the same time, however, previously known deficits have again become apparent: In Germany, too, those most affected were those whose income and education are lower, like contract workers, refugees during asylum procedures, and prison inmates – all suffering under forms of social exclusion.

*Political Implementation and Social Acceptance of Measures to Contain the Pandemic*

The political and legal measures outlined above to contain the pandemic have met with broad public approval by the German citizenry. Such legitimation is first evident in the polls which are regularly used to determine which party voters would vote for if parliamentary elections were held in the immediate or very near future. In the course of the crisis, there has been a marked increase in support for the chancellor's governing *Christlich Demokratische Union Deutschlands* (Christian Democratic Union (CDU) Party):[15] In January and February 2020, the approval rate for the CDU was constant at 27%–28%; the figures then rose to 36% by the end of March and stabilized at a level of 38%–40% by July (Forsa Institut, 2020). This is a clear indication of approval of government policy during the COVID-19 crisis. A survey carried out in July 2020 on satisfaction with the measures taken by the federal government to combat the pandemic shows that the majority of the population (65%) considered them "just right." Eighteen per cent were in favour of stricter measures, while 15% considered them "too strict" (Ifo Institut/Forsa, 2020, p. 24; see also Juhl, Lehrer, Blom, et al., 2020; see also Rees, Papendick, Rees, et al., 2020). The approval rate among supporters of the two governing parties, but also among supporters of the opposition party Die Grünen (Green Party), is approximately around 70%. Only the majority of supporters of the extreme right-wing party Alternative für Deutschland (Alternative for Germany – AfD) rejected government measures.[16]

There were repeated demonstrations by groups in various places who rejected contact bans and the obligation to wear masks. Right-wing extremist currents were represented as well as supporters of various conspiracy theories. However, only a few thousand people took part in each of these demonstrations. The relative insignificance of these protests is shown by the fact that, according to my observations and reports in the media, solidarity demonstrations with the Black Lives Matter Movement set into motion across the Atlantic in the USA, in which racial profiling by the police and extreme right-wing violence were seen, attracted noticeably higher numbers of participants in Germany during the same period.

In the course of the regionalization and the loosening up of restrictions from June 2020, differences of opinion became visible and led to controversies between the parties and the governments of the federal states (Bundesländer). However, this did not lead to serious conflicts or far-reaching social polarization. Although so-called COVID-19 parties of adolescents and young adults were occasionally observed, as well as disregard for the rules of physical distancing on weekends and holidays in areas of recreation (e.g., beaches), such forms of deviant behaviour were exponentially limited in their occurrences. Instead, at the level of party politics, far-reaching consensus emerged between the conservative CDU, Sozialdemokratische Partei Deutschlands (Social Democratic Party – SPD), Die Linke (Democratic Socialist Party), the liberal Freie Demokratische Partei (Free Democratic Party – FDP), and ecological parties such as Die Grünen, despite all the controversies in detail. Only the extreme right-wing AfD did not support the consensus. The same is also true of the media coverage of the pandemic, which validated the merits of political consensus insofar as it pertains to the management of the pandemic.[17]

The consensus-building process and the broad acceptance of the restrictions by the population need explanation, not least because they were accompanied by far-reaching interference with rights guaranteed by the constitution, as well as considerably impairing freedom and self-determination in all areas of everyday life. Indeed, Germany has seen protest movements develop even for comparatively less acute issues, leading to local or nationwide demonstrations with large numbers of participants. A first explanation for this is provided by social psychological theories, which argue that the willingness to subordinate oneself to a strong leadership increases when a strong threat is perceived from the outside, often legitimating authoritarian reactions of political actors that are perceived to be able to protect against the threat. Landau, Solomon, Greenberg, et al. (2004) summarize the state of research on this issue as follows:

> The appeal of the leader lies in his or her perceived ability to both literally and symbolically deliver the people from illness, calamity, chaos, and death as well as to demonstrate the supremacy of the worldview ... More specifically, research both in and outside the laboratory has shown that facing or even anticipating a common enemy or threat can activate superordinate identities and increase in-group solidarity. Based on this research, perhaps reminders of 9/11 and the threat of terrorism united people in a common cause and a singularly integrated identity, leading them to "rally 'round the flag" and support their current leader. (Landau, Solomon, Greenberg, et al., 2004, pp. 1138–47)

However, this socio-psychological mechanism alone is not sufficient for understanding why, even though the pandemic has led to growing support for the government in Germany, this has not been the case in other countries, such as France and the USA. From a sociological point of view, there are other aspects to be considered in this regard. In the following section, I will argue that this difference is related to the specific character of the crisis as an uncertainty of the self-understanding of modern societies and the ability of national politics to present itself as a reasonable response.

## The Pandemic as an Uncertainty in Modern Societies and Political Reactions

The fact that the COVID-19 pandemic led to the declaration of a state of emergency, the management of which would require drastic state intervention, does not appear at first glance to require further explanation. However, if one compares the COVID-19 pandemic with other infectious diseases, it becomes clear that infectious diseases that lead to serious illness and death are by no means completely exceptional, yet these do not lead to a comparable crisis response from the state. For example, the World Health Organization (WHO) recorded 1.4 to 1.6 million deaths due to tuberculosis in 2018, but this disease is far more prevalent outside developed societies of the global North (WHO, 2019,

p. 1). Therefore, a first peculiarity of the COVID-19 crisis is that it occurred in wealthy societies of the global North, confronting them with the phenomenon of a virus for which at the beginning neither vaccines nor therapies were available; that is, with a phenomenon that was previously considered a problem of underdeveloped societies presumably replete with poverty and insufficient medical care. On closer examination, however, it becomes clear that this does not sufficiently define the specifics of the COVID-19 pandemic. For even in the affluent societies of the global North, infectious diseases that can be fatal are part of the norm. In Germany, this applies to recurring waves of influenza. The most recent flu epidemic in 2017 led to around 25,000 deaths in Germany (Robert Koch Institute (RKI), 2019), without this leading to the declaration of a state of emergency or the introduction of compulsory vaccination. In contrast to COVID-19, however, influenza viruses and comparable infections are not a new phenomenon and can be adequately understood with the existing knowledge. Moreover, it is possible to protect against influenza by means of vaccination, that is, with the means offered by modern medicine. This points to a significant fundamental difference in our response to COVID-19: In the early phase of the COVID-19 pandemic, a phenomenon occurred which is not yet sufficiently understandable by means of modern science and therefore cannot be controlled by means of modern medicine.

The COVID-19 pandemic has thus led to a fundamental insecurity of the self-understanding of modern Western societies, which, as Max Weber has argued, is characterized by the rationality of calculation and controllability (Whimster & Lash, 2014). Weber thus pointed to the conviction that, in principle, all processes in nature and in society can ultimately be explained, calculated, and therefore technically controlled using methods of modern science. Subsequently, Max Horkheimer (2013) characterized modern rationality as instrumental reason, as a deceptive belief in the possibility of the comprehensive controllability of nature, society, and people. The fact that the belief in the omnipotence of instrumental reason was unsettled by the COVID-19 pandemic, because it was unclear whether or when a vaccine treatment would be found, was a central feature of the crisis until the discovery and certification of vaccines by the end of 2020. One possible reaction to such a crisis – similar to how detractors of climate change view the phenomenon – is an irrational denial and suppression of the problem, possibly in conjunction with a principled scepticism towards the sciences. Recurring tendencies towards such anti-modernist reactions to the problems of modernity have been repeatedly pointed out in sociological theories, such as in the classic ideas of Erich Fromm conveyed in his analysis of the emergence of German fascism (1969) and by Peter L. Berger, Brigitte Berger, and Hansfried Kellner in their fundamental study of anti-modern developments as they affect human consciousness in *The Homeless Mind* (1973).

In contrast to the aforementioned forms of anti-modernism, the trajectory of the German reaction can be characterized instead by a strict adherence to the orientation of political action towards modern rationality and, in connection with this, the maintaining of a very close feedback loop with political decisions informed by the scientific discourse of virology and epidemiology. In her television address on 18 March 2020, Chancellor Angela Merkel emphasized in her introduction that "Everything I am telling you about this comes from the Federal Government's ongoing consultations with experts from the Robert Koch Institute and other scientists and virologists" (Merkel, 2020, p. 1). In further political communication on the dangers of the crisis and necessary measures, she constantly referenced and deferred to the high importance of scientific expertise for political decisions. The German government thus succeeded in presenting its pandemic policies as the result of scientifically sound rational considerations that could not be reasonably criticized and for which there were no rational alternatives. Accordingly, the government appealed from the outset to the willingness of the population to orient itself on reasonable insights and defer to serious information. This is also clear from the chancellor's television address:

> I appeal to you: Stick to the rules. As a government, we will always re-examine what can be rectified, but also what may still be necessary. This is a dynamic situation, and we will continue to learn from it so that we can always rethink and react with other instruments. We will then explain that too. I therefore ask you to believe no rumors, but only the official communications, which we always have translated into many languages. (Merkel, 2020, p. 1)

An important and related point of consideration is how, according to Niklas Luhmann (2000), modern societies are envisioned through self-reflection and comparisons with other societies through their mass media. Yet even before the background comparing different countries' and leaders' approaches towards pandemic management was articulated by the media, the policy of the German government could already claim to have consistently and successfully oriented itself on reason and scientifically based assessments for pandemic management, thus being able to claim legitimation for their prudence. In this regard, another factor contributing to the acceptance of Germany's pandemic policy was how media reporting made explicitly clear that crisis management in Germany was more successful than in some other European countries, even though Germany did not impose strict curfews, as was the case in France and Spain. Since the media in Germany regularly and comprehensively reported on the political situation in the USA, it is also plausible to assume that the comparison with the USA was significant for the German citizens' self-evaluation of their government's policies and their acceptance of its crisis management approach. Indeed, the German media portrayed the former American president as an irrational and narcissistic politician who ignored scientific insights and thereby caused damage, a person who was the quintessential antipode, an extreme negative contrast, to the German chancellor.

## Physical Proximity and Distance: Sociological Lessons from the Crisis

To contain the pandemic, restrictions on social contacts and, in particular, the rule of physical distancing were important. Since the outbreak of the pandemic in Germany, the rule of keeping a distance of 1.5 metres had been practised throughout the country. In addition, schools, universities, and companies transplanted communication and collaboration through direct encounters from spatial environments to internet-based communication to the degree that it was feasible to do so. On the one hand, it became clear thereby that modern technologies allowed for a professional exchange of scientific information and professional communication with limited outlay. Moreover, not being expected to be physically present at the workplace is a considerable relief, especially if one can avoid time-consuming commuting. On the other hand, however, an uneasiness with such a mode of interaction became apparent rather quickly. Reports in the media, my own subjective experience, conversations with colleagues, and a small survey of students that I conducted indicate that communication under conditions of physical distancing is experienced as restrictive and unsatisfactory even when microphones, cameras, and the internet make audio and visual interaction possible.

Although this appears self-evident, the available repertoire of sociological theories and research results offers only limited answers to questions on why this is the case, why, even independently of the pandemic, a changeover from physical encounters to media-mediated communication is not possible without problems even in technically advanced societies. For example, in the studies on interaction order by Erving Goffman (1994) there are certainly emphases on the importance of the body in interaction. This is especially true with regard to the problematic side of physical presence and its resulting bodily vulnerability to and for violence, and the physicality of gender and its relation to the rules of a society's gender order. Furthermore, in the more recent German and French sociology of space, sociologists have accentuated that spatial arrangements and the arrangement of bodies in space express social orders and hierarchies (Löw, 2013). However, it remains largely unclear why physical presence has a positive meaning for individuals and their social coexistence even outside of physical contacts among family members and close friends, intimate sexual relationships, and physical work. This is related to the fact that the phenomenon of social modernization is understood in current sociological theories as a process that renders bodies less and less important, classically formulated in Mary

Douglas's thesis that the increasing complexity of society leads to social action resembling "intercourse between disembodied spirits" (1974, p. 110).

A starting point for understanding the reason for the positive social significance of physical presence for individuals, nevertheless, can be found in Georg Simmel (1908, p. 479ff.). In his classic chapter on the space and spatial order of society, he also addresses the question of the sociological relevance of "sensual proximity or distance between persons" (p. 479). In doing so, he distinguishes two fundamental dimensions of social relationships: on the one hand, the purely factual and impersonal dimension, and on the other hand, the emotional dimension, which concerns what he calls "das gemüt" (the "temper," the "mood") of persons (p. 480). He argues that physical proximity is not necessary when both dimensions appear in their pure form. According to Simmel, emotional relationships are possible over long distances, which he illustrates with the example of the contemporary technique of writing love letters. According to Simmel, economic and scientific communication is also possible in written form without any problems, even over a long distance. According to Simmel, physical co-presence is required when it comes to mixed forms of emotional and factual relationships; that is, when communication is aimed at interpreting factual information as well as messages from individuals. In these cases, it is not only the factual content of a statement that is important but also the intentions underlying a communication. Therefore, it is not arbitrary who communicates something, what information about their emotional state is communicated through facial expressions and gestures, and what emotional reaction this triggers in other communication partners.

This idea can be specified with the sociological communication theory of Niklas Luhmann (2001). Luhmann distinguishes three components of every communication: *information* (information), *mitteilung* (message), and *verstehen* (understanding). According to this typology, communication consists of the fact that the utterance of person A is considered relevant by person B and a distinction is made between its information content and what person A thereby expresses about himself; for example, his current emotions or his intentions. It is important to bear in mind that Luhmann is not concerned with right or wrong understanding: understanding is rather the creative achievement that consists in the fact that person B can extract meaning from both sides of A's utterance, the information and the message, and react to it in a way that enables the progress of communication. Based on this distinction, it is plausible to assume that physical presence becomes socially significant when not only is the information relevant but the message too, and facial expressions, gestures, and body postures therefore enable us to draw conclusions about the messages.

In closing, I am of the view that contemporary sociological theories of communication and interaction have not systematically followed up on this aspect of Simmel's and Luhmann's thinking. The importance of such an endeavour cannot be overemphasized even if the state apparatuses are functioning – as in this chapter's exemplar of Germany – and able to attend to the technocratic aspects of ensuring a sufficient extent of safety for its citizens. The relatively more positive experiences in the management of the COVID-19 crisis should therefore be an occasion to evolve our questioning towards a critical understanding of whether social relations in modern societies are relations between disembodied spirits, even though technological developments have fostered communication and interaction under conditions of non-spatiality. In-depth analyses of this phenomenon could also contribute to providing sociological arguments for resistance to economically motivated interests that celebrate disembodied spirits, which under current pandemic conditions can be seen to shift cooperation in industries, administrations, and, last but not least, schools and universities towards the internet for knowledge production, if only for the simple reason of cost savings. Sociology as a critical social science is challenged by the COVID-19 pandemic to develop a better understanding of the contradictions and paradoxes of the state – for example, the necessity of welfare state measures – but also the connection between statehood and exclusion and global inequalities. On the other hand, sociology should contribute to a better understanding of why human coexistence cannot be subjected to the imperatives of technology and economy without consequences, not least because human life is always also the coexistence of physical and emotional beings.

## NOTES

1 This paper looks at the development from the beginning of the COVID-19 pandemic in February 2020 to the point in time when a significant reduction in the number of new infections was stably achieved. This was the case in Germany in June 2020 and thus occurred at a time when summer academic recess began in large parts of the country.
2 This not only affected tourists but was also a consequence of the fact that international programs of German organizations (for example, international voluntary services) were suspended. In an interview with a person affected, she told me that, in the Latin American country where she was staying in March 2020, a hostile mood had arisen towards Germans and other Europeans, as locals now perceived them as potential carriers of infection.
3 See also Auswärtiges Amt updates at: https://www.auswaertiges-amt.de/de/ReiseUndSicherheit/covid-19/2296762.
4 Studies that examine the effects of the crisis on the spread of prejudice and racism are not yet available.
5 https://www.bmi.bund.de/SharedDocs/kurzmeldungen/DE/2020/06/pk-ende-grenz¬control¬len.html.

6  Unless otherwise stated, the data in the following are taken from the websites of Johns Hopkins University (https://coronavirus.jhu.edu/) and the German Robert Koch Institute (www.rki.com). The Robert Koch Institute (RKI) is an institute of the federal government engaged in the field of disease surveillance. It is also the central institution of the federal government engaged in the field of application and measure-oriented biomedical research. The recognition, prevention, and control of infectious diseases is a central task of the RKI.
7  The most recently available data from Johns Hopkins University, dated 4 February 2022, is located at https://coronavirus.jhu.edu/map.html. It recorded a total of 10,723,285 people infected and 118,508 deaths in Germany during the pandemic, giving a case-fatality rate of about 140. In the United States, the number of infected people at the same time was 75,682,712, with a case-fatality rate of about 270.
8  In Germany sex work is a legal profession.
9  This text was completed in a first version in December 2020. At that time, the pandemic was not expected to continue undiminished through 2021 and into the spring of 2022. Currently, in February 2022, it can be seen that, after a phase in which the pandemic situation worsened, it is again expected that a return to normality will be possible in the foreseeable future.
10  This includes the consumption of alcohol in public places, which is legally permitted from the age of 16.
11  In Germany, as in other countries, there is an informal labour market with illegal employment relationships. Reliable figures on the extent of this are not available. Because of relatively strict regulations, also in the area of rentals, however, living illegally in Germany is only possible to a very limited extent. In addition, rejected asylum seekers in Germany normally do not remain undocumented because they are granted temporary residence status.
12  As a rule, recipients of ALG 1 are entitled to 60% or 67% (with child) of their last net salary.
13  For single persons, this means an income of 432 euros per month, plus the assumption of limited costs for rent and health insurance. However, migrants from other European countries living in Germany who have not previously paid contributions to unemployment insurance are excluded from this basic provision.
14  The asylum procedure takes on average one year. For families with children, however, the duration of stay is limited to 6 months, for adults without children to 18 months. Following the United Nations Convention on the Right of the Child, minors without a family are accommodated in much better equipped facilities where they are looked after by social workers.
15  From 2005 to December 2021, the CDU was the strongest party in the German Parliament. Since 2013, the government consisted of a coalition of the CDU with the SPD. In addition to these two parties, three others are represented in Parliament (Die Grünen, AfD, FDP). The fact that the CDU did not emerge as the strongest party from the elections in 2021 and is no longer part of the government

is closely linked to Angela Merkel's decision not to stand for re-election after 16 years as chancellor. Since December the government has consisted of a coalition of three parties (SPD, Die Grünen, FDP). The chancellor, Olaf Schulz, belongs to the SPD. To understand the following figures, it is important to know that the chancellor in Germany is not elected directly, but by Parliament.

16 The AfD achieved 12.6% of all votes in the last election of the Bundestag in 2017. By German standards, it is considered an extreme right-wing party against which all other parties clearly distinguish themselves and with whom they avoid cooperation.

17 Since the fall of 2021, however, protests against state policy have increased and become visible in the form of numerous demonstrations. These protests were carried out by a minority that rejects vaccination on the basis of esoteric beliefs, but also by a right-wing populist spectrum that seeks to undermine the legitimacy of democracy (Frei & Nachtwey, 2021).

REFERENCES

Auswärtiges Amt. (2020). *Reisewarnungen anlässlich der COVID-19-Pandemie* [Travel warnings on the COVID-19 pandemic]. https://www.auswaertiges-amt.de/de/ReiseUndSicherheit/covid-19/2296762.

Berger, P.L., Berger, B, & Kellner, H. (1973). *The homeless mind: Modernization and consciousness.* Random House.

Bundesagentur für Arbeit. (2020a). *Arbeitslosenquote & Arbeitslosenzahlen 2020* [Unemployment rates and unemployment numbers 2020]. https://www.arbeitsagentur.de/news/arbeitsmarkt-2020.

Bundesagentur für Arbeit. (2020b). *Corona-Virus: Informationen für Unternehmen zum kurzarbeitergeld* [Corona virus: Information for companies on part-time staff budget]. https://www.arbeitsagentur.de/news/corona-virus-informationen-fuer-unternehmen-zum-kurzarbeitergeld#1478910781381.

Bundesministerium für Arbeit und Sozialordnung. (2019). *Entgeltfortzahlung bei Krankheit und an feiertagen* [Continued payment in the event of illness and public holidays]. https://www.bmas.de/SharedDocs/Downloads/DE/Publikationen/a164-entgeltfortzahlung-bei-krankheit-und-an-feiertagen.pdf?__blob=publicationFile&v=1.

Bundesregierung. (2020, June 8). *Antwort der Bundesregierung auf die kleine Anfrage der Abgeordneten Bernd reuther, Frank Sitta, Torsten Herbst, weiterer Abgeordneter und der Fraktion der FDP – Drucksache 19/19473 –* [Answer given by the federal government to the question by Bernd Reuther, Frank Sitta, Torsten Herbst, other deputies and the fdp – case 19/19473 – ]. https://dip21.bundestag.de/dip21/btd/19/198/1919821.pdf.

Douglas, M. (1974). *Ritual, Tabu und Körpersymbolik* [Ritual, taboo, and body symbolism]. S. Fischer.

Endt, C., & Zaschke, C. (2020, July 14). Wir haben einen langen Weg vor uns [We have a long way to go]. *Süddeutsche Zeitung.* https://www.sueddeutsche.de/politik/coronavirus-usa-europa-vergleich-1.4966692?reduced=true.

Forsa Institut. (2020). *Wenn am nächsten Sonntag Bundestagswahl wäre* [If there were federal elections next Sunday]. https://www.wahlrecht.de/umfragen/forsa.htm.

Frei, N., & Nachtwey, O. (2021, June 21). Wer sind die Querdenker_innen? [Who are the lateral thinkers?]. *Friedrich Ebert Stiftung E-Papierreihe: Demokratie im Ausnahmezustand. Wie verändert die Coronakrise Recht, Politik und Gesellschaft? [Friedrich Ebert Stiftung E-Paper Series: Democracy in a state of emergency. how is the corona crisis changing law, politics and society?].* http://library.fes.de/pdf-files/dialog/18030.pdf.

Fromm, E. (1969). *Escape from freedom*. Henry Holt and Company.

Giddens, A. (1979). *Central problems in social theory: Action, structure and contradiction in social analysis*. Macmillan Education.

Goffman, E. (1994). *Interaktion und Geschlecht* [Interaction and gender]. Campus Fachbuch.

Gosewinkel, Dieter. (2020). Die Rennaissance nationaler Grenzen [The renaissance of national borders]. *WZB Mitteilungen, 168*, 17–19.

Gregory, D., & Urry, J. (Eds.). (1985). *Social relations and spatial structures*. Palgrave Macmillan.

Harvey, D. (2000). *The condition of postmodernity*. Blackwell Publishers.

Horkheimer, M. (2013). *Critique of instrumental reason*. Verso.

Hurrelmann, K., & Dohmen, D. (2020, April 15). Corona-Krise verstärkt Bildungsungleichheit [Corona crisis increases educational inequality]. *Das Deutsche Schulportal*. https://deutsches-schulportal.de/expertenstimmen/das-deutsche-schulbarometer-hurrelmann-dohmen-corona-krise-verstaerkt-bildungsungleichheit/Ifo.

Institut/Forsa. (2020, July 10). *Erste Ergebnisse des Befragungsteils der BMG-"Corona-BUND-Studie"* [First results of the survey section of the BMG "corona-BUND-study"].https://www.ifo.de/DocDL/bmg-corona-bund-studie-erste-ergebnisse.pdf.

Juhl, S., Lehrer, R., Blom, A.G., Wenz, A., Rettig, T., Reifenscheid, M., Naumann, E., Möhring, K., Jrieger, U., Fridel, S., Fikel, M., & Cornesse, C. (2020, April 22). Die Mannheimer Corona-Studie: Demokratische Kontrolle in der corona-krise [The Manheim Corona Study: Democratic control in the corona crisis]. https://madoc.bib.uni-mannheim.de/55137/.

Kochhar, R. (2020, June 11). Unemployment rose higher in three months of COVID-19 than it did in two years of the Great Recession. Pew Research Center. https://www.pewresearch.org/fact-tank/2020/06/11/unemployment-rose-higher-in-three-months-of-covid-19-than-it-did-in-two-years-of-the-great-recession/.

*Koreaverband*. (2020, May 29). Antiasiatischer rassismus in der corona-krise [Anti-Asian racism in the corona crisis]. https://www.koreaverband.de/blog/2020/05/29/rassismus-corona/.

Landau, M.J., Solomon, S., Greenberg, J., et al. (2004). Deliver us from evil: The effects of mortality salience and reminders of 9/11 on support for President George W. Bush. *Personality & Social Psychology Bulletin, 30*(9), 1136–50. https://doi.org/10.1177/0146167204267988.

Landesregierung Baden-Württemberg. (2020). *Maßnahmen gegen die Ausbreitung des Coronavirus* [Measures against the spread of the coronavirus]. https://www

.baden-wuerttemberg.de/de/service/aktuelle-infos-zu-corona/aktuelle-corona-verordnung-des-landes-baden-wuerttemberg/.

Lehrer, R., Juhl, S., Blom, A.G., Wenz, A., Rettig, T., Reifenschedl, M., Naumann, E., Möhring, K., Krieger, U., Friedel, S., Fikel, M., & Cornesse, C. (2020, April 27). *Die Mannheimer Corona-Studie: Die vier Phasen des Social Distancing in Deutschland* [The Mannheim Corona Study: The four phases of social distancing in Germany]. University of Mannheim. https://madoc.bib.uni-mannheim.de/55135/.

Löw, M. (2013). *Raumsoziologie* [Sociology of space]. Suhrkamp.

Luhmann, N. (2000). *The reality of mass media*. Polity Press.

Luhmann, N. (2001). *Aufsätze und Reden* [Essays and speeches]. Reclam

Luhmann, N. (2012). *Theory of society. Volume I*. Stanford University Press.

Mauss, M. (2010). *Soziologie und Anthropologie: Band 2: Gabentausch – todesvorstellung – Körpertechniken (Klassiker der sozialwissenschaften)* [Sociology and anthropology: Volume 2: Gift exchange – death idea – body techniques (Classic of the social sciences)]. VS Verl. für Sozialwissenschaften.

Merkel, A. (2020, March 18). Fernsehansprache von Bundeskanzlerin Angela Merkel [Televised speech by Chancellor Angela Merkel]. Die Bundesregierung. https://www.bundeskanzlerin.de/bkin-de/aktuelles/fernsehansprache-von-bundeskanzlerin-angela-merkel-1732134.

Michel, B., & Stein, C. (2015). Reclaiming the European city and lobbying for privilege. *Urban Affairs Review, 51*(1), 74–98. https://doi.org/10.1177/1078087414522391.

Offe, C. (1972). Advanced capitalism and the welfare state. *Politics and Society, 2*(4), 479–88. https://doi.org/10.1177/003232927200200406.

Pintelon, O. (2012). *Welfare state decommodification: Concepts, operationalizations and long-term trends*. Working Paper No. 12/10. Herman Deleeck Centre for Social Policy, University of Antwerp.

Pogge, T. (2008). *World poverty and human rights*. Polity Press.

Rees, J., Papendick, M., Rees, Y., Wäschle, F., & Zick, A. (2020). Erste Ergebnisse einer Online-umfrage zur gesellschaftlichen Wahrnehmung des Umgangs mit der Corona-pandemie in Deutschland [First results of an online survey on how society perceives the handling of the corona pandemic in Germany]. Research report IKG. Bielefeld: Institute for Interdisciplinary Research on Conflict and Violence (IKG).

Reiss, K., Weis, M., Klieme, E., & Köller, O. (Eds.). (2019). *PISA 2018: Grundbildung im internationalen Vergleich* [Basic education in an international comparison]. Waxmann.

Robert Koch Institute (RKI). (2019). *Bericht zur Epidemiologie der Influenza in Deutschland Saison 2018/19* [Report on the epidemiology of influenza in Germany for the 2018/19 season].

Schäfers, B. (2006). *Stadtsoziologie. Stadtentwicklung und Theorien – Grundlagen und praxisfelder* [Urban sociology. Urban development and theories – Basics and fields of practice]. VS Verlag für Sozialwissenschaften.

Simmel, G. (1908). *Soziologie*. Duncker & Humblot.

Steinert, J., & Ebert, C. (2020). Gewalt an Frauen und Kindern in Deutschland während COVID-19-bedingten Ausgangsbeschränkungen. Zusammenfassung der Ergebnisse. [Violence against women and children in Germany during COVID-19 lockdown: Summary of results]. Hochschule für Politik München TUM School of Governance. https://www.kriminalpraevention.de/files/DFK/Praevention%20 haeuslicher%20Gewalt/2020_Studienergebnisse%20Covid%2019%20HGEW.pdf.

Weinkopf, C., & Hüttenhoff, F. (2017). Der Mindestlohn in der Fleischwirtschaft [The minimum wage in the meat industry]. *WSI Mitteilungen*, *7*, 533–9. https://doi.org/10.5771/0342-300X-2017-7-533.

Wendel, K. (2014). *Unterbringung von Flüchtlingen in Deutschland. Regelungen und Praxis der Bundesländer im vergleich* [Accommodation of refugees in Germany. A comparison of regulations and practice in the federal states]. Pro Asyl.

Wendling, T. (2010). Usage et abuse de la notion de fait social total [Use and abuse of the notion of total social fact]. *Revue du MAUSS*, *36*(2), 87–99. https://doi.org/10.3917/rdm.036.0087.

Whimster, S., & Lash, S. (Eds.). (2014). *Max Weber, rationality and modernity*. Routledge.

Wimmer, A., & Schiller, N.G. (2002). Methodological nationalism and beyond: Nation-state building, migration and the social sciences. *Global Networks*, *2*(4), 301–34. https://doi.org/10.1111/1471-0374.00043.

World Health Organization (WHO). (2019, November 28). Global Tuberculosis Report 2919. https://www.who.int/tb/publications/factsheet_global.pdf?ua=1.

# South Pacific

# 13 The Benefits and Drawbacks of Social Distancing: Lessons from New Zealand

BY MARIA ARMOUDIAN AND BERNARD DUNCAN

Positioned near the bottom of the world in the South Pacific with no country within 4,000 kilometres (2,500 miles), New Zealand perhaps epitomizes a geographic type of social distance from the rest of the world. During the pre-vaccination period of the COVID-19 crisis, that distance from the rest of the world provided a layer of protection from widespread illness. As in other South Pacific island nations, which fared better in protecting their citizens than most other countries, with low or no COVID infection rates, the ocean acted as a moat preventing easy access to the country and offered time to watch other countries' experiences and assess the best responses. Internally, it also exercises a type of social distance. With roughly 4.8 million people spread out across its two main islands, New Zealand has low population density, ranking 202 among 235 ranked countries in density measures. These two features factored into the success New Zealand experienced in eliminating the first variant of COVID-19 in a relatively short period of time: The global geography gave the island nation time to observe the virus ravage much of the rest of the world and to assess knowledge obtained by watching the strategies that seemed to work for managing the lethal pandemic. The country's dispersed population helped the initial four-week lockdown achieve its desired results. But it came with social, economic, political, and psychological impacts to the country. This sociopolitical history chronicles the impact of SARS-CoV-2 (COVID-19) in New Zealand and the country's response, from the initial identification of infection in early 2020 until the end of that year. It focuses on key events, arguments in New Zealand's Parliament, and coverage in the media related to social distancing, legal restrictions to mitigate the pandemic, their consequences, and their influences on the 2020 national election.

### A Miscalculation: The Early Days of the Coronavirus

The warning came in early January, when Chinese citizens in Wuhan, China, were afflicted with pneumonia symptoms. Otago University public health professor Michael Baker told Radio New Zealand (RNZ) that the virus appeared "to be severe" but did not appear highly infectious, as was the case with SARS of 2002. Nonetheless, he suggested that visitors going to Wuhan be attentive to World Health Organization (WHO) information (Todd, 2020).

By 23 January, New Zealand's government began an inquiry and determined that, at the time, the risk of large-scale infection with COVID-19 was not high for its residents. The minister of health, David Clark, introduced a few precautionary measures, including placing posters, banners, and health advice cards at international entry points and sending public health staff to the international airports to "provide information to travellers" arriving from abroad. Public health nurses were available to monitor the temperature of any passenger who reported they were unwell, he said (Clark, 2020; New Zealand Parliament, 2020). By then Baker was calling for a more precautionary approach, ominously noting, "Distance is no protection whatsoever for New Zealand" (Devlin, 2020; Woolf, 2020).

The initial sanguine public health assessment took a turn a short time later. On 2 February, New Zealand began denying entry to foreign nationals who had travelled from or through China. It followed with self-isolation orders for returning New Zealanders and a 14-day quarantine of 195 Wuhan evacuees on an army base at the Whangaparaoa peninsula north of Auckland, New Zealand's largest city, according to Clark, who noted that returnees would receive daily medical and well-being checks, and families would be kept together where possible (RNZ, 2020a). While many experts considered the decision sound, one economist, Sarah Hogan of the New Zealand Institute of Economic Research (NZIER), deemed that the travel restrictions violated international health regulations as well as being unlikely to stop the spread of the virus, both of which would have an incredible cost on the economy. By 11 February, New Zealand's Parliament began its investigation when opposition National Party member of Parliament (MP) Louise Upston asked the minister for social development about businesses calling a government-established helpline related to pandemic management, the numbers of cases reported at the Work and Income office, and the amounts paid for hardship grants. The answers were 0, 0, and NZD$882, respectively (Upston, 2020).

**A Reassessment**

As case numbers in the country rose through mid- to late March 2020, Prime Minister Jacinda Ardern announced the implementation of a four-level alert system set up to eliminate the virus from the country. The basis of the system allocated one-word descriptors to each level, beginning with "Prepare" at level 1, "Reduce" at level 2, "Restrict" at level 3, and "Lockdown" at level 4. The system further assigned risk assessment and mitigation measures for each alert level (New Zealand COVID-19 Alert Levels Summary, 2020).

A feature of the government's approach to transparency within the context of its responses to the growing threat of the coronavirus became the daily 1:00 p.m. media "stand ups" – live appearances – usually by Prime Minister Jacinda Ardern and the director-general of health, Dr. Ashley Bloomfield, on the country's three primary broadcast channels, Television New Zealand (TVNZ), TV3, and RNZ. The government also streamed on the Ministry of Health's website. At least one newspaper referred to these as "Jacinda's wartime-type briefing[s]" (van Beynen, 2020). During these briefings, Ardern outlined government decisions for managing the situation and, often, Cabinet meeting outcomes, while Bloomfield updated information about the spread of the virus, including numbers of cases, recoveries, deaths, and testing rates. Bloomfield emerged in social media as a reluctant celebrity with his image emblazoned on T-shirts (McNeilly, 2020) and, in at least one case, a woman's arm-tattoo (*NZ Herald*, 2020b). The director-general was also "humbled" to be nominated for the publicly voted TV Personality of the Year Award (TVNZ One News, 2020h). Bloomfield politely withdrew, with a statement from his office expressing his humility and noting, "Given his primary role as a public servant and not a TV personality he would like to step aside from being considered" (*Dominion Post*, 2020).

**New Zealand Locks Down**

On 23 March, the government moved the country to alert level 3, and two days later, imposed a strict level 4 lockdown. Under level 4 rules, New Zealanders were required to "stay home" and only associate with people within their "bubbles" or households, initially, with a few exceptions for those deemed to be "essential workers" in supermarkets, gas stations, and medical facilities. This meant no visitors, no in-person interactions, and no travel. New Zealanders could only legally leave their homes for groceries, medicine, or exercise, the latter of which could only take place within two kilometres of their home bases without a drive. Visiting the gym was prohibited. Violators faced fines, and in some cases, arrest and prosecution (*NZ Herald*, 2020a). Schools, offices, and parliamentary processes, including select committee meetings, took place remotely using video conferencing technology. And while Ardern called on New Zealanders to "be kind" to one another, they were also encouraged to report lockdown violators, which seemed contradictory to some (Hosking, 2020). Indeed, the New Zealand Police instituted an online facility to enable people to report guideline breaches (Leask, 2020). The volume of responses caused the site to crash at least once during its first few days of operation, while emergency call numbers were inundated (Ainge-Roy, 2020a).

For distressed children, a teddy bear hunt spread across parts of the country, in which people placed stuffed animals in their windows so that they were visible to passersby. On 25 March, a Facebook page for the "We're not scared – NZ

Bear Hunt" had 8,000 followers, according to RNZ (2020c). Children could count the bears as they went on their socially distanced family walks (Beck, 2020).

With the exception of larger centres in the country where multi-story apartments are in close proximity, and crowded households, particularly among some Māori and Pacific peoples, New Zealanders already enjoy a suburban and rural geography that, in most cases, made social distancing easier to achieve. The country's overall low population density of 18 people per square kilometre meant that walkers or joggers could generally avoid close contact. This was not the case in highly populated city centres such as central Auckland, where the population density was a thousand times greater (Statistics NZ, 2017).

**Pushbacks and Criticisms**

Even with the minimal restrictions in the early days of the virus, the economic and social costs mounted. Primary industries such as agriculture, forestry, fishing, and mining, which contributed approximately 7% to New Zealand's economy in 2018 (Statistics NZ, 2019), complained and made spending cuts (Hollyman, 2020; *Dairy News*, 2020). Furthermore, international sporting events, a social and economic mainstay of New Zealand, were cancelled. With the decision of the International Olympic Committee to postpone the 2020 Olympic Games in Tokyo for at least a year, media outlets quoted a top sports psychologist who warned of difficulties for New Zealand athletes training for the event:

> A lot of athletes will have had a four-year plan of how their lives would look and how they would set up their lives, so there is grief at the loss of those plans. Some athletes who are coming back from injury or who thought they were targeting Paris in 2024 are really thankful for another year of preparation, but at the other end, you have people who were possibly considering retiring after Tokyo. It takes a while for them to get their head around another year of commitment. (Hewson, 2020)

Government provided support in early April with an announcement from the minister of finance, Grant Robertson (who was also the minister for sport and recreation), that national sports organizations would have their current year's funding rolled over into the following year "so that current levels of investment remain through to 30, June 2021." They had "lost significant revenue streams, including broadcasting, sponsorships, sport betting, Class 4 gambling and membership fees," he said, adding:

> The Government is also working on a sports recovery package for when we get through the COVID-19 pandemic. This package will look to include some support for community organisations and our high performance and elite athletes. Sport,

recreation, and play are vital for the health and wellbeing of our communities, and we are committed to doing everything we can to support the sector, both at a grass-roots and elite level, into the future. (NZ Government Media Release, 2020)

A month later, delivering the country's national budget, Robertson announced additional support in the form of NZD$265 million government investment over the coming years to help the sports sector through the crisis. He said funding and revenue had dried up for nearly all sports organizations and that they were under "immense strain. We are providing the support needed to sports at all levels to remain viable, get stronger and adapt," he said (Reuters Staff, 2020).

Film and television productions also halted. Production managers for the *Lord of the Rings* television series suspended production indefinitely and told staff to "not report" to work, despite having already begun shooting (Keall, 2020). Movie sequels in the *Avatar* franchise were also put on hold (Frater, 2020) with Netflix, Warner Bros, and Disney cancelling work on their productions "due to keeping their village-sized crews out of the way of coronavirus," according to Newshub: The Project (2020).

## Waning Support

In the latter part of April, multi-partisan support in Parliament for the measures waned further, particularly when the government decided to extend the four-week lockdown by another five days. Despite Prime Minister Ardern's assertion that "We have stopped a wave of devastation," she went on to announce that the strictest period of national shutdown would remain in place until midnight on Monday, 27 April. At that stage, a slightly less restrictive set of level 3 measures would operate until her Cabinet reassessed the overall situation on 11 May 2020 (Graham-McLay, 2020a).

Political opponents strongly criticized the level 4 lockdown extension, with the opposition National Party leader, Simon Bridges, on 21 April, telling Parliament's Epidemic Response Committee (which he chaired) and the minister for small business, Stuart Nash, that small businesses, in particular, were suffering. A small bakery business in his electorate, for example, had complained of no income. "They're getting the wage subsidy. They say it's not enough, because they're … 'accumulating expenses, rent, rates, insurance, ACC [Accident Compensation Corporation levies], and other expenses' – the list goes on," Bridges noted, adding, "We are absolutely gutted that our livelihoods are shattered due to our business closure. We are the sacrificial lambs and are – they say – 'completely unsupported.'" He acknowledged the complainant's eligibility and gratitude for wage subsidy assistance but pressed the minister to concede that the situation was untenable and "not sufficient" for businesses forced to stop operations for up to seven weeks (Bridges, 2020). The economic downsides of the elimination strategy argued in Parliament by the government's opposition were supported by a senior epidemiology lecturer at the University of Auckland, Simon Thornley. "The costs of trying to maintain eradication are just going to be astronomical – I don't believe that it's sensible," he told Time.com; "The small-business people who have put a lot of time and effort into their livelihoods are going to have it taken away through this lockdown" (Gunia, 2020).

Perhaps the hardest hit was the tourism industry, which, in the year since March 2019, had provided 5.8% of the country's GDP and accounted for 14.4% of people employed in New Zealand, according to Tourism Industry Aotearoa (2020). In the South Island's Wanaka region, for example, tourist spending provided roughly NZD$600 million (van Beynen, 2020). Additionally, foreign students studying in New Zealand and who would not be permitted to return while the borders were locked contributed an additional NZD$3 billion annually, according to the report in Time.com (Gunia, 2020). After initial COVID-19 management protocols at campsites and other businesses, the lockdown and lockout of tourists led the government to encourage New Zealanders to vacation within their own country as well as to allocate an initial NZD$11 million package to support the industry (Cheng, 2020a). Elsewhere in New Zealand, homeowners relied on tourists renting their Airbnb accommodations to pay their mortgages. Indeed, the loss of international tourism dollars would radically change the economy and require a rethink, according to economist Brad Olsen of New Zealand economic consultancy firm Infometrics, who further noted, "The economy can survive without international tourism, but not as we know it" (Gunia, 2020).

While it may be "brutal for the economy … the huge benefit is that you have an exit strategy," countered University of Otago's Public Health professor Michael Baker. He added that since travel had stopped across the world, the damage to New Zealand's economy was going to happen no matter what. "It's not like suddenly international tourism and cruise ships are all going to start up again," Baker noted (Gunia, 2020). Furthermore, economists from the Australia and New Zealand Banking Group (ANZ) raised the prospect of a "double dip" recession that would begin from the end of 2020 and extend into 2021. They noted how the "GDP could be hit from both the absolute impact of a closed border (which we estimate is leaving around a 5% hole in the economy)" and the possibility of an additional 1% reduction due to seasonal impacts (Hargreaves, 2020).

RNZ reported that the agricultural export sector would also suffer from its inability to bring casual seasonal fruit pickers across New Zealand's closed borders. While the government agreed to extend working holiday visas for workers already in the country so they could remain for the high season, the inability of additional backpackers and casual labourers to enter the country resulted in a shortage of fruit pickers, potentially jeopardizing the NZD$10 billion horticulture industry (Chiang, 2020). Richard Palmer, chief executive of Summerfruit New Zealand, a trade body that represents the nation's apricot, cherry, nectarine, peach, and plum producers, told news media producer *Stuff* that 30% of the growers' harvest may not be able to be picked, and that shortfall could see a loss of NZD$1.25 billion in export earnings (Dunkley, 2020).

**Success and Fallout**

The lockdown measures effectively arrested community transmission of the disease by early May 2020. A study funded by New Zealand's Ministry of Business, Innovation and Employment (MBIE), the Strategic Science Investment Fund (SSIF), and the country's Ministry of Health concluded in a publication in the *Lancet*: "Within 2 weeks, lockdown was associated with a substantial reduction in daily case infection rate and improving response performance measures: the majority of cases [were] detected by contact tracing" (Jeffries et al., 2020). Referring to WHO statistics, the report added that "New Zealand experienced one of the lowest cumulative case counts, incidence, and mortality among higher-income countries in its first wave of COVID-19 following early implementation and rapid escalation of national COVID-19 suppression strategies" (Jeffries et al., 2020).

Health and safety distancing and movement restrictions enacted to halt the spread of the virus had wider sociocultural impacts. The Pasifika Festival, a cultural gathering with a Pacific Islands theme held annually in Auckland since 1993, billed as the largest of its type in the world and routinely attracting more than 200,000 visitors and participants, was cancelled (RNZ, 2020b); the previous year's festival had also been cancelled because of security issues following the mass shooting at mosques in Christchurch (RNZ, 2019). A planned Christchurch-based memorial to mark the mosque shootings one year later was also cancelled (Kronast & McRoberts, 2020). Furthermore, for the first time in 104 years, New Zealand cancelled its national day, ANZAC Day, which commemorates fallen New Zealand and Australian soldiers who fought at Gallipoli during the First World War (Wiltshire, 2020), a campaign that resulted in approximately 11,500 casualties for the Australian and New Zealand Army Corps (ANZAC). The cancellation was significant, since the dawn memorial services held in New Zealand and Australia (and internationally) became a cornerstone of New Zealanders' acknowledgment of the sacrifices suffered in wartime and the debt owed to all who contributed military service to the country. It was a "central marker of our nationhood," according to the Ministry for Culture and Heritage (2020), one that came "to mark what some saw as the foundation of a distinct New Zealand identity" (*New Zealand History – Nga korero a ipurangi o Aotearoa*, 2020).

Despite the limits imposed by the situation, public response to the government's handling of the COVID-19 crisis was generally favourable throughout

Image 13.1. COVID-19 testing facility in Central Wellington (Bernard Duncan, 10 August 2020).

the initial lockdown, as evidenced by surveys conducted by market and social research company Colmar Brunton in early April 2020. Findings revealed that survey respondents considered the government a key source of reliable information about the outbreak. High levels of trust were placed in "the judgement of our leaders": 88% of the respondents trusted their government to make the "right decisions" on COVID-19." Additionally, public trust in the New Zealand government to successfully attend to this national issue "has rocketed from 59% to 83%" (2020 Public Trust Survey conducted for the Institute for Governance and Policy Studies at Victoria University of Wellington; see also Colmar Brunton, 2020a). Indeed, three weeks later, support for the government's response to the pandemic remained as high as 87% (Colmar Brunton, 2020b).

## An Emotional Toll

While successful in halting COVID's spread, the lockdown brought social, relational, and emotional costs for some New Zealanders who grew distressed by their isolation from loved ones. Wedding plans were put on hold, and the rules surrounding hospital and retirement home visitation, especially for those with sick, elderly, and dying relatives, were considered as heartbreaking, lacking compassion, and "breaching basic human rights" by some family members (Akoorie, 2020). Those who lost loved ones had to adjust to the realization that they might never have the opportunity for a last goodbye, and would have to grieve in isolation. Funerals and *tangihanga* ("the enduring Māori ceremony to mourn the dead") (Higgins, 2011) were banned under alert level 4, and while funeral directors could continue to operate, they could only offer the options of private cremation, direct burial (with only the funeral director in attendance), or holding the departed's body until restrictions once again allowed mourners to gather (McLachlan, 2020).

The lockdown's impact on relationships was mixed. Couples were forced apart or thrown together, with some reporting an intensification of their emotions about their partners, some deciding to change a casual relationship into something more permanent, while others ended their relationships. Media reported on discoveries about infidelity, rushed cohabitation, and difficult long-distance relationships (O'Connell, 2020; Scotcher, 2020; Leahy, 2020).

Because much of the discourse about the lockdown's imposition was economically based, Dr. Susanna Every-Palmer decided to conduct a mental health study: "Clinically a number of us work with vulnerable people, and we were seeing many people struggling with the pandemic and the lockdown, and experiencing anxiety, depression and loneliness so we were interested in what was happening at a population level," she said. Academic staff from three University of Otago campuses – Dunedin, Wellington, and Christchurch – led by Every-Palmer, head of the Department of Psychological Medicine in Wellington, concluded that while most New Zealanders coped well, almost a third experienced levels of mental distress well above normal (Every-Palmer et al., 2020).

An unexpected finding was how well older people coped. "It was the 65-plus group that was doing the best," the study found. In contrast, younger adults aged between 18 and 34 were most distressed, possibly reflecting worries over work and money, or struggles with child care and other general stressors. Still, while most respondents appreciated having more time with their families and were enjoying improved family relationships, it was a very uncertain time. Every-Palmer said the study's findings emphasized the need for governments to put resources into supporting people's well-being during a pandemic and maintain a psychosocial action plan. "Messages about being kind and tolerant support mental well-being," she said (Every-Palmer et al., 2020).

The study also raised concerning aspects of lockdown, according to news outlet *Stuff* on 5 November (Martin, 2020): "About 6 percent reported having suicidal thoughts during lockdown, and 2 percent reported suicide attempts … Almost one in 10 participants reported experiencing some form of family harm … including sexual assault, physical assault, harassment and threatening behaviour." The medical director of the College of General Practitioners, Dr. Bryan Betty, added that general practitioners had "absolutely" seen an increase in patients reporting distress and anxiety, and the distress did not end when the lockdown did. "We are seeing quite a heavy workload still … the demand is definitely still there," he said (Martin, 2020).

**The Move to Level 3**

The sense of relief that could have followed the move at the end of April to a level 3 alert was muted. Business activity remained severely restricted. Public venues such as libraries, museums, cinemas, food courts, gyms, pools, playgrounds, and markets remained closed, and weddings and funerals still faced restrictions of a maximum limit of 10 people. While "controlled environments" such as some schools and workplaces could reopen, the government required them to maintain a one-metre physical distance between all who attended, and, in any other venue outside the home, the two-metre distance rule applied. This meant that retailers, except those deemed to be "essential" (which now included hardware suppliers to the trades), cafés, bars, and restaurants remained off limits, and all allowable transactions were legally mandated to be "contactless." Level 3 imposed strict limitations on regional travel and encouraged mask wearing where physical distancing was difficult. Although local travel was permitted in some instances, anyone using public transport had to, by law, wear a face covering, preferably a mask.

Still, while the emotional fallout from the lockdown was marginally alleviated, the gathering size limits continued to be a source of disquiet, particularly for people who lost someone close. Prime Minister Ardern acknowledged the pain and anguish, saying, "The thing I have found, as a human, hardest in all of this has been funerals and *tangihanga*." Her government faced strong criticism from the opposition, with National Party leader Bridges, who is Māori, calling the limit of 10 mourners a cause for "added grief and pain for families wanting to say goodbye to a loved one" (Ainge-Roy, 2020b). He contended that it was "not fair that you can have 30 people on a rugby field playing close contact sport, but you can't have more than 10 people at a funeral so they can grieve together" (the mention about contact sport was inaccurate, according to Sport New Zealand's guidance based on information from the Ministry of Health (Sport New Zealand, 2020)). The Funeral Directors Association of New Zealand added weight, calling the government decision "a cruel and heartless blow to the thousands of New Zealand families who have lost loved ones" (Ainge-Roy, 2020b).

From an economic standpoint, the government and its advisors came under increasing pressure to reduce alert levels once again, based on the impacts of level 3's strictures, according to an RNZ report (Wilson, 2020). Opposition MPs were quick to take advantage of Parliament's "Question Time" to advance their points in the debate, primarily on economic grounds. MPs asked the prime minister on 30 April 2020, if she stood by her government's actions and whether she accepted that every day, alert level 3 caused business failure and job losses. Ardern reiterated her government's support structures, including the establishment of loans, and tax and wage subsidies, saying, "We have looked at, and responded quickly to, every element of pain that COVID-19 has caused New Zealand." She added that she accepted "that any day that we move prematurely into a different alert level poses a risk to New Zealand's economy and a risk to New Zealanders' health" (New Zealand Parliament, Hansard (Debates), 2020). National Party leader Bridges pressed her, citing comments by her Australian counterpart Scott Morrison, who saw "New Zealand's approach as involving 'much more extreme economic measures'" as a reason not to follow suit in their country. Ardern responded unequivocally: "New Zealand has always responded to what is happening here with our population and our cases, and that hasn't been comparable at those times with Australia. And … Prime Minister Morrison often has had to give those responses because people have compared the results in New Zealand and Australia, often favourably around the approach that New Zealand has taken" (New Zealand Parliament, Hansard (Debates), 2020).

Doubts over the government's approach to the economic repercussions of the measures to protect public health were not confined to its Parliamentary opposition. The Labour Party's coalition partner, New Zealand First's leader (and deputy prime minister), Winston Peters, questioned Ardern about the credibility she would give "to the view that you can smash the [infection] curve whilst leaving businesses unaffected" (New Zealand Parliament, Hansard (Debates), 2020). Ardern replied, "We can see around the world there are economic consequences to an effective response to COVID-19, but there are also economic consequences if you don't get it right. You stay in lockdown for longer; the impact on business and jobs is much harsher. Our approach has always been to go hard, go early, get it right the first time." She referred to affirmative remarks of former Chief Science Advisor Sir Peter Gluckman (from 2009 to 2018). "He shared the exact same cautious approach that we have been taking. No one wants to risk the gains that we have made. No one wants a second wave of infection" (New Zealand Parliament, Hansard (Debates), 2020).

Following the 11 May Cabinet meeting, Ardern announced that New Zealand would move to alert level 2 status at 11:59 p.m. two days later. On 13 May, with no new COVID-19 cases recorded for two consecutive days, and the total number of recorded cases standing at 1,497, including 21 deaths, the government again reduced restrictions. The Ministry of Health quickly recognized "the sense of anticipation [as] both palpable and understandable. We're all looking forward to re-establishing some of the routines and rhythms of our 'normal' lives," according to its media release, which quoted Bloomfield:

> The experiences we are seeing in other countries, which have eased restrictions, increasingly emphasise[s] the need for us to be careful and not give away the

substantial gains we've made. Today's figures reinforce that we're on the right path, maintaining our gains and heading for success, but international experience shows that this is a stubborn virus. We don't want a situation where we start seeing spikes. We need to remain vigilant. (Ministry of Health, 2020b)

Level 2 restrictions, while not as stringent as those imposed in the previous two months, encouraged citizens to maintain high standards of personal hygiene and continue to vigilantly track their movements and out-of-home visits. The government guidelines about physical distancing continued to apply, and mask wearing on all public transport remained mandatory. While gatherings of up to 100 people were permitted, their organizers were legally obliged to record attendees' details to ensure that tracing was possible if necessary. Public swimming pools, gyms, libraries, and museums could reopen, but a night out at a bar or restaurant still required sitting at least one metre away from the nearest table with only one server to take and deliver each table's orders; no one else could approach the bar.

## Alert Level 1

After three and a half weeks of limitations imposed by alert level 2, Prime Minister Ardern announced on 8 June that New Zealand would shift to alert level 1 at 11:59 p.m. that night. All level 2 restrictions would be lifted, but the country's borders would remain closed, with mandatory isolation and quarantining of homecoming New Zealanders as the first line of defence against the spread of COVID-19. At the time of Ardern's announcement, there were no active COVID cases in New Zealand, and it had been 40 days since the last recorded case of community transmission. "With care and commitment our team of 5 million has united to protect New Zealanders' health and ensure we now have a head-start on our economic recovery," Ardern noted. "At Level 1 we become one of the most open economies in the world, and now we must seize our advantage of going hard and early to beat COVID-19 and use the same focus and determination we applied to our health response to rebuild our economy." She added that her government was "confident we have eliminated transmission of the virus in New Zealand for now, but elimination is not a point in time – it is a sustained effort." The Reserve Bank (New Zealand's central bank) analysis had informed the government that the economy under level 1 was expected to operate just 3.8% below normal levels, an improvement from 8.8% below normal under level 2, 19% below under level 3, and 37% below under level 4 (New Zealand Government, 2020b). The relaxation of restrictions brought a level of optimism in many quarters, including in the capital city, Wellington, with reports that "the de-escalation in alert levels could provide the adrenaline shot the struggling events sector desperately needs." Organizers could now "plan for concerts, sporting fixtures and arts performances with greater certainty" – a potential boost to the city's economy (Williams, 2020a).

With community transmission eliminated, the most likely source of any COVID-19 infection in the country could now only come from beyond New Zealand's borders, ostensibly with returning citizens and permanent residents who had been living or vacationing abroad. (Between 600,000 and one million New Zealanders live overseas, depending on the definition used (Statistics NZ, 2020)). In July, RNZ reported that job uncertainty, lack of access to welfare, and concerns over COVID were driving a flood of New Zealanders to return home (Lock, 2020). By September, more than 50,000 had returned, according to a TVNZ report (2020d). Having its borders closed to almost all travellers meant that strict rules governed entry into New Zealand, permitting only citizens and permanent residents, their partners and/or dependents, diplomats posted to the country, Australian citizens or permanent residents who normally live in New Zealand, and a few others such as residents with valid travel conditions. On a case-by-case basis, exceptions were made with a special request for those who did not meet those criteria.

After the relaxation of in-country restrictions that came with the move to level 1 status, a steady stream of eligible arrivals who were infected with COVID abroad came to the country. Strict enforcement of mandatory managed isolation and quarantine, for the most part, contained the impact of imported cases, with few short-lived isolation breaches, such as two people who sneaked out of their hotel rooms (Clent, 2020; TVNZ, 2020a, 2020e). The government's opposition, the National Party, seized on these breaches, arguing that they would not have happened under a National administration (*NZ Herald*, 2020c). The prime minister responded to the criticism of her government's handling of the situation, which included the deployment of five hundred additional defence force personnel to manage isolation and quarantine facilities, and the appointment of a new team to improve the Ministry of Health's border testing capabilities (TVNZ, 2020c). But National Party leader Judith Collins, who rose to power during this time (discussed in the next section), continued her attacks until the week before New Zealand's election day, asserting that the escapes made "a joke" of the system (Ensor, 2020).

**The Politics and the National Election**

The sociopolitical events mentioned thus far occurred during New Zealand's election year. Under the terms of the Constitution Act 1986, the date for an election is determined by the prime minister, within the requirements of the act, who in turn advises Her Majesty the Queen's representative, the governor general. The Constitution Act requires a three-year term of Parliament, "unless Parliament is sooner dissolved … for the return of the writs issued for the last preceding general election of members of the House of Representatives, and no longer" (Constitution Act 1986). Because the New Zealand Electoral Commission had returned the writ for the 2017 General Election on 12 October 2017 (New Zealand Parliament, 2017), the dissolution of Parliament was mandated to occur no later than 12 October 2020. Under these provisions, on 28 January 2020, before the onset of the pandemic crisis in New Zealand, the election date had been announced and set to take place on Saturday, 19 September (New Zealand Government, 2020).

With the introduction just weeks later of measures to control the spread of COVID-19, some political pundits such as *Stuff*'s Thomas Coughlan raised questions about whether politicians in New Zealand might postpone the election until there was less chance of spreading infection (Coughlan, 2020a). The Electoral Commission had been strategizing for a range of scenarios, including holding the election with the country being at alert level 3 or 4, and had advised the prime minister that it required a four-week lead-up to an election date to reserve venues for polling stations and to provide information to voters, according to RNZ reports (2020d).

Amidst the COVID-19-associated upheaval and the general public approval of the government's management of it, political polls continued to affirm the Labour Party's and in particular Jacinda Ardern's popularity with the electorate and the fall from favour of the National Party's then leader Simon Bridges (O'Brien, 2020; Trevett, 2020). This led to a leadership "coup" in the National Party that saw a relatively unknown MP, Todd Muller, take over as opposition leader on 22 May 2020, following a caucus vote (Trevett & Walls, 2020; Wade, 2020). Muller's service lasted a scant 53 days before he resigned, citing personal, family, and health concerns, according to RNZ (2020e). Longstanding National MP Judith Collins, first elected to Parliament in 2002, a politician with a controversial past in the party, then rose to become leader (RNZ, 2020f). By mid-August 2020 with campaigning set to begin, a new community outbreak of COVID-19 in Auckland prompted the government to return the city to alert level 3. In the Legislative Council Chamber at Parliament, Collins called for the election to be postponed. "We are calling on the prime minister … to shift out the election date to a date later in November," and if that was not possible to "defer the election to next year … probably the better alternative," she said (RNZ, 2020g).

While the Green Party co-leader James Shaw accused the National Party of "calling for the election to be delayed out of concern for their own self-interest, not democracy" and called her concerns "purely political" (RNZ, 2020h), Collins's sentiments were echoed by Deputy Prime Minister Peters during what independent New Zealand–based news and current affairs website Newsroom described as "a hastily-called press conference where he said parties would only have six days to campaign before advance voting began." Peters added, "There is now no ability to conduct a free and fair election if the prime minister decides to hold the General Election on September 19" (Newsroom Staff, 2020). Ardern relented, announcing, "the election would be postponed by four weeks" (to Saturday, 17 October). The decision came after she "had sought advice from the Electoral Commission and every political party represented in Parliament, plus Business New Zealand and the Council of Trade Unions," she said (Newsroom Staff, 2020).

## Auckland's Second COVID "Wave"

On 11 August, Prime Minister Ardern announced "four new cases of COVID-19 in one family from an unknown source" in a Ministry of Health media release (2020c). A middle-aged person in South Auckland (now referred to as the "index case"), with no overseas travel history, had tested positive after being symptomatic for a few days. Testing revealed that three other family members had also contracted the virus. Ministry officials instituted their usual protocols of enforcing self-isolation and contact tracing and testing, with the release reiterating earlier advice that this situation could arise at any time "and that time is now. The health system is well prepared for this eventuality – and the important thing now is that we don't let the virus spread in our community … we need to stamp it out." Health Director-General Bloomfield added, "This case is a wake-up call against any complacency that may have set in. We have seen how quickly it can lead to a wider resurgence in communities overseas. We are working to not let that happen here. We've done this before and we can do it again" (Ministry of Health, 2020c).

Complacency may have set in among the population because New Zealand, at this stage, had passed 102 days without any community transmission of the virus. "Many questions" remained about the source of the family's infections, said Prime Minister Ardern, adding, "Those who had tested positive did not work at the country's borders and had not traveled from overseas," according to a report in the *Guardian*. There was no known connection with managed isolation facilities where returning New Zealanders had to spend two weeks in quarantine (Graham-McLay, 2020b).

As a result of the re-emergence of the virus, Ardern returned Greater Auckland to alert level 3 from 12 August 2020, effectively imposing a partial lockdown and closing off the region to the rest of the country for at least three days with police roadblocks established around the city's borders. For the rest of New Zealand, the alert level rose to 2 (*NZ Herald*, 2020d). On 14 August, RNZ (2020i) reported that Cabinet had unanimously agreed to retain this alert level status until just before midnight, 26 August. Community cases increased within the Auckland "cluster," and on 24 August, Ardern announced that while there were promising signs that the government was managing the outbreak, Cabinet agreed to a short extension, on the director-general of health's recommendation. On 30 August, Auckland joined the rest of the country, but at a modified version of level 2, which reduced some restrictions while maintaining strict gathering-size limits. Ardern referred to this modified alert level as 2.5 (McCullough, 2020) and said that the rest of the country needed to stay at an elevated alert level 2 to mitigate the risk of cases leaving Auckland when interregional travel resumed (Coughlan, 2020b).

On 14 September Ardern announced, "On the advice of the director-general, Cabinet has decided on a short extension to the current restrictions of alert level 2.5 for Auckland and level 2 for the rest of the country." She added that while physical distancing on planes and other public transport would be immediately eased, mask use would remain compulsory. Based on further advice, she would amend restrictions on Auckland after a meeting on 21 September to "full alert level two" to begin two days later, and to remain for 14 days. The rest of New Zealand, however, would return to alert level 1 with immediate effect just before midnight on the 21st (RNZ, 2020j).

Before the end of Auckland's enforced alert level 3 partial lockdown, a new and potentially alarming trend emerged that threw the crucial importance of social distancing into stark relief. On 25 August, the Ministry of Health's daily media release linked four of seven newly reported community cases recorded that day to church gatherings in the vicinity of the outbreak of the August Auckland cluster (Ministry of Health, 2020d). Positive case numbers increased in the following days, and one suburban Auckland evangelical church was linked to higher infection rates than were found through other likely contact-infection channels. By 9 September the ministry had advised of numerous tests carried out among members of the Mt Roskill Evangelical Fellowship, and said, "As we are still seeing cases emerge in this group, we are asking all members of the congregation to be retested. We also ask anyone who may have had contact with members of the Fellowship to be tested, even if they have previously tested negative. People should be tested even if they have no symptoms." The four new cases reported that day were linked to a sub-group of the congregation, associated with bereavement activities, including visits to the funeral and the household of the bereaved (Ministry of Health, 2020f).

An *NZ Herald* report the following day suggested that the church gatherings believed to be responsible for the escalation in COVID-19 infections in the community "may have broken level-3 rules, which require people to stay within household bubbles." It quoted the level 3 rules from government, which stipulated, "Do not invite or allow social visitors, such as friends, extended family or *whanau* [Māori for extended family, a word that has become common usage in New Zealand English], to enter your home." The newspaper reported that "Mt. Roskill Evangelical Fellowship Church members were reluctant to get tested because they doubted the 'science' of the virus," thus "prompting a plea from Auckland Mayor Phil Goff for everyone to play their part" (Cheng, 2020b). Yet, the church plays a vital role in Pacific Island communities. According to *Te Ara*, the government's online Encyclopedia of New Zealand, such communities in the 2018 census accounted for 15.5% of Auckland's population (Macpherson, 2020; see also Auckland Council, 2020). It noted:

> In the Pacific Islands, community life was built around the family, the church and the village. In New Zealand, churches acted like villages. The minister was the most powerful and respected person – a role similar to the village chief. Churches became the centre of social life for many Pacific families. [They] provide health and education services, and sport, music, and social activities. (Macpherson, 2020)

On 4 September, New Zealand recorded its first COVID-19-related fatality since 28 May, per a Ministry of Health release (Ministry of Health, 2020e). The man, in his 50s, was a close contact of the Auckland cluster's "index case," and both worked at a cool store facility in the area. At that stage, 152 people had been linked to the August outbreak by genomic testing (*NZ Herald*, 2020f). That day also saw the passing of another prominent Auckland figure, 82-year-old Dr. Joseph Williams, a former Cook Islands prime minister, physician, and author, who had been admitted to a hospital just days after the city moved to alert level 3, and who had been treating patients at his suburban practice near the cool store in previous days (Ward, 2020). COVID-19 claimed its 25th victim in New Zealand on 15 September, a Waikato-based brother of the cool store worker who had passed away less than two weeks before (RNZ, 2020k). At the time of writing in November 2020, New Zealand's death toll from the virus remained at 25.

**Disbelievers and Demonstrations against the Measures**

During this time, parts of the country also witnessed a rise of disbelievers in the virus or its severity. July and August 2020, for example, featured criticisms and protests from rightist media personalities and politicians. Framing the rules as contradictory and masks as limitations on freedom, and highlighting government spending on consultancy, these public figures galvanized negative reactions and emotions as the election neared (Hosking, 2020). "We will not be silenced" was one rallying cry at an 22 August anti-lockdown protest in Auckland that attracted up to 150 people, according to press accounts (*NZ Herald*, 2020e).

Following a protest march in August, news outlet *Stuff* reported police "disappointment" at the hundreds of people who gathered, the majority of whom were not wearing masks or socially distancing. But Acting Police Inspector Chris Scott said no punishments were handed out, as police were taking an educational approach. "Police recognise people's lawful right to protest, however we also recognise the need for people to adhere to the current level three restrictions to do their part to help prevent the spread of COVID-19," he said (C. Williams, 2020). By September, *Stuff* reported more than a thousand protesters gathered for an anti-lockdown rally. Among demonstrations opposing lockdown, signs also protested 5G, vaccines, the government, and free trade agreement TPPA (Earley, 2020).

**Technological Aids for Contact and Virus Tracing**

Since the beginning of the pandemic, New Zealand's government and health authorities emphasized the importance of contact tracing as "vital at the forefront of the fight against COVID-19, with hundreds of staff mobilised to carry out the work" (Ministry of Health, 2020a). The process enabled the Ministry of Health and district health boards to "track down people who may have been exposed to COVID-19" (Ministry of Health, 2020h), test them, and provide advice about self or managed isolation, or quarantine. Tracing methods included telephone calls to those identified as having been near persons confirmed as infected, and under alert levels 1, 2, and to a lesser degree, 3, a sign-in sheet or diary at venues of permitted gatherings.

Genomic sequencing in COVID-19 tests (like other types of biological testing regimes) helped identify characteristics of the virus's DNA. Dr Joep de Ligt, the bioinformatics lead with New Zealand's Institute of Environmental Science and Research (ESR), described the process as looking in the laboratory "for a small piece of the viral genome to determine if someone carries the virus or not … [and] that is when it becomes a positive case." Positive cases are referred on from the testing laboratory to ESR to determine the complete genome structure of the virus. This helped determine the link of each viral infection to others when close contact tracing or interviews during the investigation period were inconclusive, often because of the time that had passed since notification, the asymptomatic nature of some infected persons, or the virus's rapid spread, which made connections more difficult to identify. De Ligt further noted: "We can then use the genome of the virus to identify a cluster that it is associated to. We may not always be able to ascertain that, but we can say it's from a certain area or a certain event that we can link it to and therefore know how the transmission events might have happened" (Science Media Centre, 2020).

The government also released a smartphone application, "NZ COVID Tracer," which provided users the capacity to scan QR codes displayed at places of business throughout the country to record their attendance (Kenny, 2020). In August, a public health order came into effect that provided the legal basis for the mandatory display of QR codes at all businesses and service providers, including public transportation (TVNZ, 2020b). At the time of writing, the Ministry of Health reported that the app had 2,385,800 registered users, roughly

Lessons from New Zealand 345

Image 13.2. At the entrance of an appliance store (Bernard Duncan, 10 August 2020).

half of the country's population. There had been 128,125,675 QR poster scans and 5,202,052 manual diary entries (Ministry of Health, 2020k).

Not surprisingly, some disquiet regarding the collection and storage of citizens' private information emerged. The COVID Public Health Response Act 2020 legislation that gave, "among other things, the health minister and director-general of health the power to make orders in respect of contact tracing" and required businesses to keep records of customers for that purpose was passed under urgency. Therefore, it was not sent to a Parliamentary Select Committee until after its passage, prompting the privacy commissioner, John Edwards, to submit "that the order's effect was an intrusive collection of personal information, leading to ad hoc and chaotic collection by businesses" (Kennedy, 2020).

On 23 November, the Ministry of Health announced the launch of an automatic update to the COVID-19 tracer app that would make its use easier, and also offered greater privacy protection. Its media release said, "You choose exactly what personal information and contact details you want to register – all information is optional" (Ministry of Health 2020i).

## Back to the Future

On 5 October, after 13 days with no recorded community transmission, the government decided to return Auckland again to alert level 1 status, effective at 11:59 p.m. on Wednesday, 7 October 2020. The Ministry of Health website announced, "From Thursday, people in Auckland can gather with as many people as they want, and there are no extra restrictions on businesses. Businesses, workplaces, and public transport legally must continue to display QR codes" (Ministry of Health, 2020g).

Following the landslide victory for Ardern's Labour Party in the October general election, perhaps a testament in part to its handling of the pandemic, Nick Wilson, a University of Otago epidemiologist, called for "political posturing" to be put aside to find "the best solution from a completely scientific risk management perspective. We need to look at the cost-effectiveness and the best use of people's time and effort." Michael Baker also called for an official inquiry to examine New Zealand's COVID-19 response to date (Graham-McLay, 2020c). Prominent local scientist Sir Ray Avery added his voice, saying, "You have no idea of the pandemic that could go through this country if we don't progress the real issues underpinning the transmission of the disease. New Zealand has no validated long-term COVID prevention strategies in place and no overarching agency to manage immediate and ongoing pandemic risks." Health Minister Chris Hipkins said a full review would happen but not while the country is "in the middle of dealing with" the pandemic (Williams & Mitchell, 2020).

## International Praise and Criticism

In contrast to most other countries' responses, New Zealand's government had decided on eliminating virus transmission within its borders. In their analysis, *New Zealand's COVID-19 Elimination Strategy*, a team of Otago University academics led by Baker contended, "Compared with the mitigation and suppression approaches of most Western countries, elimination can minimise direct health effects and offer an early return to social and economic activity." The report, published in August (Baker, Kvalsvig & Verrall, 2020), noted that while most "Western countries across Europe and North America were following the mitigation approach … [including] steps designed to slow entry of the pandemic, prevent initial spread and apply physical distancing measures progressively to flatten the curve and avoid overwhelming health services," it was not performing well. Some had then switched to a suppression strategy of physical distancing and travel restrictions, and a few, most notably Sweden, were pursuing a mitigation plan called "herd immunity" (Baker, Kvalsvig, & Verrall, 2020), defined by the WHO as "a concept used for vaccination, in which a population can be protected from a certain virus if a threshold of vaccination is reached" (WHO, 2020).

Some early supporters of Sweden's COVID strategy said they came to regret the approach (Vogel, 2020). The New Zealand approach appeared justified when the *NZ Herald* reported a "devastating second wave" of COVID-19 that was forcing Sweden to "a dramatic rethink" of its strategy (Smith, 2020). The *Dominion Post* reported Sweden's chief epidemiologist, Anders Tegnell, as admitting to the German newspaper *Die Zeit* that it would be futile and immoral for a state to deliberately pursue herd immunity: "There has up to now been no infectious disease whose transmission was fully halted by herd immunity without a vaccine," he said (*Dominion Post*, 2020; Patel, 2020).

Although New Zealand's approach to eliminating COVID-19 received praise from some international leaders, health experts, and media, the country also faced criticism. The prestigious scholarly journal the *Lancet* (referenced earlier) concluded that New Zealand's implementation of "graduated, risk-informed national COVID-19 suppression measures" had yielded positive results and "prevented the burden of disease experienced in other high-income countries with slower lockdown implementation, including Australia, the UK, and Italy," but said there remained "questions about the costs and sustainability of these measures." It noted, for example, that "Taiwan has shown that COVID-19 control can be achieved in the absence of complete border closure, although using advanced technological systems and ongoing strict disease suppression strategies in a society that had already normalised some of these measures from previous novel virus exposures." However, mitigating the fact that New Zealand was less prepared than Taiwan, it added, "Integral to New Zealand's response has been decisive governance, effective communication, and high population compliance – in an earthquake-prone country, [where] communities and emergency management systems are primed for disaster response" (Jeffries et al., 2020).

At the end of April, a *Time* article (Gunia, 2020) argued that New Zealand's "ambitious" elimination approach would be unlikely to work elsewhere. While reporting that the strategy seemed to be working and had "drawn admiration from all over the world," it argued that the country's combined size, wealth, geographical isolation, and low population density suggested that other countries without those features could not sufficiently emulate the strategy, and experts cautioned that the costs would be "higher than most other nations are willing to bear." The article quoted Thomas J. Bollyky, the director of the global health program at the Council on Foreign Relations, who said, "New Zealand saw its first cases on Feb. 28, at a time when the U.S. already had community spread and likely thousands of unreported cases" (Gunia, 2020). J. Stephen Morrison, the director of Global Health Policy Center at a Washington DC-based think tank, the Center for Strategic and International Studies, added that outside of the country's handful of big cities, "you have a population that is already pretty socially distant." *Time* also noted the Colmar Brunton poll findings, referred to earlier in this chapter, that "the citizens of New Zealand place a high level of trust in their government" (Gunia, 2020).

Branko Marcetic, a staff writer for *Jacobin* magazine, was strident in his criticism of Prime Minister Ardern's treatment of the crisis and her government's response to it, saying in the Democracy Project that she should "listen to her own mantra: 'Be kind'" (Marcetic, 2020). Contending that while Ardern's "decisive and effective health response to the virus has garnered both New Zealand and herself adoring headlines around the world, particularly in countries starving for basic governmental competence … it's been less good for those already on the margins, for whom weeks of lockdown has meant even greater levels of economic uncertainty and deprivation." He noted that around a thousand people a day had applied for the unemployment benefit, and that demands on food banks had risen by hundreds of percentage points – many of them from people who had never needed help before. He said that, although "some of this will be solved by moving from Level 4 to 3 of lockdown … the fact is that New Zealand's economic response to the crisis has fallen far short of the gold standard set by its health response. And with a global depression looming – and the ever-present threat of another lockdown should the virus come back – this bodes ill for the future" (Marcetic, 2020).

Citing *Captaining a Team of 5 Million: New Zealand Beats Back COVID-19, March–June 2020*, a Princeton Innovations for Successful Societies study, news outlet *Stuff* hailed "social capital" as one of the reasons behind New Zealand's successful response to the COVID-19 pandemic. Peter Crabtree, who led strategy and policy for New Zealand's government team, told researchers, "Social capital is deep. Our response was successful because social capital enabled us to adapt and work outside our systems" (Cameron, 2020).

But while much of the research commended New Zealand's response, it was not without criticism, reported *Stuff*, which outlined a lack of coordination when it came to resourcing laboratories with COVID-19 testing supplies, and gaps in the PPE supply chain and the difficulties in starting up a cloud-based platform for contact tracing. Sir David Skegg described the country's health system as being "neglected for decades" and suffering from chronic underfunding. "Public health experts said New Zealand could have been better prepared – for example, by having a better-funded health sector with more intensive care beds and better systems in place to prepare for disease outbreaks," according to the study (K. Williams, 2020b). In addition to the careful analysis of international studies, New Zealand featured in conspiratorial disinformation campaigns that discredited its handling of the coronavirus pandemic.

**Quarantine Camps and Other Disinformation**

While most international coverage of New Zealand's successful eradication of COVID-19 featured praise, some media personalities accused the country of building "quarantine camps." On 13 July, broadcast news outlet Newshub reported that controversial US author Alex Berenson claimed New Zealand was placing people in "indefinite confinement" under what he described as an "occupation" (Palmer, 2020). Newshub said the author seemed to be confused by a television interview it had conducted with public health professor Baker, in which he "warned the New Zealand system of sending returnees to managed isolation facilities for 14-day stints could continue for months – and possibly years – into the future as coronavirus continues to spread around the world." Berenson took to Twitter, writing, "Don't think developed nations with a tradition of respecting individual rights will trash them for #Covid? I have two words for you: New Zealand. Where indefinite confinement is already real, and PEOPLE IN ISOLATION ARE DESPERATELY TRYING TO BREAK FREE. Welcome to the occupation" (Palmer, 2020).

Berenson's Twitter followers, chiefly in the United States, expressed horror with comments such as "Wtf that is horrifying" and "Isolation facility sounds a lot like PRISON." New Zealanders immediately disputed his apparently confused assessment as stupid or "extremely disingenuous." One replied, "This is excellent comedy," and another, more pointedly, "If you have any intellectual honesty whatsoever, you'll delete these tweets and correct the record. People spend a maximum two weeks in a facility. Not months. Not years. That's just how long the measures will be in place, not how long people have to spend." Newshub later updated its article "to make it clearer to those reading from overseas that Kiwis will not be put in indefinite detention" (Palmer, 2020).

A few months later, Suzanne Evans, the former deputy chair of the UK Independence Party, likened New Zealand to Nazi Germany with a claim that New Zealand's managed isolation and quarantine COVID-19 protocol was "absolutely horrific," according to reports on TVNZ's One News (TVNZ, 2020f). On 27 October, she tweeted, "New Zealand now has a fascist government under @jacindaardern. Are you going to act, @amnesty?"

Across the Atlantic, Fox News presenter Laura Ingraham jumped on the bandwagon, claiming that "terrifying new" COVID-19 quarantine measures in New Zealand would end personal freedom, said One News (TVNZ, 2020g). While hosting Victor Davis Hanson of the conservative Hoover Institution think tank, Ingraham and her guest called the New Zealand government's quarantine and managed isolation rules "quarantine camps." TVNZ presenter Hilary Barry used her Twitter account to respond: "Dear Laura, You're an egg. Love from New Zealand" (TVNZ, 2020g) – in mild New Zealand slang, "an egg" is an idiot or jerk (Ohlheiser, 2016).

Media responses in New Zealand were sardonic. Alex Braae, a staff writer for the web-based magazine the *Spinoff*, reflected on these international critiques on 29 October 2020, writing, "Much as we might like to think of ourselves as the centre of the universe, these interventions aren't necessarily about New Zealand. In the UK, an intense battle is currently underway about how fiercely the country should respond to an alarming new wave of cases" (Braae, 2020).

A *Stuff* editorial published the following day was of a similar mind: "They're about a straw man projection of New Zealand as a nightmare vision of what the US might look like under Joe Biden and Kamala Harris. Quarantines and lockdowns are presented as a dystopia of oppression and tyranny to viewers wearing red hats in red states" (*Stuff*, 2020).

## Conclusion

If the year 2020 was expected to be relatively normal, the fast-moving COVID-19 virus changed perceptions and realities. Despite early miscalculations regarding the contagiousness of COVID-19, once New Zealand's political leaders realized the severity of the situation, they took decisive and relatively swift action, using the country's combined features – geography, population density, culture, relative wealth, and technology – to its advantage. The result was a quick stamping out of the virus's spread and its elimination from within the country's borders with, at this time of writing, two brief community flareups. Comparatively, that meant a considerably lower death rate than most other Western countries and widely available medical services for non-COVID-related health issues during most of 2020. At the time of writing, 25 New Zealanders had died from the virus. The country had 2,050 confirmed and probable COVID-19 cases (of which 1,956 had recovered), and had no one receiving hospital care.

While the overall commendable result is attributable to all of New Zealand's pandemic mitigation efforts, it was the leadership of Prime Minister Jacinda Ardern – her attentiveness to health developments and her compassion – that were key to the country's success. In March 2020, while other governments were struggling to contain the virus, Ardern instituted some of the earliest and toughest isolation measures in the world, including a month-long full lockdown. "We're going hard and we're going early," she told the public; "We only have 102 cases, but so did Italy once" (Redd, 2020). What followed was a lesson in solidarity. Ardern repeatedly referred to the nation as "our team of five million," urging New Zealanders to unite against COVID-19. Her government listened to and followed the advice of scientists and other health experts, making them public figures and building trust in the community. Persuasive and resolute, Ardern ended almost all her public appearances with the same message: "Be Strong. Be Kind."

While the "team of five million" mostly cooperated, the task was not simple. Naysayers, deniers, detractors, and a series of social, psychological, and economic setbacks complicated the social distancing efforts detailed in this chapter. COVID-19 meant a delayed national election and postponed plans and dreams for many in New Zealand, including those competing in the Tokyo Olympic Games, or those marrying, celebrating in annual festivals, and mourning in the nation's most important memorial event. While traversing up and down preventive alert levels with varying sets of personal and public restrictions that caused significant anxiety and economic hardship, the country's population weathered and continues to weather a health and well-being storm that rages almost unabated in other parts of the world. It is setting about a rebuilding phase as the search for a vaccine makes positive strides. Ardern, whose Labour Party government was resoundingly returned to power in the October 2020 general election, gave credit for the country's success to medical staff and the way the public supported the rules, telling them: "New Zealanders have proven themselves, and they've done so in the most incredible way."

REFERENCES

Ainge-Roy, E. (2020a, March 30). New Zealand site to report COVID-19 rule-breakers crashes amid spike in lockdown anger. *Guardian*. https://www.theguardian.com/world/2020/mar/30/new-zealand-site-to-report-covid-19-rule-breakers-crashes-amid-spike-in-lockdown-anger.

Ainge-Roy, E. (2020b, May 13). New Zealand's ban on large funerals during COVID-19 criticised as "inhumane." *Guardian*. https://www.theguardian.com/world/2020/may/13/new-zealands-ban-on-large-funerals-during-covid-19-criticised-as-inhumane.

Akoorie, N. (2020, April 10). COVID 19 coronavirus: No compassion for sick, dying elderly during lockdown, say family. *NZ Herald*. https://www.nzherald.co.nz/nz/covid-19-coronavirus-no-compassion-for-sick-dying-elderly-during-lockdown-say-family/U6Y6MLTQGPUT5EEXMPTBBBPVT4/.

Auckland Council. (2020, February 10). Pacific peoples in Auckland. https://knowledgeauckland.org.nz/media/1447/pacific-2018-census-info-sheet.pdf.

Auckland Reporters. (2020, November 10). Dr Ashley Bloomfield pulls out of running for TV personality of the year award. *Stuff*. https://www.stuff.co.nz/entertainment/tv-radio/300154165/dr-ashley-bloomfield-pulls-out-of-running-for-tv-personality-of-the-year-award.

Baker, M.G., Kvalsvig, A., & Verrall, A.J. (2020, August 13). New Zealand's COVID-19 elimination strategy. National Center for Biotechnology Information. http://www.ncbi.nlm.nih.gov/pmc/articles/PMC7436486/.

Beck, D. (2020, March 27). COVID-19 coronavirus: Teddy bear hunts shine light during lockdown. *NZ Herald*. https://www.nzherald.co.nz/rotorua-daily-post/news/covid-19-coronavirus-teddy-bear-hunts-shine-light-during-lockdown/GCYYUJT4WTKVJTKDR4FMJ6KM3M/.

Braae, A. (2020, October 29). The Bulletin: Idiots abroad infuriated by NZ's COVID response. *Spinoff*. https://thespinoff.co.nz/the-bulletin/29-10-2020/the-bulletin-idiots-abroad-infuriated-by-nzs-covid-response/.

Bridges, S. New Zealand Parliament Hansard transcripts: Epidemic Response Committee. (2020, April 21). https://www.parliament.nz/en/visit-and-learn/history

-and-buildings/special-topics/epidemic-response-committee-covid-19-2020/watch-public-meetings-of-the-epidemic-response-committee/.

Cameron, B. (2020, September). CAPTAINING A TEAM OF 5 MILLION: NEW ZEALAND BEATS BACK COVID-19, MARCH–JUNE 2020. https://successfulsocieties.princeton.edu/publications/captaining-team-5-million-new-zealand-beats-back-covid-19-march-%E2%80%93-june-2020.

Cheng, D. (2020a, February 18). Govt puts up $11m to revitalise tourism. *NZ Herald NZC C Edition – A003*. https://www.pressreader.com/new-zealand/the-new-zealand-erald/20200218/281560882808973.

Cheng, D. (2020b, September 10). COVID-19 coronavirus: Mt Roskill church cluster could push out COVID's long tail by weeks. *NZ Herald*. https://www.nzherald.co.nz/nz/politics/covid-19-coronavirus-mt-roskill-church-cluster-could-push-out-covids-long-tail-by-weeks/UVVDGHJR7KF25RURKAER5B4GSY/.

Chiang, J. (2020, September 30). Fruit picker shortage reaches new levels. RNZ: The Detail. https://www.rnz.co.nz/programmes/the-detail/story/2018765916/fruit-picker-shortage-reaches-new-levels.

Clark, D. (2020, February 11). Order paper & questions. Written questions. New Zealand Parliament. https://www.parliament.nz/en/pb/order-paper-questions/written-questions/document/WQ_00889_2020/889-2020-hon-michael-woodhouse-to-the-minister-of-health.

Clent, D. (2020, September 28). Man escapes Auckland isolation facility by "creating rope out of bedsheets." Stuff. https://www.stuff.co.nz/national/health/coronavirus/300118803/man-escapes-auckland-isolation-facility-by-creating-rope-out-of-bedsheets.

Colmar Brunton. (2020a, April 8). COVID TIMES Trust and Leadership. (2). https://static.colmarbrunton.co.nz/wp-content/uploads/2019/05/COVID-Times-8-April-2020.pdf.

Colmar Brunton. (2020b, April 24). COVID TIMES Backing New Zealand. (2). https://static.colmarbrunton.co.nz/wp-content/uploads/2019/05/COVID-Times-24-April-2020-1.pdf.

Constitution Act 1986. (1986, December 13). New Zealand Parliamentary Counsel Office/TeTari Tohutohu Pāremata. https://www.legislation.govt.nz/act/public/1986/0114/latest/whole.html#DLM94241.

Coughlan, T. (2020a, March 16). Coronavirus might postpone our election – here's how it could be done. Stuff. https://www.stuff.co.nz/national/health/coronavirus/120307598/coronavirus-might-postpone-our-election--heres-how-it-could-be-done.

Coughlan, T. (2020b, August 24). Auckland lockdown extended until Sunday, rest of country to stay at level 2. Stuff. https://www.stuff.co.nz/national/health/coronavirus/300089986/auckland-lockdown-extended-until-sunday-rest-of-country-to-stay-at-level-2.

*Dairy News*. (2020, February 5). Global dairy prices have taken a hit in the overnight auction. https://www.ruralnewsgroup.co.nz/dairy-news/dairy-general-news/coronavirus-infects-dairy-prices.

Devlin, C. (2020, January 23). Government warned NZ needs to take a precautionary approach to coronavirus. *Stuff*. https://www.stuff.co.nz/national/politics/118989349/government-warned-nz-needs-to-take-a-precautionary-approach-to-coronavirus.

*Dominion Post*. (2020, October 30). Going for herd immunity is immoral, says COVID chief.

Dunkley, D. (2020, October 22). Fruit-picking jobs going begging: Who is really to blame? *Stuff*. https://www.stuff.co.nz/business/opinion-analysis/300138256/fruitpicking-jobs-going-begging-who-is-really-to-blame.

Earley, M. (2020, September 12). Coronavirus: More than a thousand turn out for anti-lockdown rally in Auckland. *Stuff*. https://www.stuff.co.nz/national/health/coronavirus/300105763/coronavirus-more-than-a-thousand-turn-out-for-antilockdown-rally-in-auckland.

Ensor, J. (2020, October 10). Coronavirus: Judith Collins not impressed by latest managed isolation escape. *Newshub*. https://www.newshub.co.nz/home/politics/2020/10/coronavirus-judith-collins-not-impressed-by-latest-managed-isolation-escape.html.

Every-Palmer, S., Beaglehole, B., Bell, C., Gendall, P., Hoek, J., Jenkins, M., Rapsey, C., Stanley, J., & Williman, J. (2020, November 5). Mental health impacts. *He Kitenga*, University of Otago. https://www.otago.ac.nz/hekitenga/2020/otago744034.html.

Frater, P. (2020, March 17). "Avatar" sequels halt production in New Zealand due to coronavirus outbreak. *Variety*. https://variety.com/2020/film/asia/avatar-sequels-halt-production-new-zealand-coronavirus-1203536599/.

Graham-McLay, C. (2020a, April 21). Ardern accused of making New Zealand businesses "sacrificial lambs" in COVID-19 lockdown. *Guardian*. https://www.theguardian.com/world/2020/apr/21/ardern-accused-of-making-new-zealand-businesses-sacrificial-lambs-in-covid-19-lockdown.

Graham-McLay, C. (2020b, August 11). New Zealand records first new local COVID-19 cases in 102 days. *Guardian*. https://www.theguardian.com/world/2020/oct/22/ardern-urged-to-review-covid-measures-after-election-landslide.

Graham-McLay, C. (2020c, October 22). Ardern urged to review COVID measures after election landslide. *Guardian*. https://www.theguardian.com/world/2020/oct/22/ardern-urged-to-review-covid-measures-after-election-landslide.

Gunia, A. (2020, April 24). Why New Zealand's coronavirus elimination strategy is unlikely to work in most other places. *Time Magazine*. https://time.com/5824042/new-zealand-coronavirus-elimination/.

Hargreaves, D. (2020, August 4). Economists at the country's largest bank see a chance that the country will fall back into recession again from the end of this year. interest.co.nz. https://www.interest.co.nz/opinion/106356/economists-countrys-largest-bank-see-chance-country-will-fall-back-recession-again.

Hewson, S. (2020, April 18). New Zealand athletes "grieving" Olympics postponement. RNZ News. https://www.rnz.co.nz/news/sport/414543/new-zealand-athletes-grieving-olympics-postponement.

Higgins, R. (2011, May 5). Tangihanga – death customs. *Te Ara – the Encyclopedia of New Zealand*. https://teara.govt.nz/en/tangihanga-death-customs.

Hollyman, S. (2020, February 20). Coronavirus causes forestry slowdown. *Nelson Weekly*. https://nelsonweekly.co.nz/2020/02/coronavirus-causes-forestry-slowdown/.

Hosking, M. (2020, July 22). Mike's minute: Govt sucked us all in with expensive COVID spin. Newstalk ZB. https://www.newstalkzb.co.nz/on-air/mike-hosking-breakfast/video/mikes-minute-govt-sucked-us-all-in-with-expensive-covid-spin/.

Jeffries, S., French, N., Gilkerson, C., Graham, G., Hope, V., Marshall, J., McElnay, C., McNeill, A., Muellner, P., Paine, S., Prasad, N., Priest, P., Scott, J., Sherwood, J., & Yang, L. (2020, November 1). COVID-19 in New Zealand and the impact of the national response: A descriptive epidemiological study. *Lancet*, 5(11), E612–E623. https://www.thelancet.com/journals/lanpub/article/PIIS2468-2667(20)30225-5/fulltext.

Keall, C. (2020, March 15). Production suspended on *Lord of the Rings*. *NZ Herald* https://www.nzherald.co.nz/business/coronavirus-amazons-lord-of-the-rings-production-in-west-auckland-shut-down/JZE3ZEXWQQIQJXH4KYENFR3LRY/.

Kennedy, E. (2020, July 15). Exposed to privacy risk as urgent action taken to trace COVID cases' contacts. *New Zealand Doctor*. https://www.nzdoctor.co.nz/article/print-archive/exposed-privacy-risk-urgent-action-taken-trace-covid-cases-contacts.

Kenny, K. (2020, May 20). COVID tracer: What use is the government's app, really? Stuff. https://www.stuff.co.nz/technology/121575074/covid-tracer-what-use-is-the-governments-app-really.

Kronast, H., & McRoberts, M. (2020, March 14). Christchurch mosque attack memorial cancelled. Newshub. https://www.newshub.co.nz/home/new-zealand/2020/03/christchurch-mosque-attack-memorial-cancelled.html.

Leahy, B. (2020, May 3). COVID-19 coronavirus: Kiwi couples kiss and tell on their lockdown lovelife. *NZ Herald*. https://www.nzherald.co.nz/nz/covid-19-coronavirus-kiwi-couples-kiss-and-tell-on-their-lockdown-lovelife/27IC6KTPYLRCRUT6OTSZIJI6OI/.

Leask, A. (2020, March 29). COVID-19 coronavirus: New way to dob in lockdown breachers. *NZ Herald*. https://www.nzherald.co.nz/nz/covid-19-coronavirus-new-way-to-dob-in-lockdown-breachers/GZBXA64YQKFIQV7VUQUKTTYIKU?.

Lock, H. (2020, July 8). How COVID-19 changed travel home – and abroad – for Kiwis. RNZ. https://www.rnz.co.nz/news/national/420775/how-covid-19-changed-travel-home-and-abroad-for-kiwis.

Macpherson, C. (2020). Story: Pacific churches in New Zealand. *Te Ara*. https://teara.govt.nz/en/pacific-churches-in-new-zealand.

Marcetic, B. (2020, May 6). The prime minister should listen to her own mantra: "Be kind." The Democracy Project. https://democracyproject.nz/2020/05/06/branko-marcetic-the-prime-minister-should-listen-to-her-own-mantra-be-kind/0/.

Martin, H. (2020, November 5). COVID-19: Lockdown took emotional toll on Kiwis, survey finds. Stuff. https://www.stuff.co.nz/national/health/coronavirus/300149048/covid19-lockdown-took-emotional-toll-on-kiwis-survey-finds.

McCullough, Y. (2020, August 31). "Level 2.5" for Auckland comes with specific instructions. RNZ News. https://www.rnz.co.nz/news/covid-19/424823/level-2-point-5-for-auckland-comes-with-specific-instructions.

McLachlan, L. (2020, March 26). COVID-19 funeral restrictions leave families grieving in isolation. RNZ: *Te Ao Māori*. https://www.rnz.co.nz/news/te-manu-korihi/412705/covid-19-funeral-restrictions-leave-families-grieving-in-isolation.

McNeilly, H. (2020, April 16). Coronavirus: Get the Ashley Bloomfield T-shirt. Stuff. https://www.stuff.co.nz/life-style/fashion/121049868/coronavirus-get-the-ashley-bloomfield-tshirt.

Ministry for Culture and Heritage. (2020). ANZAC Day. https://mch.govt.nz/anzac-day.

Ministry of Health. (2020a, April 5). Contact tracing at forefront of fight against COVID-19. Media Release. https://www.health.govt.nz/news-media/media-releases/contact-tracing-forefront-fight-against-covid-19.

Ministry of Health. (2020b, May 13). No new cases of COVID-19. Media Release. https://www.health.govt.nz/news-media/media-releases/no-new-cases-covid-19-2.

Ministry of Health. (2020c, August 11). 4 cases of COVID-19 with unknown source. Media Release. https://www.health.govt.nz/news-media/media-releases/4-cases-covid-19-unknown-source.

Ministry of Health. (2020d, August 25). 7 cases of COVID-19. Media Release. https://www.health.govt.nz/news-media/media-releases/7-new-cases-covid-19-0.

Ministry of Health. (2020e, September 4). Ministry of Health announces COVID-19 related death in Auckland. Media Release. https://www.health.govt.nz/news-media/media-releases/ministry-health-announces-covid-19-related-death-auckland.

Ministry of Health. (2020f, September 9). 6 new cases of COVID-19. Media Release. https://www.health.govt.nz/news-media/media-releases/6-new-cases-covid-19-4.

Ministry of Health. (2020g, October 5). Auckland will move to Alert Level 1. https://covid19.govt.nz/updates-and-resources/latest-updates/auckland-will-move-to-alert-level-1/.

Ministry of Health. (2020h, October 7). COVID-19: Information for close contacts. Fact sheet. https://www.health.govt.nz/our-work/diseases-and-conditions/covid-19-novel-coronavirus/covid-19-health-advice-public/contact-tracing-covid-19.

New Zealand COVID-19 Alert Levels Summary. (2020). https://covid19.govt.nz/assets/resources/tables/COVID-19-alert-levels-summary.pdf.

New Zealand Government. (2020, January 28). PM announces election date as September 19. Media Release. https://www.beehive.govt.nz/release/pm-announces-election-date-september-19.

New Zealand History – Nga korero a ipurangi o Aotearoa. (2020). Anzac Day. (7). Modern Anzac Day. https://nzhistory.govt.nz/war/modern-anzac-day.

New Zealand Legislation. (2020). Constitution Act 1986 May 17, 2005. http://www.legislation.govt.nz/act/public/1986/0114/latest/whole.html?search=sw_096be8ed81a11be0_271_25_se&p=1#DLM94241.

New Zealand Parliament, Hansard (Debates). (2020, April 30). Hansard. Vol. 745. https://www.parliament.nz/en/pb/hansard-debates/rhr/combined/HansD_20200430_20200430.

New Zealand Parliament. (2017, October 12). 2017 General Election writ returned. Media Release. https://www.parliament.nz/en/footer/about-us/office-of-the-clerk/office-of-the-clerk-media-releases/media-release-2017-general-election-writ-returned/.

*Newshub: The Project.* (2020, March 18). New Zealand film industry feels the hit of COVID-19. https://www.newshub.co.nz/home/entertainment/2020/03/new-zealand-film-industry-feels-the-hit-of-covid-19.html.

Newsroom Staff. (2020, August 17). Election postponed four weeks to October 17. Newsroom. https://www.newsroom.co.nz/ardern-makes-call-on-election-date.

NZ Government Media Release. (2020, April 3). Funding certainty for sports through COVID-19. https://www.beehive.govt.nz/release/funding-certainty-sports-through-covid-19.

*NZ Herald.* (2020a, April 18). COVID-19 coronavirus: Man sentenced to month in prison for lockdown breaches. https://www.nzherald.co.nz/nz/covid-19-coronavirus-man-sentenced-to-month-in-prison-for-lockdown-breaches/2VWHGHUFD4CFNNZVXGMKK3PNNA/.

*NZ Herald.* (2020b, August 5). Ashley Bloomfield tattoo: Morrinsville woman's epic ink turning heads. https://www.nzherald.co.nz/lifestyle/ashley-bloomfield-tattoo-morrinsville-womans-epic-ink-turning-heads/CO23BWDUNEMHOTBYGMGH4G6WFA/.

*NZ Herald.* (2020c, August 6). COVID-19 coronavirus: Travelers would not escape quarantine under National, Judith Collins says. https://www.nzherald.co.nz/nz/covid-19-coronavirus-travellers-would-not-escape-quarantine-under-national-judith-collins-says/GZPUCZUPL6RPHQ7B6KUJN4X7EU/.

*NZ Herald.* (2020d, August 12). COVID-19 coronavirus lockdown: Jacinda Ardern says Auckland in level 3 at midday; NZ in level 2. https://www.nzherald.co.nz/nz/covid-19-coronavirus-lockdown-jacinda-ardern-says-auckland-in-level-3-at-midday-nz-in-level-2/TYGDCTPZDVYRWSKXHELV7PPX3U/.

*NZ Herald.* (2020e, August 22). COVID-19 coronavirus: "We will not be silenced" – Maskless anti-lockdown protestors march in Auckland. https://www.nzherald.co.nz/nz/covid-19-coronavirus-we-will-not-be-silenced-maskless-anti-lockdown-protestors-march-in-auckland/74HWXN2LWQI7K4BJATLROG3GP4/.

*NZ Herald.* (2020f, September 5). COVID-19 coronavirus: Man in his 50s dies in Auckland, first virus death in three months. https://www.nzherald.co.nz/nz/covid-19-coronavirus-man-in-his-50s-dies-in-auckland-first-virus-death-in-three-months/PFDWZQ2JP4QA75ZH5ZYLZCZNJI/.

O'Brien, T. (2020, May 18). Newshub-Reid research poll: Jacinda Ardern goes stratospheric, Simon Bridges is annihilated. Newshub. https://www.newshub.co.nz/home/politics/2020/05/newshub-reid-research-poll-jacinda-ardern-goes-stratospheric-simon-bridges-is-annihilated.html.

O'Connell, A. (2020, April 24). "I discovered his affair during lockdown": Is your relationship at risk in a COVID world? Stuff. https://www.stuff.co.nz/life-style/love-sex/122542549/i-discovered-his-affair-during-lockdown-is-your-relationship-at-risk-in-a-covid-world.

Ohlheiser, A. (2016, May 3). The pre-Twitter history of Twitter's famous favorite insult. *Washington Post.* https://www.washingtonpost.com/news/the-intersect/wp/2016/05/03/the-pre-twitter-history-of-twitters-favorite-insult/.

Palmer, S. (2020, July 13). COVID-19: US author Alex Berenson says New Zealand putting returnees into "indefinite confinement." Newshub. https://www.newshub.co.nz/home/new-zealand/2020/07/covid-19-us-author-alex-berenson-says-new-zealand-putting-returnees-into-indefinite-confinement.html.

Patel, B. (2020, October 29). Chasing herd immunity is both "futile and immoral", Sweden's state epidemiologist warns despite his country's no-lockdown policy. *Daily Mail.* https://www.dailymail.co.uk/news/article-8891201/Chasing-herd-immunity-futile-immoral.html.

Redd, D. (2020). An analysis of New Zealand's coronavirus response. *APAC Outlook.* https://www.apacoutlookmag.com/industry-insights/article/801-an-analysis-of-new-zealands-coronavirus-response.

Reuters Staff. (2020, May 17). NZ gives sports sector $157 million boost to get through COVID-19. Reuters. https://www.reuters.com/article/us-health-coronavirus-sport-newzealand-idUSKBN22T01B.

RNZ. (2019, March 19). Auckland's Pasifika Festival cancelled. https://www.rnz.co.nz/news/national/385075/auckland-s-pasifika-festival-cancelled.

RNZ. (2020a, February 2). New Zealanders to be evacuated from Wuhan to Whangaparaoa. https://www.rnz.co.nz/news/national/408676/new-zealanders-to-be-evacuated-from-wuhan-to-whangaparaoa.

RNZ. (2020b, March 13). Auckland's Pasifika Festival cancelled amid coronavirus fears. https://www.rnz.co.nz/international/pacific-news/411637/auckland-s-pasifika-festival-cancelled-amid-coronavirus-fears.

RNZ. (2020c, March 25). Teddy bears in windows to cheer up kids during lockdown. https://www.rnz.co.nz/news/national/412602/teddy-bears-in-windows-to-cheer-up-kids-during-lockdown.

RNZ. (2020d, May 12). New Zealand general election at level 2 or lower, PM Jacinda Ardern confirms. https://www.rnz.co.nz/news/political/416429/new-zealand-general-election-at-level-2-or-lower-pm-jacinda-ardern-confirms.

RNZ. (2020e, July 14). Todd Muller resigns as National Party leader. https://www.rnz.co.nz/news/political/421152/todd-muller-resigns-as-national-party-leader.

RNZ. (2020f, July 14). New National Party leader is Judith Collins. https://www.rnz.co.nz/news/political/421206/new-national-party-leader-is-judith-collins.

RNZ. (2020g, August 12). Collins calls for election to be pushed back to late November or 2021. https://www.rnz.co.nz/news/national/423358/collins-calls-for-election-to-be-pushed-back-to-late-november-or-2021.

RNZ. (2020h, August 13). National election's delay call purely political, says James Shaw. https://www.rnz.co.nz/news/political/423449/national-s-election-delay-call-purely-political-says-james-shaw.

RNZ. (2020i, August 14). Auckland lockdown to continue for 12 more days. https://www.rnz.co.nz/news/national/423569/auckland-lockdown-to-continue-for-12-more-days.

RNZ. (2020j, September 14). Cabinet extends current COVID-19 alert levels. https://www.rnz.co.nz/news/national/425993/cabinet-extends-current-covid-19-alert-levels.

RNZ. (2020k, September 16). Man dies from COVID-19 in Waikato Hospital, one new case at border. https://www.rnz.co.nz/news/national/426184/man-dies-from-covid-19-in-waikato-hospital-one-new-case-at-border.

Science Media Centre. (2020, April 30). Tracking COVID-19 with genome sequencing – expert briefing. https://www.sciencemediacentre.co.nz/2020/04/30/tracking-covid-19-with-genome-sequencing-expert-briefing/.

Scotcher, K. (2020, April 1). COVID-19: Love in a lockdown. RNZ. https://www.rnz.co.nz/news/national/413171/covid-19-love-in-a-lockdown.

Smith, R. (2020, October 22). COVID-19 coronavirus: Sweden's strategy takes a turn as cases spike. *NZ Herald*. https://www.nzherald.co.nz/world/covid-19-coronavirus-swedens-strategy-takes-a-turn-as-cases-spike/KP3SUPMIC7YZ5NQXEPTSIL63B4/.

Sport New Zealand. (2020, August 17). Play, active recreation and sport at Alert Level 3. *Sport New Zealand*. https://sportnz.org.nz/resources/play-active-recreation-and-sport-at-alert-level-3/.

Statistics NZ. (2017, October 26). Three in four New Zealanders live in the North Island. Stats NZ. https://www.stats.govt.nz/news/three-in-four-new-zealanders-live-in-the-north-island.

Statistics NZ. (2019, November 21). Which industries contributed to New Zealand's GDP? Stats NZ. https://www.stats.govt.nz/experimental/which-industries-contributed-to-new-zealands-gdp.

Statistics NZ. (2020, March 19). About 100,000 New Zealand residents travelling overseas. Stats NZ. https://www.stats.govt.nz/news/about-100000-new-zealand-residents-travelling-overseas.

Stuff (2020, October 30). COVID-19: New Zealand's health response is not "fascist." Stuff. https://www.stuff.co.nz/national/123236022/covid19-new-zealands-health-response-is-not-fascist.

Todd, K. (2020, January 15). Health ministry alerts hospitals to new deadly virus in China. Radio New Zealand (RNZ) News. https://www.rnz.co.nz/news/national/407390/health-ministry-alerts-hospitals-to-new-deadly-virus-in-china.

Tourism Industry Aotearoa. (2020). Quick facts and figures. https://tia.org.nz/about-the-industry/quick-facts-and-figures/.

Trevett, C. (2020, May 21). New political poll: National drops to 29%, Labour up to 59%. *NZ Herald*. https://www.nzherald.co.nz/nz/new-political-poll-national-drops-to-29-labour-up-to-59/SDY5WRCDWYPXXITR2O7Z2KRMWY/.

Trevett, C., & Walls, J. (2020, May 20). National leadership coup: Todd Muller confirms he'll challenge Simon Bridges on Friday. *NZ Herald*. https://www.nzherald.co.nz/nz/national-leadership-coup-todd-muller-confirms-hell-challenge-simon-bridges-on-friday/MAFNSJDQ5SPI5WOEISXIFZOEBI/.

TVNZ. (2020a, July 11). Returnee "in their late 60s" broke window, climbed fence to escape managed isolation – fourth to do so in a week. https://www.tvnz.co.nz/one-news/new-zealand/returnee-in-their-late-60s-broke-window-climbed-fence-escape-managed-isolation-fourth-do-so-week?gclid=EAIaIQobChMIi9b56IGB7QIV2oNLBR2x5w3yEAMYASAAEgLTXfD_BwE.

TVNZ. (2020b, August 12). Jacinda Ardern makes it mandatory for businesses to display COVID-19 tracer app QR code. https://www.tvnz.co.nz/one-news/new-zealand/jacinda-ardern-makes-mandatory-businesses-display-covid-19-tracer-app-qr-code.

TVNZ. (2020c, August 19). Ardern hits back after Collins criticises new appointments to bolster border testing as "another working group." https://www.tvnz.co.nz/one-news/new-zealand/ardern-hits-back-after-collins-criticises-new-appointments-bolster-border-testing-another-working-group.

TVNZ. (2020d, September 13). COVID-19 "brain gain" sees more than 50,000 Kiwis return home amid pandemic. https://www.tvnz.co.nz/one-news/new-zealand/covid-19-brain-gain-sees-more-than-50-000-kiwis-return-home-amid-pandemic.

TVNZ. (2020e, October 9). "Charges likely" after woman allegedly escapes from managed isolation facility in Auckland. https://www.tvnz.co.nz/one-news/new-zealand/charges-likely-after-woman-allegedly-escapes-managed-isolation-facility-in-auckland.

TVNZ. (2020f, October 28). NZ likened to Nazi Germany by far-right British politician for its COVID quarantine protocol. https://www.tvnz.co.nz/one-news/new-zealand/nz-likened-nazi-germany-far-right-british-politician-its-covid-quarantine-protocol.

TVNZ. (2020g, October 29). Fox News host goes on bizarre rant about NZ, mocks accent and details "terrifying" COVID "camps." https://www.tvnz.co.nz/one-news/new-zealand/fox-news-host-goes-bizarre-rant-nz-mocks-accent-and-details-terrifying-covid-camps?auto=6205342117001.

TVNZ. (2020h, November 3). "Not something I dreamt about" – Ashley Bloomfield nominated for TV personality of the year. https://www.tvnz.co.nz/one-news/entertainment/not-something-dreamt-ashley-bloomfield-nominated-tv-personality-year.

Upston, L. (2020, February 11). Order Paper & Questions. Written questions. New Zealand Parliament. https://www.parliament.nz/en/pb/order-paper-questions/written-questions/document/WQ_00005_2020/5-2020-hon-louise-upston-to-the-minister-for-social.

van Beynen, M. (2020, March 22). Coronavirus: Wanaka weeps as tourism industry faces mass closures and job losses. Stuff/*Sunday Star Times*. https://www.stuff.co.nz/national/health/coronavirus/120468408/coronavirus-wanaka-tourism-businesses-a-mass-of-closures-and-job-losses.

Vogel, G. (2020, October 6). SWEDEN'S GAMBLE: The country's pandemic policies came at a high price – and created painful rifts in its scientific community. *Science*. https://www.sciencemag.org/news/2020/10/it-s-been-so-so-surreal-critics-sweden-s-lax-pandemic-policies-face-fierce-backlash.

Wade, A. (2020, May 20). COVID-19 coronavirus: Simon Bridges faces National Party leadership challenge from Todd Muller, Nikki Kaye. *NZ Herald*. https://www.nzherald.co.nz/nz/covid-19-coronavirus-simon-bridges-faces-national-party-leadership-challenge-from-todd-muller-nikki-kaye/WFMOS7GIQ3VP5X5LFLIFG5BMII/.

Ward, L. (2020, September 7). COVID-19 coronavirus: Auckland's dire August – the cluster that changed everything. *NZ Herald*. https://www.nzherald.co.nz/nz/covid-19-coronavirus-aucklands-dire-august-the-cluster-that-changed-everything/TRTKQVNYPSA42NH2YZDFSORLZU/.

Williams, C. (2020, August 29). Coronavirus: Police "disappointed" but no punishment for 500 protesters breaching lockdown. *Stuff*. https://www.stuff.co.nz/national/health/coronavirus/300094774/coronavirus-police-disappointed-but-no-punishment-for-500-protesters-breaching-lockdown.

Williams, K. (2020a, June 10). Coronavirus: Wellington – it's time to reach into your pockets. *Stuff*. https://www.stuff.co.nz/national/health/coronavirus/121766075/coronavirus-wellington--its-time-to-reach-into-your-pockets?rm=a.

Williams, K. (2020b, September 29). COVID-19: Princeton University study dissects New Zealand's pandemic response. *Stuff*. https://www.stuff.co.nz/national/health/coronavirus/122917225/covid19-princeton-university-study-dissects-new-zealands-pandemic-response.

Williams, K., & Mitchell, R. (2020, October 28). Avery criticises pandemic strategy. *Dominion Post*, p. 5.

Wilson, P. (2020, May 1). Week in politics: Getting out of level 3 might not be so easy. RNZ. https://www.rnz.co.nz/news/on-the-inside/415594/week-in-politics-getting-out-of-level-3-might-not-be-so-easy.

Wiltshire, L. (2020, March 19). Coronavirus: Anzac Day services cancelled. *Stuff*. https://www.stuff.co.nz/national/health/coronavirus/120399099/anzac-day-services-cancelled-due-to-coronavirus.

Woolf, A.L. (2020, January 26). China arrivals to be checked by health staff amid coronavirus outbreak. *Stuff*. https://www.stuff.co.nz/national/health/119049195/china-arrivals-to-be-checked-by-health-staff-amid-coronavirus-outbreak.

World Health Organization (WHO). (2020, October 12). WHO Director-General's opening remarks at the media briefing on COVID-19. https://www.who.int/director-general/speeches/detail/who-director-general-s-opening-remarks-at-the-media-briefing-on-covid-19---12-october-2020#:~:text=Herd%20immunity%20is%20a%20concept,a%20population%20to%20be%20vaccinated.

# Conclusion

BY JACK FONG

In our final chapter, I attempt an assembly of considerations that points to how societies act or react in crisis situations as communities are reconfigured. With insights provided by our contributors, I hope to make visible how crisis societies generally appear when numerous stressors affect the lifeworld during pandemic distancing. From the intimate confines of the family experiencing its discontents, to emptying publics, to how the state apparatus has – effectively or ineffectively – interacted with different members of society in the name of nation, as well as to how states mete out public policies that have generated compliance or resistance, it is clear that systemic crisis creates alternative societies and privileges alternative social systems as new sites of power in the lifeworld. Systemic crisis, then, injects acute interpersonal and community challenges at every level of existence, upon all social actors as they struggle and long for a return to quotidian life. In the case of the pandemic, whose physical distancing measures in many parts of the world isolated members of society, the potential for sociopathy as well as cultural renewal and adaptation characterized a frenetic terrain where the architecture of society changed before our eyes, and in ways that affected us at the most personal level. Indeed, social changes at the end of 2020, the period in which we conclude our volume's analyses, continue as COVID-19 resurfaced in many parts of the planet when countries again became infection epicentres with the arrival of COVID-19's delta and later omicron variants.[1]

As the first known year of COVID-19's appearance drew to a close, a variety of developments related to pandemic mitigation dramatically surfaced. Chief among these was the long-awaited availability of vaccines, made possible by a variety of countries' pharmaceutical efforts to vanquish COVID-19. By the end of 2020, countries around the world were anticipating vaccines through pharmaceutical corporations like Pfizer and Moderna, or China's Sinopharm, its state-owned pharmaceutical enterprise, to name but a few. In the North American context, Canada had already secured the most doses per capita by December

2020 (Mullard, 2020), while large-scale (Phase 3) testing was scheduled for the British-Swedish AstraZeneca corporation, as well as Janssen Pharmaceutica, headquartered in Beerse, Belgium. Russia's Gamaleya Research Institute of Epidemiology and Microbiology had also involved itself in pandemic mitigation, offering its own vaccine, not unlike Germany's BioNTech SE biotechnology company, which had already applied to the United States Food and Drug Administration (FDA) to distribute its vaccines. The United Kingdom, Bahrain, the United Arab Emirates, Canada, Saudi Arabia, the United States, Mexico, Kuwait, Singapore, Jordan, and Oman had approved emergency vaccine measures to be implemented, with the United Kingdom being the first country to offer its COVID-19 vaccines in early December 2020. Pfizer and BioNTech distributed 800,000 vaccines for UK citizens according to Gavi, The Vaccine Alliance, an organization that aims to "protect people's health by increasing equitable and sustainable use of vaccines," especially in poor countries (Gavi, 2021b). By mid-December 2020 alone, 138,000 people in the UK had already received their vaccines (*Guardian*, 2020b).[2] Other companies were immersing themselves in pandemic mitigation with alacrity: The United States' Novavax, France's Sanofi, the German pharmaceutical company CureVac, and China's Sinovac were not far behind in vaccine development at the time of this writing in late 2020. By October 2021, 21 vaccines were available, while another 128 vaccine candidates were undergoing trials (Gavi, 2021c). One month later, Finland began distribution of the AstraZeneca vaccine to Côte d'Ivoire, with 98,400 doses out of a promised 3.65 million doses reaching the country by 4 November 2021 (Gavi, 2021a). Yet global inequalities persist, as vast populations of people in low-income countries would have to wait until 2023 or 2024 for available vaccinations, according to the forecast of the Duke Global Health Innovation Center based in Durham, North Carolina, USA (cited in Mullard, 2020).

Despite continuing challenges, a sense of optimism conveyed through media channels about the *healing and repairing of society* has yet to find traction. Instead, public officials convey narratives that point to a return to "normalcy" as one determined by economic and technocratic activity, growth, and employment. Yet what COVID-19 has taught us is that global humanity did express its collaborative potential in ways that were not entirely cheapened by political climate, misinformation, and ideology. Efforts by charitable organizations and different actors around the world did engage in activities for those who were negatively affected. We can appreciate a few examples, such as when Ireland sent more than US$1.8 million to the Navajo Nation and Hopi Reservation as a return gesture to the Chocktaw Nation – the "first tribe to be relocated during the Trail of Tears" (O'Loughlin & Zaveri, 2020). During the Irish potato famine in 1847, the Choctaw sent US$170 to starving Irish families after hearing about their struggles, feeling great empathy for them after they heard "such a similar tale [to their own] coming from across the ocean" (O'Loughlin

& Zaveri, 2020). The fundraiser provided resources to secure "clean water, food and health supplies" to Navajo and Hopi communities in the American Southwest. Scenes of community rebuilding were also observed in Siena, Italy, not far from Florence, as people sang traditional folk songs from their balconies for neighbours, inspiring people in other cities such as Naples to engage in similar actions (Porterfield, 2020). In Vietnam, organizations such as the Centre for Sustainable Rural Development and the Centre for Supporting Community Development Initiatives engaged in activities that provided loans and foodstuffs to farmers and rural workers (Ravelo, 2020). During June 2020 in Thailand, a 77-year-old grandmother and her associates in the Village Health Volunteers program in rural Ang Thong province travelled on scooters and modified rickshaws to remind arrivals to the province of the importance of self-quarantine and physical distancing (*South China Morning Post*, 2020). In India, police inspector Rajesh Babu in Chennai gained fame when photographed donning a coronavirus helmet to convey to commuters the importance of remaining at home. The helmet, designed by local artist B. Gowtham, projected spikes akin to how the COVID-19 virus appears under a microscope. When talking to drivers of automobiles, Babu would approach the driver's side window of commuters as the virus and say, "If you come out, I will come in" (Yeung, 2020). In many areas of the United States, local charity groups organized food drives and distributed masks to members of their communities in frequently televised events. Also in the United States, however, varying degrees of state responses to citizens as well as citizens' responses to the state, and to one another, exhibited a degree of volatility and sociopathy that was heavily influenced by the incompetence and cultural warmongering of Donald Trump. For many political analysts, his botched response to the arrival of COVID-19 further contributed to the negative outcomes of pandemic mitigation in the country. Indeed, Peter Wehner, a fellow at the Ethics and Public Policy Center, a conservative think tank, already noted in the *Atlantic* early in March 2020 as COVID-19 arrived in the United States:

> Mr. Trump's virulent combination of ignorance, emotional instability, demagogy, solipsism and vindictiveness would do more than result in a failed presidency; it could very well lead to national catastrophe. The prospect of Donald Trump as commander in chief should send a chill down the spine of every American. (Wehner, 2020)

Wehner highlighted how Trump, "a habitual liar," managed to ignore "early warnings of the severity of the virus" and "dismantled the National Security Council's global-health office, whose purpose was to address global pandemics." He further emphasized, "We're now paying the price for that" (Wehner, 2020). The last sentence in the article presciently stated, seven months before

2020 elections, that as a result of the president's drivel and incompetence, "the Trump presidency is over" (Wehner, 2020).

Less than two months after Wehner's article, developing pandemic restrictions and the stresses they caused communities in the United States were further amplified and exacerbated by the indignation felt by many following the death of George Floyd on 25 May 2020, an African American who died extra-judicially by the actions of police officers in the state of Minnesota. The event set into motion a series of massive social protests, involving tens of thousands, against police violence and systematic racism that took place across many urban environments in the United States and in some cities around the world. Such developments require us to see the prevalence of protest during health crises set into motion by acute morbidity events like COVID-19.

Floyd's murder ultimately incited the anger of many Americans who did not forget the loss of Breonna Taylor – another African American shot in her home by the police on 13 March 2020. Such developments reminded the American public about how social inequalities in health were further intersected by the country's perennially unresolved issues on ethnic and race relations. The massive gatherings of protesters donning masks suggested that a pandemic consciousness had taken root in how Americans configured their lives during the pre-vaccination phase: by mixing politics with public health. By early 2021, continued violent attacks and verbal abuses directed at primarily Asian American senior citizens and/or women compelled the new president, Joe Biden, to engage in nationally televised public service announcements to condemn such acts. He also threw his support behind many legal and community groups that gathered in urban environments to voice their anger against what was seen as misplaced violence directed against a people simply because of their phenotypical appearance of looking "Oriental" or Chinese. That Trump, while holding office, recklessly played the blame game by calling COVID-19 the "Chinese Flu" further inspired his sycophants to unleash their vitriol against Asian Americans. Additionally, much footage across the United States and in many countries around the world revealed the blatant hostilities many have against government mandates for pandemic mitigation, an anger that those who subscribe to the merits of public health and its best practices were made to bear. Verbal abuse of and physical confrontations with those who ask others to don masks can be found in the multitude on social media, acts committed by belligerents who forget that freedom is as much about individual responsibility to others as it is about individual rights for the self.

The blueprint of protest against the state apparatus and citizens protesting against one another was thus replicated repeatedly in 2020 during the pandemic in the United States, even well into 2022. The striking difference, apart from the aforementioned issues related to America's never-ending culture wars, is how later protests involved those who were vehemently against lockdown measures and/or mask wearing. By the time vaccinations became available, an added third dimension of angst, that of being anti-vaccine, or being an "anti-vaxxer," emerged as a new battle cry for those disenchanted with or fear-ridden about institutional approaches towards pandemic mitigation. Indeed, in many countries, as we shall examine, the catalysts informing such groups were their distorted views on freedom and how their rights were being taken away, a view further textured with misinformation about COVID-19 stemming from cultic conspiracy theories floating through social media that viewed government mandates as a means of nefarious social control.

The historian Paul Slack (2020, 2021) has written about plagues and their social consequences. Slack argues that responses to plagues in the form of increased institutional control and the protest it often generates are not new. From as far back as the plague in the fourteenth century to later pandemics of the nineteenth and twentieth centuries, governments have always attempted to secure some form of social control to mitigate the spread of infection. Preindustrial societies had already set the precedent for what we today experience as pandemic lockdowns, physical distancing, and/or quarantine in developing social mechanisms designed to control the population so as to prevent further spread of infections. During the bubonic plague of the sixteenth century:

> Plague victims were isolated and their contacts traced and incarcerated. There were restrictions on movement ... [and] quarantine regulations for travelers and shipping. Bedding and houses were fumigated. All this necessitated the growth of local administrative machines and an expansion of state power, the invention of "medical police" in fact. It also implied serious restrictions on individual liberty and provoked opposition for that reason, among others. (Slack, 2020, p. 409)

The many examples of government-imposed measures to stem infections from pandemics are beyond the scope of this work, but Slack's *Plague* (2021) provides some notable examples: During the Black Death of the mid-fourteenth century, when a third of the European population perished, Venice and Florence "set up special health commissions of leading citizens to deal with the crisis," with their main duty being the "enforcement of existing sanitary laws" (Slack, 2021, p. 75). The dukes of then Tuscan and Venetian republics "ordered local hospitals to take in the infected, had temporary shacks and cabins thrown up for them ... and then had a special plague hospital built for them" (Slack, 2021, p. 76). In early seventeenth-century England, the monarchy ensured that watchmen had "legal authority to use 'violence' to keep the infected shut up in their houses"; any individual with plague sores "found wandering in the company of others ... was guilty of felony and might be hanged," and "anyone else going out could be whipped as a rogue" (Slack, 2021, p. 78). The Great Plague of Marseille in France (circa 1720) resulted in its government establishing cordons that were enforced by "one-quarter of the French standing army" stationed on "the frontiers of Provence" (Slack, 2021, p. 82). By the late eighteenth century, the Austrian-Hungarian and Ottoman empires similarly employed their military to enforce plague mitigation measures. When the pneumonic plague struck China in 1910–11, officials of the Qing Dynasty set up a Plague Prevention Bureau that enforced compliance through its sanitary police, who inspected conditions of the home environment and, if need be, ensured that the sick were sent to isolation stations. Furthermore, Japan at the time had presciently adopted "German concepts of state medicine and public health" (Slack, 2021, p. 89). For Slack, "most of what we understand about public health, its basic rationale and ideology, was first formulated in the context of the plague, and in the first decades of the second pandemic" (Slack, 2021, p. 74). In spite of his careful detailing of state responses to epidemics/pandemics, Slack emphasized that none of the mitigation attempts were "achieved painlessly" (Slack, 2021, p. 74).

The protests that surfaced in many countries during the pre-vaccination phase of COVID-19, then, should not be seen as unusual. Indeed, Slack notes how restrictions on social interaction generated "conflicts between public and private interests and between the dictates of medically informed prudence and the imperatives of popular morality, which arose in the case of later epidemics from cholera to AIDS [but] can first be truly documented in Europe in the age of the plague" (Slack, 2020, pp. 409–10). Citizens through time have found offensive "those practices of close surveillance and compulsory segregation that shattered community cohesion and violated the moral norms which sustained it" (Slack, 2021, p. 91). Such actions demand a deeper examination of their consistency simply because, in the long chronology of plagues, epidemics, and pandemics, citizens in their reconfigured communities and societies were "daily faced with conflicting obligations to families, friends, and neighbors" (Slack, 2021, p. 70). For Slack, the glaring "reality for everyone involved in an epidemic was personal stress. They had to weigh concern about their own safety and that of their families against the pull of other obligations and loyalties [of] everyday ties of friendship, business, and neighborhood" (Slack, 2021, p. 68). Therefore, those who violently lash out in the era of COVID-19 in protest against the state, and against scapegoated populations and/or people who try to remind others about the importance of wearing masks, exhibit but atavistic outbursts from yesteryear, as during the Black Death when over a thousand of Strasbourg's Jews were "burned on an island in the Rhine in 1349" while "large Jewish communities elsewhere … were almost wholly exterminated" (Slack, 2021, p. 48). Although such horrific outcomes did not befall those being scapegoated or who followed state mandates for pandemic mitigation in 2020, social and community dynamics did deteriorate owing to state-mandated policies to reduce its most current morbidity. What then followed in reconfigured community and society is the act of blaming, be it of Jews or witches from yesteryear, or today of foreigners (or those perceived to be foreigners), and/or those perceived to be willing or unwilling conformists to state mandates (Slack, 2020, p. 414). Additionally, with the development of the plague in Europe over many centuries, a class reading of outcomes evolved to disenfranchise its most disadvantaged. Campaigns against vagrants, prostitutes, and beggars followed, and "the plague-infected poor were to be incarcerated … and if necessary punished in the interest both of public health and of public order, broadly defined" (Slack, 2020, p. 420). The most disadvantaged, the lumpenproletariat of the day, became "a conspicuous target," to be held in contempt, since they "threatened the health, social position, and peace of mind of elites" (Slack, 2020, pp. 420–1).

When another plague struck England during the mid-1600s, criticism was more civil, with publications that nonetheless misinformed communities that "'alterations in the air' were far more important causes of infection than 'contagion' between people," while in Königsberg, Prussia, "local protests … killed more people than the epidemic itself" (Slack, 2021, p. 83). In the case of England, social regulations did not always prevent "unlawful assemblies" where neighbours were caught "drinking together" or amassing at funerals (Slack, 2020, p. 418). Some aspects of resistance took on bizarre expressions, as in the epidemic of Avignon where dancers were seen performing among graves in the late 1300s, while in 1720 others played leapfrog over corpses. In the city of York in 1632, a man was seen "'dancing and fiddling' among infected houses" while another sarcastically remarked to "an inquiring constable that all in his house 'were in health but his cat was sick'" (Slack, 2021, p. 64). During the late nineteenth century when the plague struck India and China, India's enforcement of the Epidemic Diseases Act in 1897 through plague committees met with resistance. In the previous year, over 100,000 Indians had already migrated out of Bombay (now Mumbai), fully aware that British-mandated forms of plague mitigation, as acknowledged by an officer in the Indian Army Medical Service, were designed for "riveting our [British] rule in India" and also for "showing the superiority of our western science and thoroughness" (Slack, 2021, p. 86). Not surprisingly, during this same period in Bombay, "crowds attacked ambulances and stormed the plague hospital," while in Kolkata (formerly Calcutta) riots occurred alongside the population's "mass exodus" from the city (Slack, 2021, p. 88). For Slack, racialization of the plague combined with colonial hubris informed public health policies during Western expansion into different regions of Asia. The director-general of the Indian Medical Department at the time noted that "the bubonic plague was a 'disease of filth, a disease of dirt, and a disease of poverty'" (Slack, 2021, p. 89). A European in Hong Kong attending to the plague during the same period remarked, "no Sanitary Board … no sanitary or preventive measures … no isolation of cases'. All of these were introduced by the British government of Hong Kong, much to the dismay of the Chinese community" (Slack, 2021, p. 86).

We thus cannot be complacent about the variability seen in outcomes through time insofar as pandemic mitigation is concerned – even if modern citizens are seeing comparatively less bloodletting directed against those from the disadvantaged classes and/or minorities – since atavisms of COVID-19 can be expected to remain with humanity for the long term. And even if our virus can be eliminated or softened due to endemicity, future epidemics/pandemics will usher in a revival of new calls for compliance as well as new hostilities directed towards compliance. Cues for the consistency of resistance in the twenty-first century can already be seen throughout 2020 (fortunately in the minority of populations), during the pre-vaccination phase of COVID-19's dispersion – arguably with a greater infusion of aberrant sociopathy that expressed itself most vociferously in Western democracies, creating ironically numerous "democracy deserts" (Fong, 2025, forthcoming).

By early April 2020, the British, and Americans in many US states, were engaged in protests against imminent vaccinations and stay-at-home orders. Taking place in major urban centres of their respective countries, these signalled more activities to come, although the size of protests varied from small groups to a few thousand. Many conveyed sociopathic and hyper-paranoid views exported by the United States that were and remain shaped by Q-Anon proponents, who saw a malevolent "deep state" staffed by a cabal of satanic child-trafficking pedophiles at work, according to Ian Haimowitz (2020) of the United States' most prestigious think tank, the Center for Strategic and International Studies. For example, a protester in Michigan at the time proclaimed, "We're more afraid of the government than we are of the virus at this point," while others saw the imminent use of vaccines as a sinister scheme to enrich pharmaceutical companies (Sommer & Kucinich, 2020). By May, anti-vaccine advocates, conspiracy theorists, and other anti-state activists were already engaged in protests against Germany's lockdown measures, with some protestors claiming that Bill Gates was responsible for the COVID-19 pandemic while others accused the German government of imposing "dictatorship-like conditions" in the country (Ankel, 2020). Canada jumped into the fray during this time, with its Toronto and Vancouver protestors proclaiming that their protests were for the sake of small businesses, even though some protestors carried elaborate signs that said "Stop 5G" (these were part of the 5G Truthers group who believe COVID-19 is linked to 5G towers), while another had a sign that proclaimed "Bill Gates Vaccine Patent is 060606, Mark of the Beast, Wake Up" (Press Progress, 2020). By September of 2020, a few hundred 5G and child-trafficking conspiracy theorists and right wing adherents emanating from "extreme private Facebook groups," along with anti-mask and anti-vaccine advocates, had again gathered in Vancouver, with one of the organizers proclaiming, "This is all due to a virus that has spread [to] the world. It's not called COVID, it's called socialism and communism" (Little, 2020).

Between August and September 2020, New Zealanders gathered in Auckland, first by the hundreds, later with over a thousand participants, to promote a "freedom rally" that condemned the lockdown, 5G, vaccines, and the Trans Pacific Partnership Agreement (TPPA) (Earley, 2020). Australian police made dozens of arrests during protests in Melbourne, Sydney, Brisbane, Adelaide, and Perth (BBC, 2020a).[3] In August, thousands in South Korea protested against Moon Jae-in's policies as the country experienced a spike in infections, with some protestors carrying signs that read "Expel Moon Jae-in" (Cha, 2020), while opponents of vaccination in London gathered in Trafalgar Square to protest lockdown regulations, with many railing against 5G as well as expressing their scepticism about COVID-19. These events coincided with another protest in Berlin that drew thousands of COVID-19 sceptics, with one holding a sign that read "Back Then Witch Craze, Today Corona Craze" (Gayle & Blackall, 2020; WION, 2020). In Japan, a small group of protestors supporting a fringe political party gathered to protest mask wearing in Tokyo's Shibuya Station. Its leader encouraged the group of 100 protestors to ride the Yamanote Line to confront passengers on the train and make them "feel stupid" for donning masks; only "a dozen or so protestors went along for the ride" (*Japan Today*, 2020).[4] During the fall of 2020, there were anti-mask, anti-lockdown, and anti-vax protests in Italy and France. Hundreds protested in the poorer areas of Madrid, Spain, which during September 2020 had the highest COVID-19 infection rate in the EU. Some held signs proclaiming "No to Segregation" and "No to a Class-based Lockdown" (BBC, 2020b).

In October, the Italian cities of Genoa, Milan, Palermo, Rome, Trieste, and Turin saw protests erupt against the strict curfew and closure of restaurants, bars, gyms, and cinemas – not unlike the many protests seen across the United States during the same period (BBC, 2020b). The protests in Italy during this period were followed by a small group who demonstrated in Mumbai against the mandatory use of masks and imminent vaccinations. In the case of India, Anant Bhan, researcher of Bioethics and Health Policy at Global Health, responded sensibly, noting that "while wearing a mask may seem like a cumbersome exercise for many, it has to be worn because of the protection it offers. Just like wearing a helmet while driving isn't particularly convenient … but people wear it, for safety reasons" (Dey, 2020). In Thailand, protests during the COVID-19 period were due to what Thais perceived as their government's tepid efforts to mitigate the negative economic effects of the pandemic during a period of severe economic downturn. Thousands voiced their grievances against the government of Prayuth Chan-ocha, former military man and now prime minister, and indirectly against the country's unpopular monarch, Maha Vajiralongkorn (Robinson, 2020; Macan-Markar, 2020; *Guardian*, 2020a). By the conclusion of 2020, 26 countries, including Mexico, Brazil, and Nigeria, saw over 30 major protests erupt across their urban centres (Carothers & Press, 2020).

In spite of the trying circumstances experienced by many states and societies around the world, vaccine developments by the end of 2020 offer us reasons for cautious optimism. We can also find sombre optimism in Slack's observation of plague dynamics and the responses they elicited across time: "Wherever they were enforced, quarantine regulations" and their imposition of "segregation and isolation on small families" were nonetheless "as often supported by people who wished to protect themselves and their children as they were opposed by those who found themselves shut up" (2021, p. 71). Yet regardless of differing views, one must remember that the vast majority of people in the countries examined in our work accepted the need for compliance, if grudgingly, even as the beginning of the pandemic saeculum amplified our exhaustion and stress. Al Jazeera provides a comprehensive listing of countries around the world that mandated different degrees of lockdown duration between 16 January 2020 and 15 January 2021 (see Table 14.1) (Haddad, 2021).

As our work draws to a close at the time of this writing in 2022, public officials, media personalities, and news soundbites around the world continue to express their hopes of vanquishing COVID-19 amidst a frenetic terrain of vaccination dynamics. In the United States, even as early as the end of 2020, one heard during nightly news that we were "seeing the light at the end of the tunnel," a common refrain, one superimposed upon news imagery of people receiving COVID-19 tests and/or vaccines. Even with the arrival of the delta variant in 2021, the presence of vaccines for more fortunate countries changed the dynamics of how people managed their bodies in their communities. Many parts of the globe certainly developed a more humbled understanding of the human condition, seen historically by Slack as one where "people somehow learned to live and die with such privations" (Slack, 2021, p. 72). Yet numerous unresolved problematics remain about the crisis environments that are our communities and societies. That is, and in hindsight, the body *was* sick – but what about the social forces of community in the lifeworld as they reconfigured, if not deformed, during the pandemic? No vaccines exist for such community repair or citizens' fears of their public spaces.

Our work highlighted a variety of diacritica from which this chapter offers some open-ended considerations, many of which refer to the significance of crisis contexts as epistemic and ontologic, regardless of whether we are here responding to micro- or macro-level liminalities and environments. As can be seen in Adur and Narayan's globally oriented examination of the discontents associated with the pandemic, as well as Petersson Hjelm's chapter on Sweden's response: where private contexts are not accessible by some authoritative apparatus of the state insofar as providing resources for pandemic mitigation are concerned, manifestations of domestic violence or negligence towards our elderly surface, reflecting how crises, although systemic in nature, can be demographic in terms of their consequences. Systemic crisis also lays bare the

Table 14.1. Lockdown durations around the world between 16 January 2020 and 15 January 2021; adapted map from Al Jazeera and the Oxford Coronavirus Response Tracker (Haddad, 2021). Note: Italicized countries are those examined in the volume.

| 300–366 days (countries ordered by days of lockdown) | | | |
| --- | --- | --- | --- |
| Country | Days of lockdown | Country | Days of lockdown |
| Bolivia | 320 | Argentina | 303 |
| Peru | 307 | Bahamas, Jamaica, Venezuela | 302 |
| Honduras | 306 | El Salvador | 301 |
| Paraguay | 305 | *India* | 300 |

| 250–299 days (countries ordered by days of lockdown) | | | |
| --- | --- | --- | --- |
| Country | Days of lockdown | Country | Days of lockdown |
| Algeria | 299 | Suriname | 273 |
| Chile | 297 | Guyana | 272 |
| Congo | 294 | Bangladesh | 271 |
| Nigeria | 293 | Rwanda | 270 |
| Guinea | 292 | Uganda | 263 |
| Tonga | 291 | Morocco | 260 |
| China | 289 | Fiji | 257 |
| Kosovo | 287 | Brazil | 256 |
| Mexico | 285 | Libya | 252 |
| Palestine | 278 | Panama | 252 |
| Myanmar | 277 | Dominican Republic | 251 |
| Puerto Rico, Chad | 276 | | |

| 200–249 days (countries ordered by days of lockdown) | | | |
| --- | --- | --- | --- |
| Country | Days of lockdown | Country | Days of lockdown |
| Philippines | 247 | Pakistan | 229 |
| Kenya | 246 | South Africa | 220 |
| Zimbabwe | 245 | Kazakhstan | 213 |
| Eritrea | 263 | Haiti | 211 |
| Israel | 233 | *United States* | 201 |

| 150–199 Days (countries ordered by days of lockdown) | | | |
| --- | --- | --- | --- |
| Country | Days of lockdown | Country | Days of lockdown |
| Nepal | 199 | Jordan, Sri Lanka | 177 |
| Belgium, Ecuador, Iraq | 195 | Oman | 173 |
| *Vietnam* | 194 | Ireland | 171 |
| Guatemala | 193 | Colombia, Russia | 169 |
| Lebanon, Kyrgyz Republic, Portugal | 192 | Liberia | 163 |

*(Continued)*

Table 14.1. Lockdown durations around the world between 16 January 2020 and 15 January 2021; adapted map from Al Jazeera and the Oxford Coronavirus Response Tracker (Haddad, 2021). (*Continued*)

| Country | Days of lockdown | Country | Days of lockdown |
|---|---|---|---|
| Australia, Indonesia | 185 | Italy, Tunisia | 156 |
| Spain | 183 | Eswatini | 155 |
| Azerbaijan | 181 | Afghanistan | 154 |
| Türkiye | 180 | | |
| **100–149 Days (countries ordered by days of lockdown)** | | | |
| Country | Days of lockdown | Country | Days of lockdown |
| France, Georgia | 146 | Madagascar, South Sudan | 122 |
| Albania | 145 | Romania | 120 |
| Uzbekistan | 141 | Cape Verde | 117 |
| Greece | 138 | Cuba | 116 |
| Austria | 137 | Yemen | 109 |
| Bhutan | 132 | Gabon | 108 |
| Cyprus | 131 | Democratic Republic of Congo, Senegal | 107 |
| Angola | 129 | Barbados | 105 |
| Aruba | 128 | Bermuda | 102 |
| Hungary, Mauritania | 124 | | |
| **50–99 Days (countries ordered by days of lockdown)** | | | |
| Country | Days of lockdown | Country | Days of lockdown |
| Namibia | 96 | Mongolia, Mauritius | 68 |
| United Kingdom, Malaysia | 94 | Gambia | 67 |
| Sudan, Slovakia | 92 | Belize | 65 |
| Egypt, Togo | 91 | Syria, United States Virgin Islands | 62 |
| Botswana | 90 | East Timor | 60 |
| Bahrain, *Germany* | 88 | Lesotho | 59 |
| Czech Republic, Luxembourg | 79 | Dominica | 56 |
| Burkina Faso, *Singapore* | 72 | Djibouti | 55 |
| *New Zealand* | 71 | Moldova, Serbia | 53 |
| Trinidad and Tobago | 70 | San Marino | 52 |
| Malta, Saudi Arabia | 69 | Estonia | 50 |
| **1–49 Days (countries ordered by days of lockdown)** | | | |
| Country | Days of lockdown | Country | Day/s of lockdown |
| Croatia, Netherlands | 49 | Zambia | 23 |
| Guam | 41 | Cambodia | 21 |
| United Arab Emirates, Papua New Guinea | 40 | Ghana | 20 |
| Bulgaria | 37 | Kiribati | 16 |

| Country | Days of lockdown | Country | Days of lockdown |
|---|---|---|---|
| Seychelles | 36 | Iran | 13 |
| Laos | 35 | Uruguay | 11 |
| Central African Republic | 34 | Poland | 9 |
| Lithuania | 31 | Canada | 7 |
| Somalia, *South Korea* | 28 | Solomon Islands, Tajikistan | 3 |
| Sierra Leone | 24 | Mozambique | 1 |
| **No nationwide lockdown enforced, only recommended or isolated lockdowns** | | | |
| Andorra, Cote d'Ivoire, Costa Rica, Ethiopia, Finland, Greenland, Hong Kong, *Japan*, Latvia, Mali, Malawi, Niger, Norway, Qatar, Slovenia, *Sweden*, Switzerland, *Thailand*, Turkmenistan, Ukraine, and Vanuatu | | | |
| **No nationwide restrictions or recommendations to stay at home** | | | |
| Burundi, Benin, Belarus, Brunei, Cameroon, Faroe Islands, Iceland, Macao, Nicaragua, Taiwan, Tanzania | | | |

propensity for survival through social class and social networks that are functions of group access to resources, as notes McCaffree and Saide's chapter on social forces influencing pandemic distancing outcomes in the United States.

In scenarios where more stable outcomes related to pandemic mitigation were realized by certain countries during early 2020, the initial systemic crisis experienced by the state had consequences that affected political outcomes – a dynamic that renders social crises a political issue and catalyst for social change, as documented in Armoudian and Duncan's chapter on New Zealand. Indeed, politics frequently underlie the framing of health care. Yet as noted by Pangsapa in her examination of Thailand, sociocultural forces also shape a country's pandemic mitigation efforts when they continue to inspire respect for authority within the population. Additionally, the assumption that the state can, by default, serve as an agent of crisis management, even if in its function there is a temporary curtailment of individual rights that is incurred for the greater good, is a theme addressed by Huh and Yoon in their discussion of South Korea as well as in Scherr's examination of Germany during the pandemic's pre-vaccination phase, with both countries' efforts resulting in overall population compliance with their governments' emergency measures. Another key manifestation of state actions towards the pandemic is how health figures – when conveyed by countries not traditionally seen as belonging to the core or semi-periphery, or to developed or metropole nations, however one chooses to employ the terms – often incur Orientalist dismissiveness from Western observers towards indigenous conceptions of stress and these countries' communicating of positive outcomes, a situation seen in Dao's chapter on Vietnam and its efficacious pandemic mitigation strategies.

Understanding pandemic mitigation and systemic crisis, however, also set into motion a process of critically assessing how research can be conducted for the betterment of community, as explored by Sen and Tan in experiencing the rupture of researcher-subject in India's West Bengal and Singapore, and Tan and Tran's problematizing of doctor-patient interaction at a hospital in New York City when it was the epicentre of COVID-19 infections in the United States. Our work attempted to demonstrate how many of the aforementioned dynamics related to pandemic mitigation outcomes are tied to the health of the lifeworld, one that has seen many publics emptied from society, disrupting healthy bonds and relationships, while reanimating destructive impulses and actions, as seen in my chapter examining the greater Los Angeles area. Such manifestations often surface as law-breaking activities, which require us to consider alternative spatial approaches towards assessing, measuring, and mapping crime, according to Plickert and Cooper. Indeed, if one were to move our level of analysis from cities and return to the level of the state, lacklustre public policies and poor distributional logistics that are unable to fulfil the social contract can point to an inefficacious state, a condition extrapolated from Japan's early pandemic response in Fukurai's chapter, which was seen to privilege recreational revenue over health concerns. The various discourses that have emerged in this volume thus suggest how a complex interplay of state regulations – enforceable or unenforceable – and actors contesting or supporting state mandates, as well as forces related to urban disassembly and denudation of different publics, served to globally reconfigure community and society during the pre-vaccination phase of the pandemic.

Extrapolations derived from our volume point to some important considerations for future address of crisis research. Chief among these is that societies experiencing systemic crisis of a non-warfare and prolonged nature should be conceptualized as bona fide entities with their own social systems and institutional manoeuvres, albeit ones experiencing delimitations and malfunctions. Systems in crisis can be seen as alternative societies that have reordered social priorities to respond to existential threats to their population while social forces of recreational culture are constrained, if not prevented, from operating. Yet such systems experiencing crisis, especially in decentralized social environments, can enable sectarian indignations and hostilities to surface in a variety of different spatialities. Such discontents appear in the most intimate confines of the micro-level as well as across major social institutions constituting the macro-level terrain; social discontents become a venting mechanism for many of society's unresolved issues, issues whose gravity is amplified if systemic crisis is accompanied by a Habermasian legitimation crisis, seen when public opinion rejects the legitimacy of respective governments and their dictates for compliance. Thus, a cautious extrapolation that can be seen as an undercurrent flowing through the chapters of this volume is that the state exhibits varying degrees of success in fulfilling its social contract, with fulfilment dependent not only on public systems and their logistical efficiencies but also on the sense of legitimation citizens are willing to acknowledge in their states' actions towards resolving medical morbidities *and* social crises. However, the chapters herein suggest that the efficacy of pandemic mitigation across different countries tends to be due to increasing centralization for the sake of fulfilling the social contract, even within democracies, and frequently at the expense of community dynamics. Such a process also made visible abrupt shifts towards emphasizing occupational roles and legal statutes that attend to our existential human condition, the degree of which should be seen as indicative of the legitimacy of the state's social contract with its population. In this regard, future examination of the social consequences of pandemics will need to encapsulate more cultures and countries across more continents, a blind spot still apparent in our volume's lack of coverage of states and cultures in South America, Africa, and the Middle East.

Another important consideration not yet addressed in explicit detail is the notion of how pandemic fatigue is an expression of one's inability to attend to a solitude that forces individuals to confront themselves. I am of the view that social scientists will need to embark on conducting research in this particular area so as to equip our armamentarium for future responses to pandemic-generated crises. That is, how do our aforementioned dynamics cascade towards the existential self, one that for Kotarba and Fontana "is the product of both experience and the language used to render that experience understandable" (1984, p. xii)? Such language – be it through law and/or state mandates – invokes Durkheim's view that "We cannot live without representing to ourselves the world and the objects which fill it" – yet by virtue of relying on the representations, "we get attached to the world at the same time that we get attached to ourselves" (cited in Tiryakian, 1962, p. 48; see also Fong, 2014). Although voluntary solitude is ideal for regeneration and enhancing mindfulness, systemic crisis of a pandemic nature and its accompanying physical distancing mandates that *force solitude upon the actor* are an entirely different matter (Coplan & Bowker, 2014; Fong, 2014). The distinction between a healthy, voluntary solitude and its unhealthy counterpart in the guise of mandated isolation remains a less examined topic in the context of systemic crisis. That is, what needs address, perhaps with a philosophical discourse, is who we *have* become, or perhaps more importantly, who we *can* become, under pandemic distancing, as our communities and societies experienced and continue to experience the consequences of COVID-19. In an age when embedded materialist culture has indoctrinated us to be externally oriented creatures dependent on outside validation for our sense of self-worth, we need to explore who we have become and who we can become when forced to experience the fearsome process of knowing the self without superficial and/or misinformational stimuli. Such an undertaking is important, for if states, societies, and communities fail in their dispatch of well-being, only selves and in-groups will remain as arbiters of existence in the lifeworld.

This concluding chapter, having considered some of the blind spots in this volume, nonetheless argues that the primary impetus influencing the vast majority of social forces in operation during the pandemic is focused on redeeming community and health institutions of society as well as reanimating the raison d'être of the state – even if it requires actors to first expose the state's inefficacious "insides." Our conclusion argues that during systemic crises brought forth by the pandemic, social discontents and angsts expressed by actors are but different clarion calls for the social contract to be realized, with many countries' governments rising to the occasion through their emergency responses, while others faltered. Our volume also attempted to demonstrate how the social contract itself became a variable during the pandemic: Some social environs saw their citizens exhibit compliance that upheld the raison d'être of the state, while others articulated anti-systemic narratives that suggest how these states are experiencing the pathos of a bona fide legitimation crisis of a Habermasian nature.

Our contributors remind us that the manner in which states, communities, and individuals have responded during the first year of the COVID-19 pandemic crisis, therefore, reflects how the social contract for members of society becomes a variable shaped by economics, political climate, history, and social forces that have manifested concomitantly during urban *disassembly*. These social dynamics, our authors emphasize, have in turn reconfigured large swathes of society. With full acknowledgment of society's liminalities, our scholars demonstrated how a state's competence or incompetence can generate alternative communities and practices that can, in the context of emptied and reconfigured publics, impact and shape all aspects of the population's ability to thrive in the lifeworld, rendering the COVID-19 crisis as much a political, cultural, and philosophical issue as it is a health issue. For better or worse, systemic crisis brought about by the pandemic, then, is epistemic and ontologic – it offers up new social knowledge about what it means for a human *being* to *become* as their communities experience crisis conditions. How we can write the next chapter in the saga about the human condition as it churns through its reconfigured communities and societies will depend on what we have learned when humbled by our vulnerabilities, for civilization is dependent on its citizens expressing its sociality in planned, regulated spatialities, a dependency that can, however, take us towards or away from cultures, towards or away from knowledge, towards or away from empowerment. With many communities and societies being compelled to reimagine life under existential urgencies that have tested people daily in 2020, we can only hope the information, insights, and narratives contained herein can presage social transformations that will herald humanity's comeback with a new will to power.

NOTES

1 When I was beginning the volume's final chapter in late December 2020, California – the state where I live – had announced a 0% capacity in intensive care units (ICUs). To get a sense of scale, in a state with close to 40 million people and with a land area larger than the country of Japan, *no* hospitals in the state at the time in which we concluded our work had available ICUs. This situation *preceded* the arrival of COVID-19's more pernicious delta variant, which later affected countries thought to have escaped the pandemic's effects because of the introduction of vaccines. Thailand and India are notable examples, as the delta variant wreaked havoc on their populations through much of 2021.
2 Yet progression was followed by regression, as in the United States when it registered over 330,000 deaths from COVID-19 by the conclusion of 2020.
3 Melbourne, located in the state of Victoria, accounted for 75% of Australia's total infections and 90% of all deaths (BBC, 2020a).
4 Between January 2020 and January 2021, Japan and Thailand did not experience enforced but only recommended (or isolated) lockdowns (Haddad, 2021).

REFERENCES

Ankel, S. (2020, May 24). Germany is at the forefront of a global movement of anti-vaxxers obsessed with Bill Gates and it could mean the coronavirus is never defeated. *Business Insider.* https://www.businessinsider.com/germany-becomes-forefront-of-a-global-movement-of-anti-vaxxers-2020-5.

BBC. (2020a, September 5). Coronavirus: Arrests at Australia anti-lockdown protests. https://www.bbc.com/news/world-australia-54040278.

BBC. (2020b, September 20). COVID-19: Hundreds protest against localised Madrid lockdowns. https://www.bbc.com/news/world-europe-54227057.

BBC. (2020c, October 27). COVID: Protests take place across Italy over anti-virus measures. https://www.bbc.com/news/world-europe-54701042.

Carothers, T., & Press, B. (2020, October 15). The global rise of anti-lockdown protests – and what to do about it. *World Politics Review.* https://www.worldpoliticsreview.com/articles/29137/amid-the-covid-19-pandemic-protest-movements-challenge-lockdowns-worldwide.

Cha, S. (2020, August 14). Thousands protest against Moon as Seoul scrambles to curb virus resurgence. Reuters. https://www.reuters.com/article/us-health-coronavirus-southkorea/thousands-protest-against-moon-as-seoul-scrambles-to-curb-virus-resurgence-idUSKCN25B0A0.

Coplan, R.J., & Bowker, J.C. (2014). All alone: Multiple perspectives on the study of solitude. In Coplan, R., & Bowker, J. (Eds.), *A Handbook of Solitude: Psychological Perspectives on Social Isolation* (pp. 3–13). Wiley-Blackwell Publishers.

Dey, S. (2020, October 6). Inside India's anti-mask movements: Why these people don't believe in masks, vaccine for coronavirus. News18. https://www.news18.com/news/buzz/inside-indias-anti-mask-movements-why-these-people-dont-believe-in-masks-vaccine-for-coronavirus-2937147.html.

Earley, M. (2020, September 12). Coronavirus: More than a thousand turn out for anti-lockdown rally in Auckland. Stuff. https://www.stuff.co.nz/national/health/coronavirus/300105763/coronavirus-more-than-a-thousand-turn-out-for-antilockdown-rally-in-auckland.

Fong, J. (2014). The role of solitude in transcending social crises – new possibilities for existential sociology. In Coplan, R., & Bowker, J. (Eds.), *A handbook of solitude: Psychological perspectives on social isolation* (pp. 499–516). Wiley-Blackwell Publishers.

Fong, J. (2025, forthcoming). The American experience of democracy deserts during the pandemic. In Bowker, M. (Ed.), *Severe social withdrawal and the Covidian era: Psychoanalytic perspectives*. Phoenix Publishing House.

Gavi: The Vaccine Alliance. (2021a). Finnish COVAX donation kicks off with shipment to Côte d'Ivoire. Retrieved 8 November 2021, from https://www.gavi.org/news/media-room/finnish-covax-donation-kicks-shipment-cote-divoire.

Gavi: The Vaccine Alliance. (2021b). Gavi, the vaccine alliance helps vaccinate almost half the world's children against deadly and debilitating infectious diseases. Retrieved 19 February 2021, from https://www.gavi.org/our-alliance/about.

Gavi: The Vaccine Alliance. (2021c). The COVID-19 vaccine race – weekly update. Retrieved 30 October 2021, from https://www.gavi.org/vaccineswork/covid-19-vaccine-race#:~:text=In%20December%202020%2C%20the,start%20of%20the%20month.

Gayle, D., & Blackall, M. (2020, August 29). Coronavirus sceptics, conspiracy theorists and anti-vaxxers protest in London. *Guardian*. https://www.the*guardian*.com/world/2020/aug/29/coronavirus-sceptics-conspiracy-theorists-anti-vaxxers-protest-london.

*Guardian*. (2020a, July 18). Thousands of anti-government protesters rally in Thailand's capital. https://www.the*guardian*.com/world/2020/jul/18/thousands-of-anti-government-protesters-rally-in-thailands-capital.

*Guardian*. (2020b, December 16). 138,000 people in UK receive COVID vaccine in first week. https://www.the*guardian*.com/world/2020/dec/16/138000-people-in-uk-have-received-covid-vaccine.

Haddad, M. (2021, March 28). Mapping coronavirus anti-lockdown protests around the world. Al Jazeera. https://www.aljazeera.com/news/2021/2/2/mapping-coronavirus-anti-lockdown-protests-around-the-world.

Haimowitz, I. (2020, December 17). No one is immune: The spread of Q-anon through social media and the pandemic. Center for Strategic & International Studies. https://www.csis.org/blogs/technology-policy-blog/no-one-immune-spread-q-anon-through-social-media-and-pandemic.

*Japan Today*. (2020, August 11). Anti-mask group in Tokyo slammed for "cluster festival." https://japantoday.com/category/national/anti-mask-group-in-tokyo-slammed-for-cluster-festival.

Kotarba, J.A., & Fontana, A. (1984). *The existential self in society*. University of Chicago Press.
Little, S. (2020, September 13). QAnon conspiracy theorists, far-right group join Vancouver anti-mask rally. Global News. https://globalnews.ca/news/7332529/vancouver-anti-mask-rally-qanon/.
Mullard, A. (2020, November 30). How COVID vaccines are being divvied up around the world. *Nature*. https://www.nature.com/articles/d41586-020-03370-6.
O'Loughlin, E., & Zaveri, M. (2020, May 5). Irish return an old favor, helping Native Americans battling the virus. *New York Times*. https://www.nytimes.com/2020/05/05/world/coronavirus-ireland-native-american-tribes.html.
Porterfield, C. (2020, March 13). Italians serenade each other to fight loneliness during coronavirus lockdown. *Forbes*. https://www.forbes.com/sites/carlieporterfield/2020/03/13/italians-serenade-each-other-to-fight-loneliness-during-coronavirus-lockdown/.
Press Progress. (2020, May 8). Canada's anti-lockdown protests are a ragtag coalition of anti-vaccine activists, conspiracy theorists and the far-right. https://pressprogress.ca/canadas-anti-lockdown-protests-are-a-ragtag-coalition-of-anti-vaccine-activists-conspiracy-theorists-and-the-far-right/.
Ravelo, J.L. (2020, October 9). Behind Vietnam's COVID-19 success story. *Devex*. https://www.devex.com/news/behind-vietnam-s-covid-19-success-story-98257.
Robinson, G., & Markar, M. (2020, October 21). Thai protests build as pandemic fuels unrest across Southeast Asia. *Nikkei Asia*. https://asia.nikkei.com/Spotlight/The-Big-Story/Thai-protests-build-as-pandemic-fuels-unrest-across-Southeast-Asia.
Slack, P. (2020). Responses to plague in early modern Europe: The implications of public health. *Social Research*, *87*(2), 409–28. https://doi.org/10.1353/sor.2020.0046.
Slack, P. (2021). *Plague: A very short introduction*. Oxford University Press.
Sommer, W., & Kucinich, J. (2020, April 22). Anti-vaxxers and lockdown protesters form an unholy alliance. *Daily Beast*. https://www.thedailybeast.com/anti-vaxxers-are-forming-an-unholy-alliance-with-shelter-in-place-protesters.
*South China Morning Post*. (2020, June 8). From Cold War to coronavirus, Thailand's hero grandma fights to protect villagers from COVID-19. https://www.youtube.com/watch?v=d2_lWTDU5Gc&ab_channel=SouthChinaMorningPost.
Tiryakian, E. (1962). *Sociologism and existentialism: Two perspectives on the individual and society*. Prentice Hall.
Wehner, P. (2020, March 13). The Trump presidency is over. *Atlantic*. https://www.theatlantic.com/ideas/archive/2020/03/peter-wehner-trump-presidency-over/607969/.
WION. (2020, September 6). No-mask, anti-lockdown protests gather pace across the globe. https://www.wionews.com/photos/no-mask-anti-lockdown-protests-gather-pace-across-the-globe-325774#no-mask-movements-in-italy-325768.
Yeung, J. (2020, March 30). A police officer is wearing a coronavirus helmet to warn people to stay inside during India's lockdown. CNN. https://www.cnn.com/2020/03/30/asia/coronavirus-helmet-chennai-intl-hnk-scli/index.html.

# Contributors

**Shweta Adur**, California State University, Los Angeles

**Maria Armoudian**, University of Auckland

**Emily Cooper**, George Mason University

**Amy Dao**, California State Polytechnic University, Pomona

**Bernard Duncan**, Ministry of Education of New Zealand

**Jack Fong**, California State Polytechnic University, Pomona

**Hiroshi Fukurai**, University of California, Santa Cruz

**Kelly Huh**, California State Polytechnic University, Pomona

**Kevin McCaffree**, University of North Texas

**Anjana Narayan**, California State Polytechnic University, Pomona

**Piya Pangsapa**, Thammasat University

**Ann-Christine Petersson Hjelm**, Uppsala University

**Gabriele Plickert**, California State Polytechnic University, Pomona

**Anondah Saide**, University of North Texas

**Albert Scherr**, Freiburg University of Education

**Amritorupa Sen**, National University of Singapore

**Junbin Tan**, Princeton University

**Phu Tran**, New York Presbyterian-Queens Hospital

**Hyejin Yoon**, University of Wisconsin–Milwaukee

# Index

A3PCON. *See* Asian Pacific Policy & Planning Council
Abe, Shinzo, 12, 13, 56; criticism of, 69; on herd immunity, 70; resignation of, 73; on Tokyo Summer Olympics, 59–60
Advanced Cardiac Life Support (ACLS), 205
Afghanistan, 372
age: pandemic concern and, 192; well-being and, 184. *See also* elderly population
Albania, 372
Alexander, Paul, 73
Algeria, 371
alienation, xix
Alternative für Deutschland, 313
alternative world-making projects, 89
Althusser, Louis, 208–9
Andorra, 373
Angola, 372
anonymity, 248
anti-fragility thesis, 6
anti-modernism, 315–16
anxiety, 179
ANZ. *See* Australia and New Zealand Banking Group
ANZAC Day, 333
appearational ordering, 251, 265

Ardern, Jacinda, 25, 328–9, 350; criticism of, 336–7, 348
Argentina, 371
Aruba, 372
Asian Americans: hostility against, 3, 264–5, 267; protests, 266; Trump and, 364
Asian Pacific Policy & Planning Council (A3PCON), 264
AstraZeneca, 362
audience role prominence, 249
*Austerlitz* (Sebald), 92
Australia: lockdown in, 372; protests, 369
Australia and New Zealand Banking Group (ANZ), 332
Austria, 372
authoritarianism: in South Korea, 34–5; third places and, 257
authority: social distance and respect for, 34; South Korean respect for, 34–5; in Sweden, 278, 281, 289–91
automobile: theft, 226, 232, 234; urban experience and, 248
Avery, Ray, 346
Azerbaijan, 372

Babu, Rajesh, 363
Bach, Thomas, 56

Bahamas, 371
Bahrain, 372
Baker, Michael, 328, 332, 346
Bangladesh, 371
Barbados, 372
Barry, Hilary, 349
*Behavior in Public Spaces* (Goffman), 249
Belarus, 373
Belgium, 371
Belize, 372
Benin, 373
Berenson, Alex, 348–9
Berger, Brigitte, 315
Berger, Peter L., 315
Bermuda, 372
Betty, Bryan, 335–6
Bhan, Anant, 369
Bhutan, 372
Biden, Joe, 364
biological essentialism, 141–3
BioNTech SE, 362
Black Death, 18, 177, 365
Black Lives Matter Movement, 313. See also Floyd, George
Black Swan event, xv
Bloomfield, Ashley, 329
Bolivar, Simon, 66
Bolivarian Revolution, 66
Bolivia, 371
Bollyky, Thomas J., 347
Bolsonaro, Jair, 67; crime against humanity charge, 58, 71–4
Botswana, 372
Braae, Alex, 349
Brazil: crime against humanity in, 58, 71–4; Cuban doctors in, 67; lockdown in, 371
breaching experiment, 101–2
break-of-bulk, 21, 262–4
Bridges, Simon, 331, 340
Brunei, 373
Bulgaria, 372

Burkina Faso, 372
Burundi, 373
Butler, Judith: on distance, 14–15, 88, 91–2; on suffering, 88

CAA. *See* Chinese for Affirmative Action
calculation rationality, 315
California, 19–21; crime trends, 227–37; crime trends 10 months before and 10 months after, 234–7; crime trends March–June 2020, 231–4; emergency declared in, 258; hospital capacity, 377n1; interventions, 229; justice reforms, 228; social distance in, 20. *See also* Los Angeles
Cambodia, 372
Cameroon, 373
Canada: lockdown in, 373; protests, 368; vaccine in, 361–2
Cape Verde, 372
*Captaining a Team of 5 Million*, 348
Caraccio, David, 267
cardiopulmonary resuscitation (CPR), 206
Castillo, Maria Luisa, 267
Castro, Fidel, 66
Castro, Raul, 66
catastrophe, 26n6
*Catastrophe and Social Change* (Prince), 8
CCSA. *See* Centre for COVID-19 Situation Administration
CDU. See *Christlich Demokratische Union Deutschlands*
celebrity, 247–8
Center for Strategic and International Studies, 368
Center of Juvenile and Criminal Justice, 227, 228
Central African Republic, 373
Central Disaster and Safety Countermeasures Headquarters, 40

Centre for COVID-19 Situation
  Administration (CCSA), 110
Chad, 371
Chang, Elbert, 267
Changchun Heber, 65
child abuse, 158; intimate partner
  violence and, 167
childcare, 41
Chile, 371
China: Cuba collaborating with, 12,
  65–6; Cuban doctors in, 65–6; GDP
  per capita, 59; lockdown in, 371; test
  ranking, 63
Chinese for Affirmative Action (CAA),
  264
Chocktaw Nation, 362–3
Christchurch shooting, 333
*Christlich Demokratische Union
  Deutschlands* (CDU), 312, 320n15;
  approval rate, 313
Chun Doo-hwan, 35, 36
Chung Sye-kyun, 40
The Circuit Breaker, 93
civil inattention, 249–50
civility towards diversity, 249–50
Clark, David, 328
collectivism, 34, 50
Collins, Judith, 339, 340
Colombia, 371
colonialism, 16–17; exoticization and,
  149; numerical representation and,
  138; Vietnam and, 142–3. *See also*
  decolonization
commercial burglary, 226, 232
Committee on Disaster Studies,
  4–5
Committee Terms of Reference, 281
common nationalism, 23, 305; defining,
  303; responsibility and, 303; visibility
  of, 304
Communicable Diseases Act, 112, 282–3;
  changes to, 284–5

communication: components of, 318;
  internet-based, 317; Luhmann on, 318;
  media and, 317; sociological, 318; in
  Vietnam, 140
community recovery resources, 7
compliance: political identity and, 181; in
  South Korea, 11; well-being and, 182
Compulsory Psychiatric Care Act, 282
Congo, 371
Constitution Act 1986, 339–40
contact, unsettling, 215–17
contact tracing: genomic sequencing
  and, 344; in New Zealand, 344–5;
  privacy and, 43, 345; in Singapore, 93;
  in South Korea, 42–9; technology for,
  42; in Vietnam, 139. *See also* disease
  surveillance
contracting, 203
control, intimate partner violence and,
  158–9
controllability, 315
conversational entanglements, 249
cooperative motility, 248–9
coping strategy: restrained helpfulness
  as, 249; territorializing as, 39
corporate crime, 70
Costa Rica, 373
Cote d'Ivoire, 373
Coughlan, Thomas, 340
CPR. *See* cardiopulmonary resuscitation
Crabtree, Peter, 348
Crenshaw, Kimberlé, 159
crime: of aggression, 72; assumptions,
  222; automobile, 226, 232, 234;
  California 10 months before and
  10 months after, 234–7; California
  March–June 2020, 231–4; California
  trends, 227–37; child abuse, 158,
  167; commercial burglary, 226, 232;
  corporate, 70; decline in, 227–8; event,
  226; financial security and, 221; in
  Florida, 224; future considerations,

crime (*continued*)
237–9; geographic boundaries and, 20; government, 71; hurricanes and, 223–4; Influenza Pandemic 1918 and, 223; in Japan, 68–74; occupational, 70; poverty and, 224–5; property, 226; residential burglary, 226, 232; social distance and, 19–20; state-corporate, 71; theoretical explanations for, 225–7; trends, 223–37; unemployment and, 224–5; vandalism, 226, 232–3; variation in, 221; war, 72; white collar, 68–74; women and, 39

crime against humanity, 57; Bolsonaro charged with, 58, 71–4; in Brazil, 58, 71–4; herd immunity as, 70; in Japan, 68–74; in United States, 71–4

Crime Pattern Theory, 225–7
critical situations, xix
Croatia, 372; test ranking, 63
Cuba: China collaborating with, 12, 65–6; lockdown in, 372
Cuban doctors, 68; in Brazil, 67; in China, 65–6; in Venezuela, 66–7; in Vietnam, 65
cultural diversity, 4; in South Korea, 37
cultural norms, 34
culture shock, xvi–xvii
culture wars, 364–5
Cummings, Dominic, 69
Cupboard of Happiness, 123–5
CureVac, 362
Cyprus, 372
Czech Republic, 372

Dam, Vu Duc, 139
data acquisition process, 9
death: foreseeable, 212; helplessness and, 211; medical workers and, 210–15; passivity and, 211; untimely, 212
death toll: in Germany, 305–6; in Italy, 68; in Japan, 57; in New Zealand, 343;
South Korea, 45; in Sweden, 276; in Taiwan, 57; tuberculosis, 314–15; in United States, 3; in Venezuela, 66–7; in Vietnam, 16–17, 65, 141, 143–7
Declaration of a New Coronavirus Emergency, 62
decolonization steps, 136
deep state, 368
Defense Production Act, 72–3
Delaware, 25n2
de Ligt, Joep, 344
Democratic Republic of Congo, 372
Democrats (USA): pandemic concern and, 187, 189, 191; psychological well-being and, 189; trust in media and, 187, 189, 191
depression, 179
dignified life, 286
Disaster Impact Model, 7; diagram, 8
Disaster Research Center (DRC), 5; founding of, 25n2; funding for, 25n2
disaster studies, 4; early iterations of, 5; geography and, 6; internalization of, 5; Lindell and, 7; mental health and, 5; military and, 5; positive foundational consequences and, 5; sociology and, 6
disease surveillance, 320n6
disinformation: in New Zealand, 348–50; vaccine, 365
distance: Butler on, 14–15, 88, 91–2; emotional, 19, 210–15; ethical obligations and, 14–15; fieldwork and, 103; gaze from, 14; mediating, 215–17; New Zealand geographic, 327; privilege and, 88; structural, 88; suffering and, 14–15, 88; witnessing and, 90–1. *See also* physical distancing; physical proximity; sensual proximity; social distance
diversity: civility towards, 249–50; cultural, 4, 37; in South Korea, 37
Djibouti, 372

Index 387

domestic violence: defining, 158; protests, 165; shelters, 166. *See also* child abuse; elder abuse; intimate partner violence
Dominica, 372
Dominican Republic, 371
donations, 26n5
Douglas, Andrew, 148–9
Douglas, Mary, 212, 318
DRC. *See* Disaster Research Center
Durkheim, Emile, 26n4, 375
Dutta, Mohan, 96
Dynes, Russell, 25n2

earthquakes, 6; Haiti, 7; Tōhoku, 7, 58–9
Eastern representations, 135
East Timor, 372
Ebola pandemic, 67
economic impact: New York, 162; New Zealand, 330–2, 336–7; political identity and, 183; Tokyo Summer Olympic Games 2020, 60; United States, 178; World Rugby Cup 2019, 60
Ecuador, 371
Edwards, John, 345
Egypt, 372
elder abuse, 158
elderly population: framework of care, 285–7; in New Zealand, 335; restrictions on, 292–4; in Sweden, 21–3, 275–8, 285–7, 292–4
elective surgeries, 205
Elimination of Violence against Women, 160
El Salvador, 371
emergency declaration: in California, 258; in Japan, 12, 56, 60; in Thailand, 115
emergency infrastructure, 207
emotion: dimensions of, 318; rules of, 207; in Thailand, 120; toll in New Zealand, 334–6. *See also* feeling rules

emotional distance: collapse of, 210–15; threats to, 19
empathy, 208
Epidemic Diseases Act of 1897, 367
Eritrea, 371
Estonia, 372
Eswatini, 372
ethical obligations, 14–15
Ethiopia, 373
European Union, 21–4; border closures, 304; nationalistic hostilities in, 304–5; xenophobia in, 305. *See also specific countries*
Evans, Suzanne, 349
Every-Palmer, Susanna, 335
exceptionalism, 126
exclusion, 37–9
executions, public, 251–2
Exemplars in Global Health, 142
exoticization, 149
exposure, 215; Merleau-Ponty on, 216
extra-community assistance, 7

face, 215
facilitating, 203
factual dimensions, 318
family separation immigration policy, 73
Faroe Islands, 373
fear: of hospitals, 206; in Thailand, 114–19; urban experience and, 249–50
feeling rules, 207; modifying, 208
feminist scholarship, 158–61
Ferencz, Benjamin, 73
field data, 9; distance and, 103; privilege and, 103
field rats, 149
Fiji, 371
financial security: crime and, 221; mitigation strategies and, 18
financial well-being, 182, 184–5; pandemic concern and, 187, 192–3; Republicans and, 187

Finland, 276, 373
5G protests, 368–9
flatten the curve, 137
Florida, 224
Floyd, George, 180, 364; physical distancing and, 3–4
Foo Chow, 259
Formosa, 150n1
France, 372
freedom: Germany and, 23; societal imperatives and individual, 22
Freie Demokratische Partei, 313
French Flu, 27n8
Fritz, Charles, 5
Fromm, Erich, 315
*Fukushima* (Lochbaum), 7
Fukushima Nuclear Power Plant, 58–9, 71
funerals, New Zealand ban on, 335–6
future considerations, 374

Gabon, 372
Gaillard, J.C., 8, 9
Gamaleya Research Institute of Epidemiology and Microbiology, 362
Gambia, 372
Garcetti, Eric, 258
Gates, Bill, 368
gaze: from distance, 14; Lacan on, 15, 90–1; politics of, 88–92; researcher's, 87, 102; Vietnam and, 17; Western, 8, 17. *See also* witnessing
gender: explanations for intimate partner violence, 158–60; health precautions and, 179; pandemic concern and, 192; physicality of, 317; well-being and, 184
General Strain Theory, 225–7
genocide, 72
genomic sequencing, 344
Genovese, Kitty, 249
geography: crime and, 20; disaster studies and, 6; distance, 327

Georgia, 372
Germany, 24; approval rate, 313; asylum procedure, 320n14; death toll in, 305–6; early phase in, 305–9; everyday life impact in, 308–9; excess capacity in, 306; feedback loops in, 316; first cases in, 305; freedom and, 23; government measures, 307–8; international air shutdown in, 303–4; international comparisons, 306; lockdown in, 372; masks in, 313; media in, 316–17; overview, 305–7; physical distance in, 308, 317–19; physical proximity in, 317–19; political implementation in, 312–14; political reactions in, 314–17; protests in, 321n17; public healthcare in, 23, 310; public space in, 309; reactions in, 309; reception centres, 312; repatriation operation, 303–4; responsibilities in, 308; schools in, 311–12; short-time work relief, 310; sickness benefit, 310; social acceptance in, 312–14; social impact in, 309–14; sociological lessons, 317–19; socio-spatial differentiation in, 309; success of, 306–7; unemployment in, 309–10; United States compared with, 306; urban experience in, 309; welfare state importance in, 310–11; welfare state limits in, 311–12
germaphobia, 21
Geyer, Felix, xix, 26n4
Ghana, 372
ghost workers, 8–9
GHSA. *See* Global Health Security Agenda
Giddens, Anthony, xix
gig workers, 260–1
Global COVID-19 Index, 26n7
Global Health Security Agenda (GHSA), 144
globalization, 303

Global North, 57, 314–15
global positioning systems (GPS), 34
Global South, 57
global village, xviii
Gluckman, Peter, 337
Goff, Phil, 342
Goffman, Erving, 249, 317
*goshiwon*, 37, 51
Go To Campaign programs (GTC), 61
Government Bill 2019/20, 155, 283
government crime, 71
Gowtham, B., 363
GPS. *See* global positioning systems
Great Global Recession 2008, 58
*The Great Good Place* (Oldenburg), 253
Great Plague of Marseille, 365–6
Greece, 372
Greenland, 373
Green Party, 282
*greng jai*, 121
grief, 216–17
Die Grünen, 313
GTC. *See* Go To Campaign programs
Guam, 372
Guatemala, 371
Guevara, Che, 66
Guiberson, Brenda Z., 6
Guinea, 371
guns, 164
Guterres, António, 17
Guyana, 371

H1N1. *See* Influenza A virus
Haas, J. Eugene, 25n2
Habermas, Jürgen, xv, 256–7
Haimowitz, Ian, 368
Haiti: earthquake, 7; lockdown in, 371
Halifax maritime explosion, 8
Hall, Daniel R., 267
Han Bat Shul Lung Tang, 259–60
Hanke, Steve: on field rats, 149; reactions to, 148; on Vietnam, 147–9

Hanshin Pocha, 260
Hanson, Victor Davis, 349
healing and repairing society, 362
healthcare: in Germany, 23, 310; in Thailand, 125–6; in Vietnam, 65
healthcare workers. *See* medical workers
Heisei Era, 58
helplessness, 211
herd immunity: Abe on, 70; Alexander on, 73; as crime against humanity, 70; Cummings on, 69; defining, 346; in Japan, 69–70; in Sweden, 69–70; Tegnell on, 347; Trump on, 69, 73
Higginbotham, Adam, 7
Hipkins, Chris, 346
Hochschild, Arlie, 202, 208; on medical workers, 207
Hogan, Sarah, 328
*The Homeless Mind* (Berger, Berger, & Kellner), 315
home nursing, 286
Honduras, 371
Honey Pig, 260
Hong Kong, 373
Hopi Reservation, 362–3
Horkheimer, Max, 315
hospitals: California capacity, 377n1; fear of, 206; in New York, 19, 203–7
household income, 184
human rights: intimate partner violence and, 160; Thailand and, 115. *See also* crime against humanity
Human Rights Watch, 115
Hungary, 372; test ranking, 63
hurricanes: crime and, 223–4; Katrina, 6, 224; Rita, 224
Husserl, Edmund, xv
hydroxychloroquine, 3

ICC. *See* International Criminal Court
Iceland, 373
illegal employment relationships, 320n11

IMF. *See* International Monetary Fund
Immanuel, Stella, 3
India, 13, 15; lockdown in, 97–101, 371; migrant workers in, 14, 97–101; protests in, 369
Indonesia, 372
Industrial Revolution, 253–4
Influenza A virus (H1N1), 34
Influenza Pandemic 1918, 18, 27n8; crime and, 223
information, 318. *See also* disinformation
Ingraham, Laura, 349
instrumental reason, 315
Instrument of Government, 282, 283
interaction order, 317
Interferon *alfa-2b*, 65, 67
International Criminal Court (ICC), 58
international mobility, 303
International Monetary Fund (IMF), 37
International Olympic Committee (IOC), 56
internet-based communication, 317
intersubjective encounters, 207–10
intimate partner violence (IPV): child abuse and, 167; conditions for, 164; connectedness and, 164; control and, 158–9; Crenshaw on, 159; economic instability and, 161; feminist scholarship and, 158–61; first responders, 165–6; gender explanations for, 158–60; guns and, 164; Guterres on, 17; as human rights issue, 160; increase in, 163–7; individual experiences of, 168; intersectional approach to, 159, 168; Mlambo-Ngcuka on, 163, 167; national response plans and, 167; opportunities for, 164; pandemics and, 164; power and, 158–9; as public problem, 159; social support and, 164; stimulus plans, 167; structural inequality and, 161; structural oppression and, 159–60; theoretical underpinnings of, 158–61; under-reporting, 163
IOC. *See* International Olympic Committee
IPV. *See* intimate partner violence
Iran, 373
Iraq, 371
Ireland, 371
Israel, 371
Italy: death toll in, 68; lockdown in, 372; protests in, 369

Jackson, Steve, 141
*jai yen yen*, 121
JAL. *See* Japan Airlines
Jamaica, 371
Janssen Pharmaceutica, 362
Japan, 12, 373; Anti-Coronavirus National Task Force, 62; crime against humanity in, 68–74; death toll in, 57; decentralizing responsibility, 61–4; delayed responses, 58–64, 69; economic recovery priorities of, 59–61; emergency declaration in, 12, 56, 60; Fukushima Nuclear Power Plant and economic cost to, 58–9; GDP per capita, 59; Go To Campaign programs, 61; Great Global Recession 2008 and, 58; Heisei Era, 58; herd immunity in, 69–70; masks distributed in, 62; polymerase chain reaction tests in, 63; protests in, 369; schools in, 62; South Korea exploited by, 35; testing strategy, 63; test ranking, 63; white collar crime in, 68–74
Japan Airlines (JAL), 57
Japanese Olympic Committee (JOC), 59
Jeung, Russell, 264
JOC. *See* Japanese Olympic Committee

Jordan, 371
Jun Eun-kyeong, 50
justice reforms, 228

Kami, Masahiro, 63
Kazakhstan, 371
KCS. *See* Korea Customs Service
KDCA. *See* Korea Disease Control and Prevention Agency
Keenan, Margaret, xxn2
Kellner, Hansfried, 315
Kenya, 371
Kerbeck, Robert, 7
Khanh, Truong Huu, 145
Khokan, Zakir Hossain, 96
Killian, Lewis, 5
Kim, John, 260
Kim Dae-Jung, 36–7
Kiribati, 372
Koh, Tommy, 96
Koike, Yuriko, 62
Korea Customs Service (KCS), 42
Korea Disease Control and Prevention Agency (KDCA), 33
Kosovo, 371
Kyrgyz Republic, 371

Lacan, Jacques, 89; on gaze, 15, 90–1
Laos, 373
Latvia, 373
Lebanon, 371
Lefebvre, Kristine, 260
legitimation crisis, 374
Lesotho, 372
Levinas, Emmanuel, 202; on empathy, 208; on vulnerability, 208–9
LG Corporation, 40–1
Liberia, 371
Libya, 371
lifeworlds, xv
Lindell, Michael, 7
Die Linke, 313

Lithuania, 373
Little, Daniel, 26n4
livestock, 150n2
local knowledge, xvii
Lochbaum, David, 7
lockdown: in Australia, 372; in Brazil, 371; in Canada, 373; in China, 371; in Cuba, 372; duration around world, 371–3; in Germany, 372; in Haiti, 371; in India, 97–101, 371; in Italy, 372; in New Zealand, 64, 329–30, 372; none enforced, 373; peak, 162; physical impact of, 179; privilege and, 100–1; psychological impact of, 179; in Singapore, 93, 372; in South Korea, 373; in Spain, 372; in United Kingdom, 372; in United States, 371; in Venezuela, 371; in Vietnam, 65, 371
Lofland, Lyn, 21, 245–6; on celebrity, 247–8; relevance of, 247–53; on strangers, 247
Loneliness Scale, 185
Los Angeles: Chinatown, 259–60; Koreatown, 259–60; Malibu wildfires, 7; social angst in, 21; social distancing in, 257–68; as warehouse, 262–4
Louisiana Uniform Crime Report, 224
Luhmann, Niklas, xix; on communication, 318; on self-reflection, 316
Luxembourg, 372

Macao, 373
MacArthur, John, 146
Madagascar, 372
Maha Vajiralongkorn, 369
*mai phen rai*, 121
Malawi, 373
Malaysia, 372
Mali, 373
*Malibu Burning* (Kerbeck), 7
Malta, 372

Marcetic, Branko, 348
Marcuse, Herbert, 256
masks: contaminated, 62; in Germany, 313; Japan distribution of, 62; political identity and, 182; rationing, 42; refusal to wear, 3; South Korea regulations for, 42; in Thailand, 116–18, 120; in Vietnam, 139
materialism, 375
Mauritania, 372
Mauritius, 372
Mauss, Marcel, 301
Mayer, Johanna, 27n8
MBIE. *See* Ministry of Business, Innovation and Employment
McLuhan, Marshall, xviii
media: communication mediated by, 317; in Germany, 316–17; on healing and repairing society, 362; on migrant workers, 85; modernism envisioned through, 316; political identity and, 184
media, trust in, 180–4, 186, 188, 190; Democrats and, 187, 189, 191; pandemic concern and, 187, 191–2; psychological well-being and, 187, 189; Republicans and, 185, 187; social well-being and, 189
medical workers: backlash against, 267–8; complicity and, 213; death and, 210–15; grief and, 216–17; Hochschild on, 207; movement of, 205; pay cuts for, 205; responsibility and, 213; risks to, 177–8; thinning out of, 213–14. *See also* Cuban doctors
Memorandum of Understanding (MOU), 66
Meng, Ng Chee, 95
mental health: anxiety, 179; depression, 179; disaster studies and, 5
Merkel, Angela, 307; television address of, 316

Merleau-Ponty, Maurice, 202; on exposure, 216; on touch, 209–10
MERS-CoV. *See* Middle East respiratory syndrome coronavirus
message, 318
Mexico, 371
Middle East respiratory syndrome coronavirus (MERS-CoV): public official responsibility and, 34; South Korea reactions to, 33
*Midnight in Chernobyl* (Higginbotham), 7
migrant workers: exodus of, 98–9, 102; housing conditions, 94–5; in India, 14, 97–101; interviews with, 99–100; media reports on, 85; quarantine of, 93–4; in Singapore, 13, 92–7; supporting, 104n5
Ministry of Business, Innovation and Employment (MBIE), 333
mitigation strategies: financial security and, 18; global expanse of, 12; social connections and, 18; Thailand, 15–16; willingness to comply with, 18. *See also specific countries and strategies*
Mlambo-Ngcuka, Phumzile, 157; on intimate partner violence, 163, 167
Moderna, 361
modernism: anti-, 315–16; media and, 316; South Korea, 35
modern rationality, 315–16
Modi, Narendra, 97
Moldova, 372
Mongolia, 372
Moon Jae-in, 34
Moore, Harry Estill, 7
Moore, Matthew, 146
Morales, Evo, 67
Mori, Yoshio, 59
Morocco, 371
Morrison, J. Stephen, 347
Morrison, Scott, 337

MOU. *See* Memorandum of Understanding
Mozambique, 373
Muller, Todd, 340
*The Mushroom at the End of the World* (Tsing), 89
Myanmar, 371
Myung-Bak, Lee, 37

Namibia, 372
Nash, Stuart, 331
National Coalition against Domestic Violence, 166
nationalism: common, 23, 303–5; in European Union, 304–5. *See also* regionalism; territorializing
National Opinion Research Center (NORC), 4
National Restaurant Association (NRA), 260
natural barriers, 74
natural experiment, 101–2
nature, technology and, 6
Navajo Nation, 362–3
negative pressure isolation devices, 206
neoliberalism: International Monetary Fund and, 37; in South Korea, 36–9
Nepal, 371
the Netherlands, 372
new normal, 122
Newsom, Gavin, 258
New York: economic impact on, 162; respiratory specialist shortage, 204–5
New York hospitals, 19; best practices lapses, 205–6; disruption of, 203–7; fear of, 206; intubation cases, 204
New Zealand, 24–5; alert system, 328; children in, 329–30; church infections, 342–3; contact tracing in, 344–5; criticisms of, 330–3, 343–4, 346–8; death toll in, 343; disbelievers in, 343–4; disinformation in, 348–50; early miscalculations in, 327–8; economic impact on, 330–2, 336–7; elderly population in, 335; election, 339–41; emotional toll in, 334–6; fallout in, 333–4; film and television production in, 331; foreign nationals blocked entry into, 328; funerals banned in, 335–6; geographic distance of, 327; index case, 341; international praise for, 346–8; level 1, 338–9, 346; level 2, 337–8, 342; level 3, 336–8, 341–2; lockdown in, 64, 329–30, 372; October 2020 election, 25; population density of, 327; privacy in, 345; protests in, 343–4, 369; pushbacks in, 330–1; quarantine camps, 348–50; reassessment, 328–9; relationship impact in, 335; residents returning to, 339; second wave, 341–3; social distancing in, 327; sports programs, 330–1; success of, 24–5, 64, 333–4; Tokyo Summer Olympic Games and, 330; tourism in, 332–3; transparency in, 329; warning, 328
*New Zealand's COVID-19 Elimination Strategy*, 346
Nicaragua, 373
Nietzsche, Friedrich, xv, 10
Niger, 373
Nigeria, 371
Nishiura, Hiroshi, 63
Noda, Yoshihito, 59
non-strangers, 247–8
NORC. *See* National Opinion Research Center
North Korea, 11; as brother country, 36–7; South Korea ceasefire with, 34
Norway, 373
noticing, 87, 103; senses activated through, 89
Novavax, 362
Novel Coronavirus Expert Meeting, 13

NRA. *See* National Restaurant Association
numerical representation, 135–6; colonialism and, 138; distrust in, 147–9; interpretation of, 137–8; production of, 137; qualitative methods and, 137; truth value of, 137–8
*nyepi*, xvii, xxn3
NZ COVID Tracer, 344–5

occupational crime, 70
Oldenburg, Ray, 21, 245–6; relevance of, 253–7; on third places, 253–5; on urban experience, 255–6
Olsen, Brad, 332
Oman, 371
one-dimensional beings, 256
one-penetrated-by-the-other, 209
ordering: appearational, 251, 265; spatial, 250–1, 262, 318
*Orientalism* (Said), 16–17, 135
other: in South Korea, 34; trust in, 180. *See also* strangers
*Otherwise Than Being* (Levinas), 202, 208
overpsychologization, xix

Pakistan, 371
Palmer, Richard, 333
Panama, 371
pandemic: defining, 161; fatigue, 375; historicizing, xxn1; of inequality, 94; intimate partner violence and, 164. *See also specific pandemics*
pandemic concern: age and, 192; Democrats and, 187, 189, 191; financial well-being and, 187, 192–3; gender and, 192; political identity and, 190–1; psychological well-being and, 187, 189; of Republicans, 185, 187; social well-being and, 189, 192–3; trust in media and, 187, 191–2

Papua New Guinea, 372
Paraguay, 371
Parker, Chris, 117–18
Park Guen-hye, 37
Park Jung-hee, 35–6
Park Neung-hoo, 43
Parliamentary Ombudsmen, 287–8
partisan attitudes, 180–2
Pasifika Festival, 333
passivity, 211
PCR. *See* polymerase chain reaction tests
*pecalang*, xxn3
people centrality, 111
People's Republic of China (PRC). *See* China
perpetual foreigner, 265
Peru, 371
Peters, Winston, 337, 340–1
Pfizer, 361–2
*The Phenomenology of Perception* (Merleau-Ponty), 202
Philippines, 371
Phillips, Brenda, 8
Phuc, Nguyen Xuan, 142
physical distancing, 258; Floyd murder and, 3–4; in Germany, 308, 317–19; in Thailand, 116
physical proximity, 317–19
Piot, Peter, 142
Pirtle, Laster, 163
Pittinsky, Todd, 250
*Plague* (Slack), 365, xxn1
Plague Prevention Bureau, 366
Pogge, Thomas, 23, 303
Poland, 373
political identity: compliance and, 181; economic impact and, 183; masks and, 182; media and, 184; pandemic concern and, 190–1; religion and, 183; schools and, 181–2; self-reported, 183; study limitations, 193–4. *See also* tribal partisanship models

*Political Life in the Wake of the Plantation* (Thomas), 89
polymerase chain reaction tests (PCR), 63
Portugal, 371
positive foundational consequences, 5
poverty, 224–5
power: dynamics in Sweden, 295; intimate partner violence and, 158–9
Prayut Chan-o-cha, 111, 369
PRC. *See* China
President's Coronavirus Guidelines for America, 257–8
Prince, Samuel Henry, 8
privacy: bypassing, 44; contact tracing and, 43, 345; in New Zealand, 345; in South Korea, 43, 50
privilege: distance and, 88; fieldwork and, 103; lockdown and, 100–1
Proposition 47, 228
Proposition 57, 228
Protestantism, South Korean, 34–7
protests: against hostility to Asian Americans, 266; in Australia, 369; in Canada, 368; domestic violence, 165; 5G, 368–9; Floyd murder, 180, 364; in Germany, 321n17; historical, 365–7; in India, 369; in Italy, 369; in Japan, 369; in New Zealand, 343–4, 369; Slack on, 365–7; in South Korea, 46, 369; in Spain, 369; in Thailand, 369; in United Kingdom, 369; in United States, 368; vaccine, 368–9
pseudoscience: Trump enabling, 3. *See also* disinformation
psychological well-being: Democrats and, 189; pandemic concern and, 187, 189; Republicans and, 187; trust in media and, 187, 189
public policy, 7
*The Public Realm* (Lofland), 247, 248
Public Safety Realignment, 228

public space: in Germany, 309; in South Korea, 37–9. *See also* third places; urban experience
public sphere, 256–7
Puerto Rico, 371

Q-Anon, 368
Qatar, 373
qualitative methods, 8; numerical representation and, 137
Quarantelli, E.L., 4, 5, 6, 25n2
quarantine: camps in New Zealand, 348–50; inspections, 56; of migrant workers, 93–4; public notice of, 44; in South Korea, 43–4

racial capitalism, 163
reconstructive imagination, 203
regionalism, 34–7
*Relations in Public* (Goffman), 249
religion: political identity and, 183; in South Korea, 37; well-being and, 185
remote work ability, 178
Republicans (USA): defined in opposition, 18–19; financial well-being and, 187; pandemic concern and, 185, 187; psychological well-being and, 187; social well-being and, 189; trust in media and, 185, 187
researcher's gaze, 87, 102
residential burglary, 226, 232
response-ability, 89
restrained helpfulness: as coping strategy, 249; malfunction of, 249; urban experience and, 249–50
Rhee, Syngman, 35
Robert Koch Institute (RKI), 320n6
Robertson, Grant, 330–1
Roh Moo-hyun, 36–7
Romania, 372
Rome Statute 1998, 72
Rousseff, Dilma, 67

Routine Activity Theory, 225–7, 232, 238
routines, 178
Russia, 371
Rwanda, 371

*sabai sabai*, 121
Safer at Home Order for Control of COVID-19, 258
Said, Edward, 16–17, 135
Samsung, 40–1
San Marino, 372
Sanofi, 362
SARS-1, xvi
Saudi Arabia, 372
Sayed, Zakiya, 267
schools: in Germany, 311–12; in Japan, 62; political identity and, 181–2; South Korea changes in, 41, 49
Schweitzer, David, xix, 26n4
Scott, Chris, 343–4
Sebald, W.G., 92
self: Luhmann on, 316; openness of, 202; reflection, 316
Senegal, 372
sensual proximity, 318
Serbia, 372
sex work, 320n8
Seychelles, 373
shadow pandemic. *See* intimate partner violence
Shamdasani, Ravina, 73
shared humanity, xvi, 4
Shaw, James, 340
SHELDUS. *See* Spatial Hazards Events and Losses Database for the United States
Shibuya, Kenji, 63
*Shincheonji*, 36, 39
Sierra Leone, 373
silenced assistants, 8–9
Simmel, Georg, 318

Singapore, 14–15; contact tracing in, 93; lockdown in, 93, 372; migrant workers in, 13, 92–7; slogans, 93
Sinopharm, 361
Sinophobia: of Trump, 3. *See also* Asian American hostility
Sinovac, 362
situated encounters, 207–10
Skegg, David, 348
Slack, Paul, 370, xxn1; on protests, 365–7
Slovakia, 372
Slovenia, 373
smallpox, 6
Smart-City Innovation and Development Project, 43
social connections, 18
social contract variability, 376
Social Democratic Party, 282
social distance: in California, 20; crime and, 19–20; in Los Angeles, 257–68; multicultural consequences of, 264–8; in New Zealand, 327; no options for, 162–3; respect for authority and, 34; in South Korea, 40–2, 45–6, 48–9; in Thailand, 115–16
social impacts, 7; in Germany, 309–14
Social Services Act, 282, 286
social well-being: pandemic concern and, 189, 192–3; Republicans and, 189; third places and, 255; trust in media and, 189
societal complexity, 318
sociological communication theory, 318
sociology: disaster studies and, 6; German lessons in, 317–19
solitude, voluntary, 375
Solomon Islands, 373
Somalia, 373
South Africa, 371
South Korea, 12; authoritarianism in, 35; call centres, 39–40; cell-phone data, 45, 47; childcare, 41; church

hierarchies, 36; collectivism in, 34, 50; contact tracing in, 42–9; contract-based housing, 37; credit card spending data, 45, 47; cultural diversity in, 37; cultural norms, 34; death toll, 45; democratization of, 36; deregulation in, 37–8; digital infrastructure of, 43; economic plans, 35; exclusion in, 37–9; first reports from, 39; industrialization of, 36; Japan exploitation of, 35; lockdown in, 373; mask regulations, 42; MERS-CoV reactions, 33; missionaries in, 35; modernization in, 35; neoliberalism in, 36–9; North Korea ceasefire with, 34; othering, 34; privacy in, 43, 50; Protestantism, 34–7; protests in, 46, 369; public compliance in, 11; public space in, 37–9; public transportation data in, 45, 46; quarantine in, 43–4; regionalism in, 34–7; religious pluralism in, 37; respect for authority in, 34; school changes in, 41, 49; Shamanic and Confucian traditions in, 35–6; social distance in, 40–2, 45–6, 48–9; socio-economic polarization in, 36–7; territorializing politics in, 34–7; test ranking, 63; third places in, 38–9, 50–1; workplace changes, 40–1
South Sudan, 372
Sozialdemokratische Partei Deutschlands, 313
Spain: lockdown in, 372; protests in, 369
Spanish Flu, origins of, 27n8
Spatial Hazards Events and Losses Database for the United States (SHELDUS), 224
spatial ordering, 318; inability to access, 262; in industrial society, 250; in preindustrial society, 251; strangers and, 250; urban experience and, 250
Special Immigration Procedure, 42–3

Sri Lanka, 371
SSIF. *See* Strategic Science Investment Fund
Stallings, Robert, xix
Standing Committee on Social Questions, 288
state-corporate crime, 71
state security, 5
Stop AAPI Hate, 264
strangers, 248; defining, 247; Lofland on, 247; spatial ordering and, 250; urban experience and, 247, 252
Strategic Science Investment Fund (SSIF), 333
stress, indigenous conceptions of, 11
structural distance, 88
structural inequalities, 101–4; intimate partner violence and, 161
structural oppression, 159–60
structural paranoia, 211
Sturgis Motorcycle rally, 179–80
subjectivity, xix
Sudan, 372
suffering: Butler on, 88; distance and, 14–15, 88
Suga, Yoshihide, 61
Summerfruit New Zealand, 333
Suriname, 371
Sutherland, Edwin, 70
Suzuki, Naomichi, 62
Sweden, 373; administrative structures, 281; authority command in, 289–91; authority confidence in, 278, 281; case law in first phase, 287–8; Corona Commission evaluation directives, 291–2; crisis preparedness, 281–4; criticism of, 276; death toll in, 276; do no harm strategy, 275; early phase strategy, 294–7; elderly population in, 21–3, 275–8, 285–7; elderly restrictions in, 292–4; guiding principles, 283–4;

Sweden (*continued*)
  Health and Social Care Inspectorate, 290–1; herd immunity in, 69–70; infection penalties in, 277–8; legal preconditions, 277–8, 281–8; motto of, 275; National Board of Health and Welfare, 279, 290; power dynamics in, 295; principle of responsibility, 278; Public Health Agency, 279, 289–90; responsibility principle, 281; theoretical approaches in, 281; welfare state ideology of, 281; well-being and, 286
Sweden Public Health Agency, 22
Switzerland, 373
SydeKick for ThaiFightCOVID, 113
Syria, 372
systemic crises, xix, 25n1

Taiwan, 373; Center for Disease Control, 56–7; death toll in, 57; success of, 64, 347; test ranking, 63
Tajikistan, 373
Takahashi, Tai, 70
Taleb, Nassim Nicholas, xv, 6
Tanzania, 373
Tarde, Gabriel, 26n4
Taweesilp Visanuyothin, 113–14
Taylor, Breonna, 364
technology: for contact tracing, 42; nature and, 6. *See also specific technology*
Tegnell, Anders, 347
telemetry units, 204
Tep, Josephine, 95
TEPCO. *See* Tokyo Electric Power Plant
territorializing: as coping strategy, 39; in South Korea, 34–7
Thai Chana, 114
Thailand, 373; air quality in, 118; curfew, 15; economic cost on, 125–6; emergency declared in, 115; emotion in, 120; exceptionalism, 126; fear mongering in, 114–19; first report from, 15, 110; frequent phrases in, 121; government response in, 111–13, 126; handling ranking, 15; Human Rights Watch condemnation of, 115; interview responses, 116–18; lessons learned in, 115; masks in, 116–18, 120; mitigation strategy, 15–16; penalties in, 122; physical distancing in, 116; protests in, 369; public health system in, 125–6; public response in, 115–19; social distance in, 115–16; success of, 15, 110, 126–8; success summary, 127; traditional greeting in, 118; travellers arriving in, 112; Universal Precautions adopted by, 112

*Theories of Alienation* (Geyer & Schweitzer), xix
third places, 21, 245; authoritarianism and, 257; neutrality of, 254; Oldenburg on, 253–5; public sphere and, 256–7; as social liabilities, 263; social status and, 255; in South Korea, 38–9, 50–1; therapeutic effect of, 254; in United States, 38, 254; women and privatization of, 38–9
Thomas, Deborah, 89
Thompson, Vetta L. Sanders, 10
Thornley, Simon, 332
Thu, Huong Le, 147
*Thus Spoke Zarathustra* (Nietzsche), 10
Thwaites, Guy, 146
tightness-looseness theory (TL), 110, 119–25; components of, 122
time perception, 179
TL. *See* tightness-looseness theory
Togo, 372
Tōhoku earthquake, 7, 58–9
Tokyo 2020 Bid Committee, 59
Tokyo Electric Power Plant (TEPCO), 71
Tokyo Summer Olympic Games 2020, 12, 56; Abe on, 59–60; economic

cost of, 58; economic impact of, 60; investments in, 60; New Zealand and, 330; postponement of, 61; prioritizing, 58, 59–61; promotion of, 58, 59; sponsorships, 60; tourism expectations of, 60

Tonga, 371

*Tornadoes over Texas* (Moore), 7

*Totality and Infinity* (Levinas), 202, 208

touch, 209–10

toxic masculinity, 158–9

toxic substances, 7

transparency, 138; justifications for, 145; in New Zealand, 329; Vietnam and, 143–7

tribal partisanship models, 18, 181, 194

Trinidad and Tobago, 372

Trump, Donald: Asian American hostility of, 364; criticisms of, 181; on herd immunity, 69, 73; negligence of, 72–3; pseudoscience enabled by, 3; Sinophobia of, 3; Wehner on, 363–4

trust: in government, 180; in media, 180–92; in others, 180; well-being and, 180, 182–3, 185. *See also* numerical representation

Tsing, Anna, 89

Tunisia, 372

Tuohy, Laurel, 120, 121

Türkiye, 372

Turkmenistan, 373

Turner, Sarah, 8–9

Uganda, 371

Ukraine, 373

uncertainty, 314–17

understanding, 318

unemployment: crime and, 224–5; in Germany, 309–10

United Arab Emirates, 372

United Kingdom: lockdown in, 372; protests in, 369

United States, 18–21; crime against humanity in, 71–4; culture wars in, 364–5; death toll, 3; economic impact on, 178; Germany compared with, 306; lockdown in, 371; protests, 368; third places in, 38, 254. *See also specific states*

United States Virgin Islands, 372

Universal Precautions, 15; Thailand adopting, 112

University of Chicago, 4–5

University of Maryland, 4–5

University of Oklahoma, 4–5

Upston, Louise, 328

urban disassembly, 376

urban experience: anonymity and, 248; audience role prominence and, 249; automobile and, 248; civil inattention and, 249–50; civility towards diversity and, 249–50; consistency in, 248; cooperative motility and, 248–9; fear and, 249–50; in Germany, 309; through material consequences, 246–7; Oldenburg on, 255–6; restrained helpfulness and, 249–50; spatial ordering and, 250; strangers and, 247, 252

*Urban Ills* (Yeakey, Thompson, & Wells), 10

urban politics, 10

Uruguay, 373

US CARES, 167

Uzbekistan, 372

vaccine: anti-, 365, 368; availability of, 361, 362; in Canada, 361–2; disinformation, 365; distribution of, 362; emergency measures, 362; first person to receive, xxn2; optimism from, 370; protests, 368–9. *See also specific vaccines*

Vaccine Alliance, 362

vandalism, 226, 232–3

Vanuatu, 373

Venezuela: Cuban doctors in, 66–7; death toll in, 66–7; lockdown in, 371

Vietnam: auditing, 143–7; biological essentialism and, 141–3; colonialism and, 142–3; communication in, 140; contact tracing in, 139; Cuban doctors in, 65; death toll in, 16–17, 65, 141, 143–7; erasing success of, 141–3; first wave, 139; flights suspended to, 139; Hanke on, 147–9; health infrastructure investments in, 65; Jackson on, 141; lockdown in, 65, 371; MacArthur on, 146; marine life disaster in, 150n1; masks in, 139; Moore on, 146; National Steering Committee, 138–9; Piot on, 142; reporting controversies in, 17, 135, 143–9; strategy overview, 138–40; Thwaites on, 146; transparency and, 143–7; Western gaze and, 17

violence: against Asian Americans, 3, 264–7, 364; domestic, 158, 165–6; Guterres on, 17; private sphere, 17; against women, 157. *See also* crime; intimate partner violence; *specific forms of violence*

vulnerability, 215, 216; Levinas on, 208–9

Waco Tornado, 7
war crimes, 72
WCC. *See* white collar crime
Weber, Max, 315
Wehner, Peter, 363–4
well-being: age and, 184; compliance and, 182; demographic characteristics, 184; financial, 182, 184–5; gender and, 184; household income and, 184; hypotheses, 182–4; impacts on, 177–80; mediation analysis, 187–9; pandemic prioritization and, 185; psychological, 182, 185; religious concern and, 185; research objectives, 182–4; social, 183, 185; study limitations, 193–4; study participants, 184; study results, 185–90; survey methods, 184; Sweden and, 286; trust and, 180, 182–3, 185

Wells, Anjanette, 10
*werkverträge*, 312
Westerhäll, Vahlne, 293
Western gaze, 8; Vietnam and, 17
white collar crime (WCC), 58; in Japan, 68–74; prosecuting, 71; taxonomies of, 70–1
WHO. *See* World Health Organization
Williams, Joseph, 343
Wilson, Nick, 346
witnessing, 87, 103; boundaries and, 89–90; as co-performative, 89; distance and, 90–1; response-ability demanded by, 89; Thomas on, 89
women: crime targeting, 39; privatization of third places by, 38–9; violence against, 157
Woodward, Bob, 72
workplace changes, 40–1. *See also* remote work ability
World Health Organization (WHO), 12
*The World of Perception* (Merleau-Ponty), 202, 209–10
*A World of Strangers* (Lofland), 247, 250
World Rugby Cup 2019, 60
writing, politics of, 88–92

xenophobia: in European Union, 305; germaphobia and, 21

Yang Chow, 259
Yeakey, Carol Camp, 10
Yemen, 372
Yoshimura, Hirofumi, 69

Zalo, 134
Zambia, 372
Zimbabwe, 371